Textbook of DAIRY ENGINEERING

THE AUTHORS

Dr. S. Ravi Kumar, had obtained his B.Sc. (Dairy Technology) in 1976, M.Sc. (Dairy Engg.) in 1981 and Ph.D in 1999 from NDRI, Karnal. He had joined APDDCF Ltd, in 1976. He continued to work as Dairy Manager, Marketing Manager at various places in Andhra Pradesh. He joined in 1988 as Associate Professor (Dairy Engg.) in the then APAU, (later changed to ANGRAU and SVVU), and was promoted as Professor of Dairy Engg., in 2001, after acquiring his Ph.D from NDRI. He was head of Dairy Technology Programme and later College of Dairy Technology from 2005 to 2012. During his career he held the posts as Assoc. Dean, Dean (Faculty of Dairy Science), Estate Officer and Member, Board of Management in the SV Veterinary University.

He was project in-charge for various projects on use of non-conventional energy use in dairy equipment and cold storages (for dairy and fisheries). Conducted training programmes for technical officers and technicians in various dairies for plant operations. He retired as Professor, in the year 2017.

He continues to teach various dairy engineering subjects as Guest faculty, and also incharge of IGNOU, study center at Heritage Foods Ltd.

Dr. B.M Manjunatha, Asst. Professor and Head, Dept of Dairy Engineering, College of Dairy Technology, Sri Venkateswara Veterinary University, Tirupati. He had his graduation in B.Tech (Dairy Technology) from UAS, Bengaluru and master's and Ph.D degrees in Dairy Engineering from National Dairy Research Institute, Karnal.

He has gained a rich experience in execution of around 14 Dairy and Food industry projects in India and overseas. He was involved in execution of Dairy and Food industrial projects which includes milk processing, milk products, milk powders, coffee powders and Egg powders, *etc.*

His Doctoral research was on innovative subject of Nano technology application in surface engineering of Dairy Equipment. He has got academic excellence award for his Ph.D from National Dairy Research Institute. His path breaking research adventure has cleared way for many other researchers to follow.

He was co-principle investigator of departmental research projects on Vapor absorption refrigeration and solar energy-based bulk cooler. He is life member of Indian Dairy Association and Indian Dairy Engineers Association.

Textbook of
DAIRY ENGINEERING

— Authors —

Dr. S. Ravi Kumar

Professor of Dairy Engineering (Retd.)
Former Dean, Faculty of Dairy Science
Sri Venkateswara Veterinary University, Tirupati
Former Dairy Manager, APDDCF Ltd

Dr. B.M. Manjunatha

Head, Department of Dairy Engineering
College of Dairy Technology
SVVU, Tirupati
Former Senior Process Engineer, GEA ltd.

2025

Daya Publishing House®
A Division of
Astral International Pvt. Ltd.
New Delhi – 110 002

ISBN : 978-93-5919-823-1 (HB)

Published by : **Daya Publishing House®**
A Division of
Astral International Pvt. Ltd.
– ISO 9001:2015 Certified Company –
4736/23, Ansari Road, Darya Ganj,
New Delhi-110 002
Ph. 011-43549197, 8130496929
E-mail: info@astralint.com
Website: www.astralint.com

Dedicated to our families

Sandhya

Parikshith, Ruthu, Yathi Ramanujam

&

Swapna

Shreshta

Foreword

The growth and development of dairy industry in India in past two decades is phenomenal. The dairy equipment to meet this growth in terms of capacity and technological demands for manufacture of a wide variety of dairy products, especially the indigenous dairy products on such a large scale is unprecedented. Explaining the design .and working principles of the processing equipment is a truly challenging task for any author of text book on dairy engineering. Both the authors Dr. Ravi Kumar and Dr. Manjunath have passion for teaching. They have years of experience in dairy plant operations and teaching students in dairy technology and are uniquely qualified to write this book. The book covers the fundamentals along with the latest development in the dairy equipment manufacture. The book includes well labelled diagrams that explicitly display the working principles of the equipment. The authors have done well in explaining the selection of materials for equipment manufacture, the processes undergoing in the equipment and also the maintenance aspects of the equipment. The book covers the syllabus of engineering topics of a degree program in dairy technology.

I know the authors of the book, Dr. Ravi Kumar and Dr. Manjunath for many years. Both are distinguished teachers of Dairy engineering and technology. They have extensive experience of milk procurement in rural areas, operation of small dairies, milk product factories and installation of milk powder plants One of the authors (Dr. Manjunath) erected multiple effect milk evaporating equipment for a multinational company.

Both the authors received advanced degrees in dairy engineering from National Dairy Research Institute where I had the privilege of working close to three decades. I am confident that this book will be useful to the students and teachers of dairy engineering and also as a reference to dairy plant managers.

S.C. Sarma

B.Sc. (Hons), B.E. (IISc), M.S., Ph.D (Utah State)

Formerly, Visiting Professor of Mechanical Engineering

Utah State University, Logan, Utah

Visiting Professor of Food Process Engineering,

University of Missouri, Columbia,

and Project Engineer Emeritus,

Desert Aire Corporation, Milwaukee, Wisconsin, USA

Preface

The subject of Dairy Engineering being taught for the under graduate students of Dairy Technology, is a unique combination of both basic and applied engineering subjects. Further, it has to deal with certain aspects of microbiology, food and hygiene and Chemistry. The 5th Dean's committee of Indian Council of Agricultural Research (ICAR), has included Energy Conservation too as one of the subjects. The developments in any of the basic and applied engineering subjects, find its resonance in the subject of the Dairy Engineering and its application in dairy equipment design, manufacture and use.

As the number of colleges offering degree in Dairy Technology as well as post-graduation degree are increasing, a requirement of text book in dairy engineering grew to keep up with these developments. An e-course content was developed, under aegis of ICAR to normalize and serve the teaching requirements, in number of colleges, which are in the process of strengthening their faculty and hence needed a knowledge base, that aids in teaching. A series of consultation with various college faculties resulted in the above, in which the author had been active in developing dairy engineering course content. It was felt by the author that a book bringing in update to these contents would be useful.

The present effort is to meet the requirement of syllabus of two courses of Under graduation in particular, the Dairy Engineering and Dairy Process Engineering. The focus had been on explaining the principles and design requirement of dairy equipment that are commonly used in established dairy factories as well as in small-scale dairy manufacture establishments of indigenous dairy products. The authors in writing the book, have used liberally their lecture notes and drawings for teaching over the years.

The book has been divided into 25 chapters, explaining the basic features of materials used in manufacture of dairy equipment, the equipment for processing liquid milk, milk powder, western and indigenous dairy products.

The chapters 1 to 12 deal with basics of liquid milk processing equipment. This broadly covers the syllabus for the Dairy Engineering I course. Chapter 11 gives methods of milk packing and storing, which include aseptic packaging of milk for long life. Types of agitation in the dairy processing has been introduced in Chapter 12.

From Chapter 13 to 23, catering to syllabus of Dairy Process Engineering, cover the equipment involved in manufacture of concentrated milk, drying of milk powder, butter, cheese and Ice cream. The growing interest in mechanization of manufacture of indigenous dairy products has resulted in research and development equipment for continuous manufacture of these products. They have been introduced in Chapter 21. Certain technologies which have found recent applications in dairy industry *viz.* Membrane Technology, has been adequately explained in Chapters 24 and 25.

The author acknowledges the authors, and publishers of various books, periodicals and industry manuals from which some of the information presented here is gathered. In particular we have consulted the following:

1. Milk Powder Technology: Evaporation and Spray Drying by GEA Niro, Denmark (2010)
2. Spray Drying Hand book by K. Masters, (1979) 3rd Edn.
3. Outlines of Dairy Technology, by Prof. Sukumar De
4. Handbook of Food Engineering Edited by D.R. Heldman and D.B. Lund (2007)

Thanks are due to all our students, friends and well-wishers from both academia and industry for their support and encouragement to write the text book.

The authors express their deepest gratitude to Dr. S.C Sarma, former Director (Officiating), NDRI, for not only writing the Forward, but going through the draft and making important improvements in the basic structure of the book. The author acknowledges the helps of Mr. Sravan Surineni, USA, in final stages in speeding up computers work and information accessing. The authors record their grateful thanks to the Astral International Pvt. Ltd., New Delhi for their co-operation and patience during the developing of this text book.

Dr. S. Ravi Kumar

Dr. B.M. Manjunatha

Contents

Foreword *vii*

Preface *ix*

List of Symbols Used in the Text *xiii*

1. Materials used and Sanitary Features of Dairy Equipment 1

2. Sanitary Pipes and Fittings, Standard Glass Piping, Plastic Tubing and Gaskets 11

3. Milk Storage Tanks, Silos, Road and Rail Tankers 19

4. Can Washing, Bottle Washing and CIP Systems 27

5. Separation Equipment 37

6. Centrifugal Separation Equipment 51

7. Homogenization 71

8. Pasteurizers 87

9. Sterilization Equipment 103

10. UHT Processing Equipment 119

11. Filling Equipment 127

12. Mixing and Agitation Equipment 135

13. Evaporation	**145**
14. Multiple Effect Evaporators	**161**
15. Auxillary Equipment for Evaporators	**189**
16. Principles of Drying	**203**
17. Classification of Dryers	**213**
18. Spray Drying	**221**
19. Theory of Solid-Gas Separation	**245**
20. Butter Making Machines	**261**
21. Mechanization of Indian Dairy Products Manufacture	**275**
22. Ice Cream Making Equipment	**285**
23. Cheese making Equipment	**293**
24. Membrane Processing	**305**
25. Applications of Membrane Processes	**315**
Annexures	**329**
Index	***341***

List of Symbols Used in the Text

The following SI unit symbols used throughout the text, unless otherwise stated wherever it is introduced.

Basic Units

Quantity	*Unit*	*Symbol*
Length	Meter	m
Mass	Kilogram	kg
Temperature	Kelvin	K
Time	Second	s
Electric current	Ampere	A
Amount of substance	Mole	Mol
Frequency	Hertz	Hz
Force	Newton	N
Pressure	Pascal	Pa
Energy, Work, quantity of heat	Joule	J
Power	Watt	W
Quantity of electricity, electric charge	Coulomb	C
Electric potential, potential difference	Volt	V
Capacitance	Farad	F
Electric resistance	Ohm	Ω

Derived Units

Quantity	*Unit*	*Symbol*
Linear velocity	m/s	V
Linear acceleration	m/s^2	A
Angular velocity	Rad/s	ω
Angular Acceleration	Rad/s^2	∞
Mass Density	Kg/m^3	ρ
Force, Weight	$Kg - m/s^2$	N
Pressure	N/m^2	p
Work, Energy, Enthalpy	N – m	W,E,H
Power	J/s	W
Absolute or dynamic viscosity	$N - s/m^2$	μ
Kinematic viscosity	M^2/s	ϒ
Frequency	1/s	*F*
Gas constant	J/kg K	R
Thermal conductance	$W/m^2 K$	h
Thermal conductivity	W/m K	k
Specific Heat	J/kg K	C
Molar mass or Molecular Mass	kg/mol	M
Thermal diffusivity	m^2/s	α

Some Non-conventional Terms

mm : Milli meter

SWG : Standard Wire gauge

GI: Galvanized Iron

MS: Mild Steel

SS : Stainless Steel

PPM: Parts per million

min: Minute

BMC : Bulk Milk Cooler

L : Litres

h : Hour

UHT : Ultra High Treatment

BOD : Biological Oxygen Demand

g/cc : Grams per Centimeter cube

Chapter 1

Materials used and Sanitary Features of Dairy Equipment

Introduction

Dairy processing industry needs various types of equipment for handling, processing and storing milk and milk products. Dairy Equipment needs special materials, methods of design and operation in comparison to the general engineering equipment. As these equipments are designed to handle milk and milk products which are highly perishable, hygienic aspect being an essential factor in the design. The special features applicable to the equipment used in dairy industry are listed below:

Sanitary Features

- Equipment is easily accessible and readily cleanable either in an assembled position or when disassembled. Removable parts are readily demountable.
- All product contact surfaces are self-draining.
- When assembled, no pipe or fitting threads are exposed to the product. (For example Milk storage tank supporting legs, pasteurizer frame *etc.* Further, SMS type easily cleanable threads are preferred.)
- All permanent joints are welded and grounded to smooth finish.
- Interior surfaces are smooth and even, especially at joints, thus eliminating crevices and projections.

- Gaskets are eliminated wherever possible. If used, they should be smooth, non-toxic and product resistant. They should not contribute to any additional flavours to the product.

  In addition to the above, following operational aspects are also required to be considered in the overall sanitary design of the equipment.
- Possibility of processing under vacuum, which make the equipment susceptible to external contaminants.
- Effective velocity of flow and optimum agitation to prevent formation of surface film
- Pre-heating of the product with external devices
- Providing large vapour space from where water from condensing vapours to be removed
- Rapid removal of water vapour
- Avoiding air leaks

Specific Requirements for Manufacture of Dairy Products

- Continuous operation throughout the year causing more wear and tear
- Wide temperature variations during operations
- Withstand the effects from water, steam, brine, refrigerants, cleaning and sanitizing solutions.
- Disassembled more frequently which may increase damage to the components
- Dust proof and corrosion resistant external surface
- Smooth polished internal surface
- Minimum clearance of 100 mm between equipment base and floor with ball foot
- Absence of sharp corner and edges
- Elimination of vertical dead end or passage and capped tees
- Proper slope towards drain outlets
- Properly ground and polished welds
- Raised edge opening to prevent surface drainage in to container
- Sanitary type inlet and outlet fittings

As sanitary aspects are of paramount importance in the design of dairy equipment, the equipment must be designed and manufactured with the correct materials to achieve the above requirements.

The following diagrams illustrate the special requirements of dairy equipment which may differ from design of general purpose equipment. The Figure 1.1 enunciates the important aspect of self-draining of the vessel, so that no liquid is left in it.

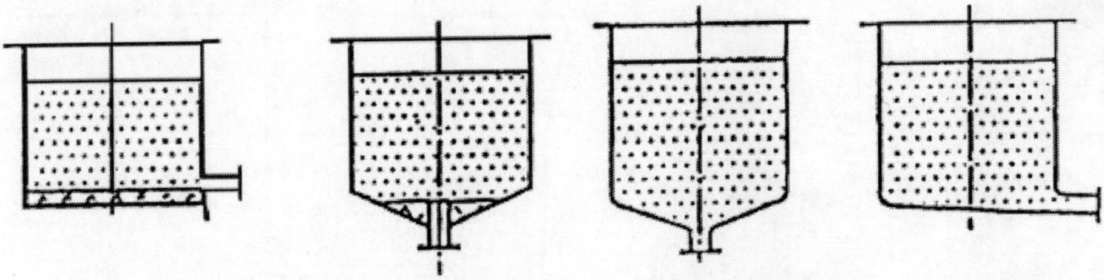

Not Drainable Design (unacceptable) Drainable design (acceptable)

Figure 1.1. Sanitary Design Aspects

Even supports and other attachments that do not coming directly in contact with milk and milk products should be designed to avoid accumulation of dust and soil.

The Figure 1.2 (a) to (h), (adapted from Anon,2005) illustrate several principles required to be observed in the design of the equipment. Figure 1.2 (a) shows, how the pipeline design should avoid dead spaces, to prevent trapping of air or dirt accumulation. It is possible that these "T" joints may be left for future extension or for change in pipe connections. However, if it is unavoidable, then a flow of fluid can be directed, so that the turbulence, does not allow settling of material at the dead ends.

Figure 1.2a-h. Additional Principles for Sanitary Design

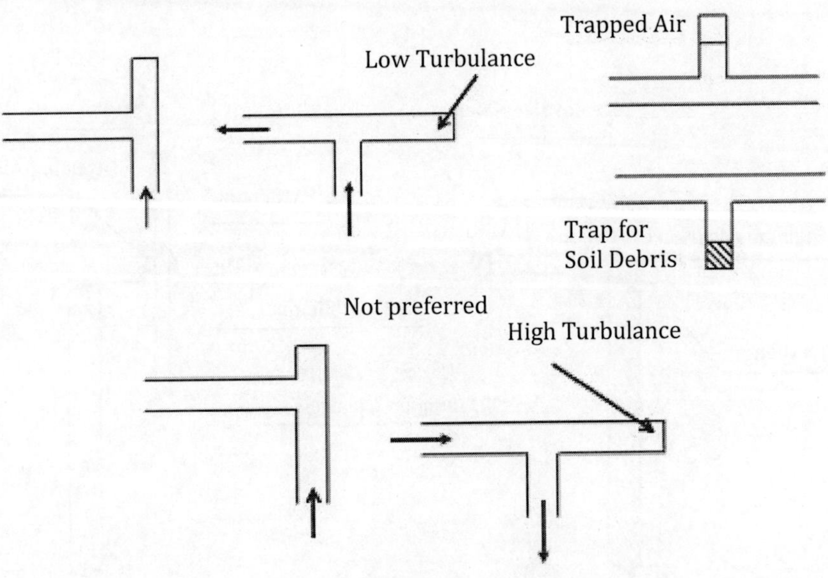

(a)

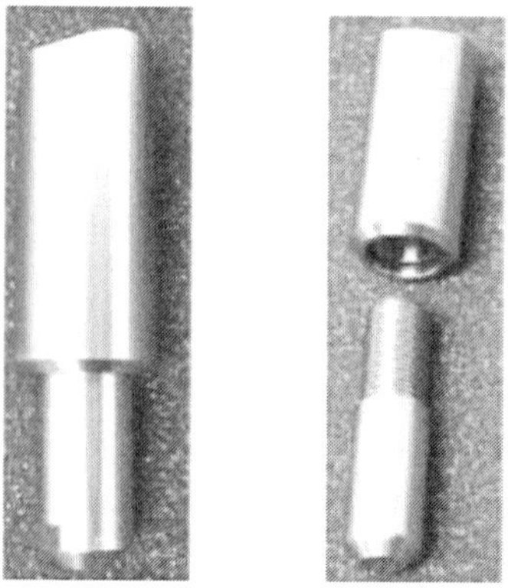

(b)

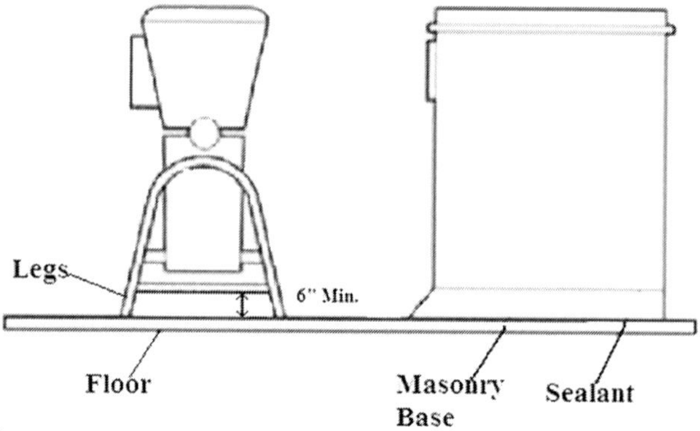
Legs
6" Min.
Floor
Masonry Base
Sealant

(c)

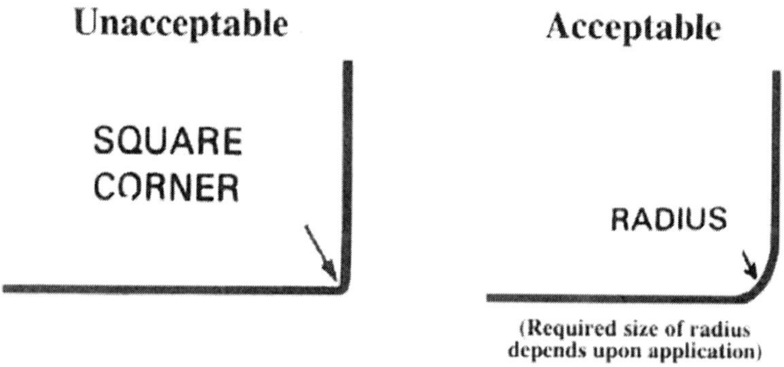
Unacceptable
SQUARE CORNER
Acceptable
RADIUS
(Required size of radius depends upon application)

(d)

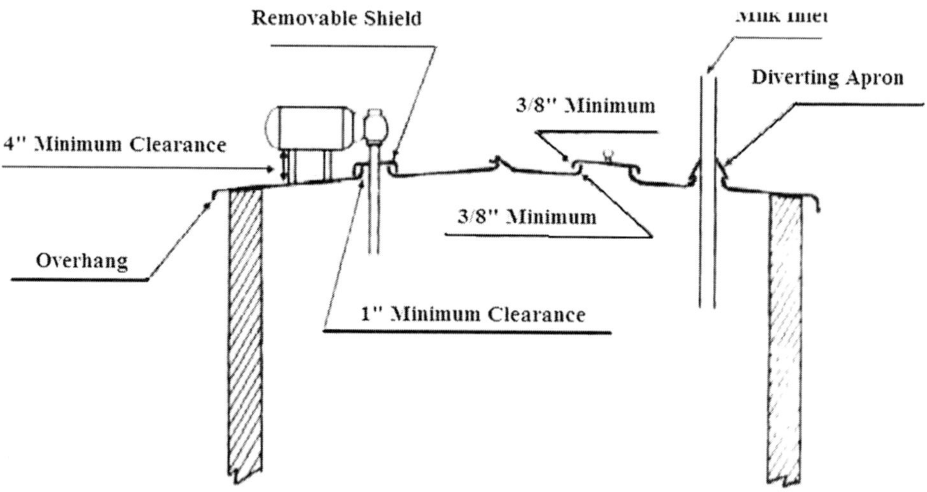
Removable Shield
Diverting Apron
3/8" Minimum
4" Minimum Clearance
3/8" Minimum
Overhang
1" Minimum Clearance

(e)

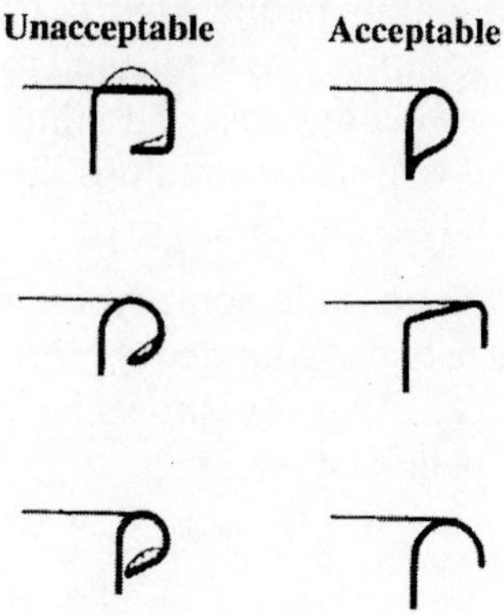

(f)

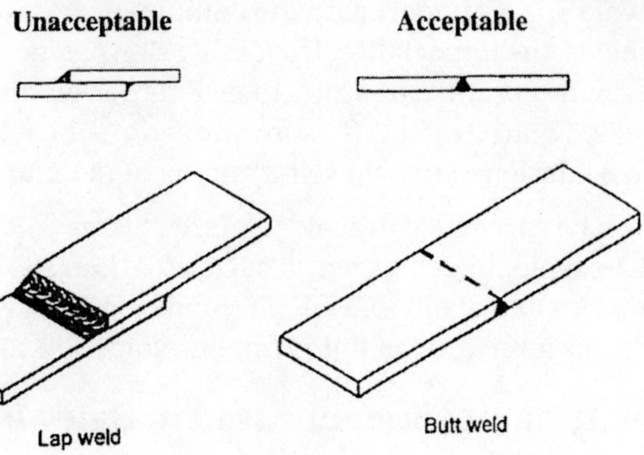

(g)

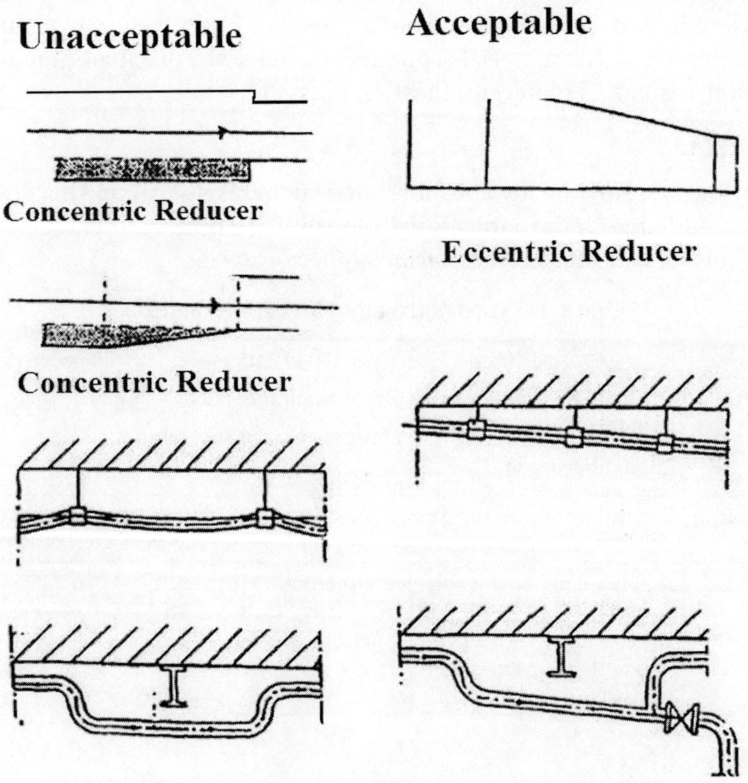

(h)

Figure 1.2a-h. Additional Principles for Sanitary Design

Figure (b) shows, how the adjustable legs of storage tanks are designed to avoid exposure of threads to the surrounding, so that splashing of water during cleaning of floors or accumulation of dirt is prevented. Similarly, Figure (c) shows, how to avoid spaces underneath the equipment from the accumulation of dirt. Figure (d) shows, the importance of avoiding sharp edges, which are difficult to clean, and to provide certain curvature, to facilitate hygienic conditions inside a vessel holding milk. Figure (e) and (f) show how the lid covers for the vessels or openings are to be designed to help self-draining of moisture and avoid accumulation of dirt.

It is preferred to avoid joints, welds, gaskets as much as possible, some cases, it is unavoidable. However, in such situations the welds have to be ground to smooth surface and avoid unhygienic conditions (ref Figure 1.2 g). Figure (h)shows how the accumulation of fluids can be avoided in the milk pipeline layout.

Materials commonly used in Dairy Industry

Materials used for the construction of dairy equipment also play an important role in the sanitary design and maintenance of hygiene. They are broadly divided into those parts of equipment that come in contact with the milk and milk products and those which do not come in contact and only act as supports, and other service equipment. The materials commonly used are: Iron, Steel, Stainless steel, Aluminium, Insulation materials, Plastics, Glass *etc.*

Iron and Steel

Iron and steel are mostly used for material that is not coming in contact with milk. Iron added with Carbon is Steel. Depending on the percentage of carbon present, it is further divided into various types of steels.

Table 1.1. Types of Iron and Steel and their Uses

Metal	*Characteristics*	*Application*	*Carbon per cent*
Cast Iron	Difficult to weld, brittle in nature Cannot be forged	Heavy equipment beds Main castings of engines Gear box cases	2 to 4.5 per cent
Wrought Iron	Easily welded, tough, ductile, malleable	Chains, Hooks, couplings, rivets, shock loading parts	0.06 to 0.08 per cent
Mild Steel	Highly ductile, malleable, tough, strong, easily welded, possesses magnetic properties, can be cast and forged	Structural elements, building equipment, conveyor frames	0.1 to 0.25 per cent

Stainless Steels

Steel containing more than 10.5 per cent Chromium is called Stainless Steel. It is particularly used where corrosion should be kept minimum and where the material is coming in contact with milk and milk products. Different proportions of Nickel are also added to further increase resistance to corrosion.

The Stainless Steels are broadly divided into Austenitic, Ferritic, Martensitic. Among Austenitic steels, of particular importance is the American Iron and Steel Institute's grades, AISI 304 and AISI 316. These grades are especially used in parts of equipment making contact with the milk and milk products. The martensitic and ferritic are designated with 400 series, while Austenitic are designated with 300 series. Both martensitic and ferritic are magnetic and have carbon up to 1 percent or higher. They can be hardened by quenching. Austenitic has little or no carbon, and cannot be hardened by quenching, but can be work hardened (will turn magnetic then). Figure 1.3 shows graphically the various combinations of Steel with chromium, nickel and molybdenum. When hardened stainless steel is required, then the grades of steel are used.

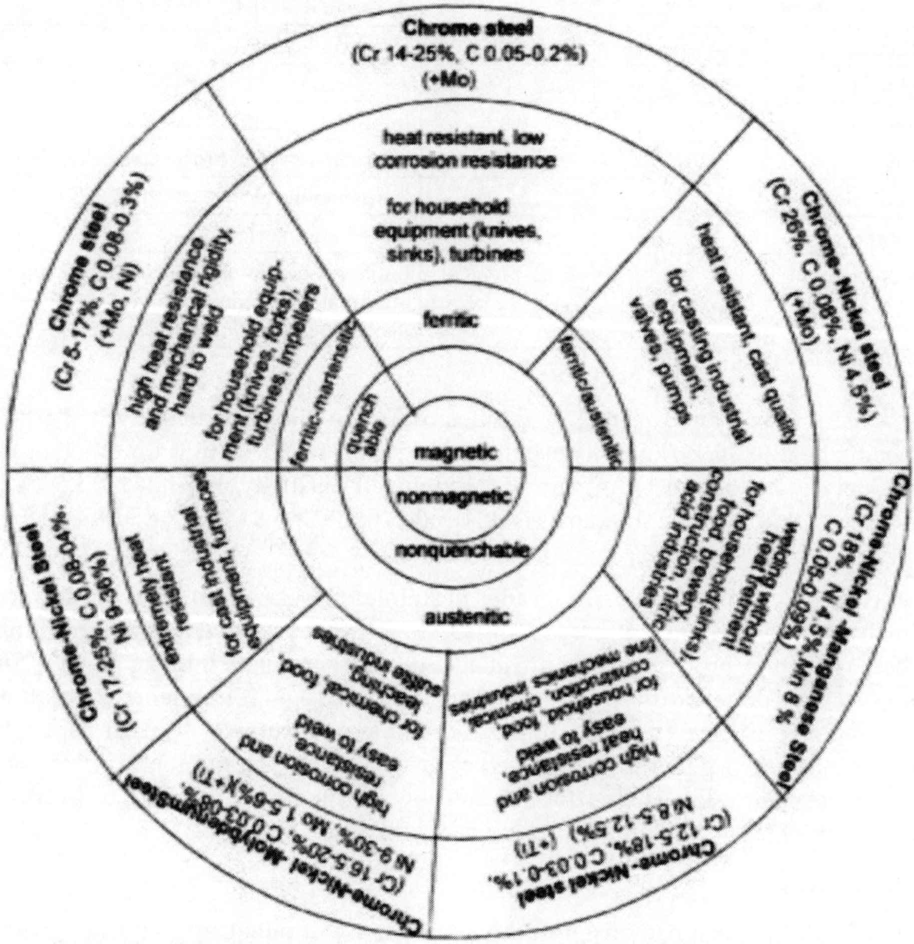

Figure 1.3. Characteristics and Uses of different Alloys

Table 1.2 gives the details of the composition of AISI 304 and AISI 316, explaining the role of each component in their overall utility. The role of Molybdenum is the important difference between the two. Consequently, AISI 316 is comparatively more expensive than AISI 304. However, components of the dairy equipment and those not coming in contact with milk are also made of other grades of stainless steel, to maintain hygiene and to meet sanitary design.

Table 1.2. Chemical Composition of AISI 304 and AISI 316

Component	*AISI 304 (Per cent max.)*	*AISI 316 (Per cent max.)*	*Remarks*
Carbon	0.08	0.10	0.04 per cent for Low carbon grades, designated as 304L
Manganese	2.0	2.0	Improves wear resistance
Phosphorous	0.04	0.04	
Sulphur	0.03	0.03	Improves machining properties
Silicon	1.0	1.0	Improves resistance to heat and chemicals
Chromium	18 – 20	16 – 18	Hardenability
Nickel	8 – 10	10 – 14	Gives better surface finish
Molybdenum	—	2 – 3 per cent	Improves corrosion resistance at higher temperatures; reduces chances of "weld-decay" in fabricated stainless steels

The physical and thermal properties of AISI 304 are: Density, 7750 kg/m^3; Specific heat, 0.50 kJ/kg K; Thermal conductivity, 16.28 W/mK; Co-efficient of thermal expansion, 10^{-6} mm/mm °C; Modulus of Elasticity, 1.940 x 10^6 kg/cm^2. Its designation in other standards is: IS Grade 10; BS 304 S15; DIN X5 CR Ni 18 9; SIS 2332.

Though, the use of selective grades of stainless steel is important, the surface finish and maintenance of corrosion resistance property is equally important. The passivation process (using nitric acid or other strong oxidizing agents) for the surface is done for this purpose, initially and at a regular frequency thereafter, to maintain a passive (non-reactive) oxide film on the surface. Another aspect is the surface finish, generally expressed as grit finish (as grit number), or average roughness factor, Ra. The surface should be as smooth as possible to prevent formation of bio-films, initiated by microorganisms.

Aluminium

Aluminium, is mostly used in the manufacture of milk cans, but the current trend is to shift to stainless steel for their manufacture for better hygiene. Aluminium is light weight (2.7 gm/cm^3), insoluble in milk, does not impart any flavour and can be spun into various shapes. But it is porous in nature, making it difficult to weld. Its resistance to abrasion is poor, and gives grey stain. The surface is generally given Anodizing treatment, which involves part to be treated is made anode and dipped in electrolytic bath (of sulphuric, chromic or oxalic acid), to form a colour less and stable film of aluminium oxide on the surface by passing direct current through an electrolytic solution. The aluminium oxide is formed by release of hydrogen at the cathode (the negative electrode) and oxygen at the surface of the aluminium anode, creating a build-up of aluminium oxide. The anodizing current varies with the area of aluminium surface being anodized, typically in the ranges from 30 to 300 amperes/square meter. The recyclable nature of aluminium is an advantage in use of this metal.

Polymers

Polymers, have wide range of applications, and are broadly classified into Thermoplastics and Thermosetting plastics. Thermoplastics are manufactured by a process called addition polymerization. They are liquids at higher temperature and crystallize readily at low temperatures because of their simpler molecular structure. They generally show poor resistance to organics. Structurally, they are straight chain polymers, and have no cross links.

Those used in dairy industry of the above category are Poly Tetra Flouro Ethylene (PTFE), Poly Ethylene (HDPE and LDPE), Poly Vinyl Chloride (PVC), Acrylonitrile butadiene styrene (ABS), Cellulose acetate butyrate (CAB) *etc.*

The PTFE, is practically unaffected by all alkalis and acids except fluorine and chlorine gas. It retains its properties up to 260 °C. In the form of PTFE clothes, it is used as barriers between heat-sealing rods and milk pouch films. This is also useful in manufacture of shaft seals, agitator attachments. The High Density polyethylene (HDPE) is used for milk crates and cans. ABS has good resistance to nonoxidizing and weak acids but is not satisfactory with oxidizing acids. The upper temperature limit is about 65°C. Its resistance to weak alkali solution is excellent. CAB is not affected by dilute acids and alkalis, but chlorinated solvents cause swelling.

Thermo-stetting plastics on the other hand have cross links in their structure, and undergo permanent changes at higher temperatures. Hence, they cannot generally be recycled. Thermo-setting plastics used in dairy industry are Fibre Reinforced Plastic (FRP) storage tanks, Epoxy glass, polyester reinforced glass, plastic gears, and Elastomers like Polybutadiene and Polychloroprene (the commercial name is Neoprene). Neoprene is resistant to ozone, sunlight, oils and halogenated solvents. Some of these are used as gasket material in the pipe line joints and similar applications. Silicon rubbers have high resistance to high or low temperatures as well as oils and greases. They are used in dairy industry for higher temperature applications.

Glass

Among the various types of glasses, Borosilicate or Pyrex glass is used for pipe lines handling corrosive liquids, gauge glasses, laboratory apparatus. The Safety glasses are used as light and sight glasses for milk storage tanks, pipelines where transparency is required. These glasses on breaking, do not fly into pieces. They are made by making two or three layers of flat sheets of glass and in between them a thin layer of vinyl plastic is placed and subjected to slight pressure. It is then heated, till the glass layers ad plastic layers merge into one another to make it as a sandwich.

Insulation Material

The insulation materials in dairy industry, are mainly for the chilled water, steam, chilled milk pipe lines, milk storage tanks and spray drier chamber. The materials commonly used are glass wool, mineral wool, polystyrene, polyurethane. Use of asbestos material was discontinued when it was found to be carcinogenic. There are three types of insulations in general. These are:

a. Fibrous insulations (*e.g.* Ceramic fibre, glass mineral wool, rock mineral wool, *etc.*)
b. Cellular insulations (*e.g.* Polyurethane, polystyrene and polypropylene, *etc.*)
c. Granular insulations (*e.g.* Calcium silicate, perlite expanded and vermiculate)

For the insulation of hot surface, the insulation materials used are: Glass mineral wool, Calcium silicate, Ceramic fibre and Rock mineral wool. For insulation of cold surfaces, the insulation materials used are: Cork, Expanded nitrile rubber, polyethylene foam, polypropylene foam and polyurethane foam (PUF). For the insulation to be effective, the material should be correctly applied, and provided with a vapour barrier. The water vapour barriers like bituminous, epoxy and aluminium foils of minimum 0.06 mm thickness are required on the hot side of the insulation. Polyurethane Foam (PUF) is the most commonly used insulation material in dairy and food industry. For in-situ insulation it can be applied using blocks of material or by spray foam which hardens instantly.

REFERENCES

Anonymous, "Sanitary design and construction of food equipment", Extension bulletin FSHN0409, IFAS Extension, University of Florida, 2005.

Thermal Insulation Materials: A Tool for Energy Conservation Gajanan Deshmukh *et al.*, J. of Food Process Technology, 2017, 8: 4.

Chapter 2

Sanitary Pipes and Fittings, Standard Glass Piping, Plastic Tubing and Gaskets

Introduction

Design and selection of pipes for handling milk and fluid milk products has to be done with same care and attention to sanitary features as is done for the dairy equipment, where the surfaces are coming in contact with the product. The type of material has to be Stainless Steel of grade AISI 304 or AISI 316. The fluid, which is milk, is also flowing, and hence, the corrosion due to erosion as well as pitting has to be factored in while designing. In some of the instances, instrumentation for measuring the properties of the flowing fluid will be required, and hence need special type of pipe fittings. Care also has to be taken regarding the surface finish on both inside and outside the pipe lines. In general, the following aspects are important in the design of sanitary pipe lines for a dairy plant.

General points to be considered while selecting sanitary pipe lines are given below:

1. Pipes selected should be cold drawn seamless tube
2. They should be softened, descaled, and pickled
3. Pipes should be polished internally or externally or both sides (as required)
4. Thickness generally adopted is given below in Table 2.1

Table 2.1. Outer Diameter and Thickness of Stainless Steel Pipes

O.D	*Thickness*
25.4 mm	1.25 mm
38.1 mm	1.25 mm
50.8 mm	1.25 mm
63.5 mm	1.5 mm
76.2 mm	1.5 mm

5. Pipeline should be clean, smooth and free from surface defects, straight and free from longitudinal grooving both inside and outside.
6. Material for fitting to be same as that of pipe.
7. The condensate from the surface of milk pipe lines should not be allowed to fall on places where unhygienic conditions or milk contamination is likely to occur.
8. The capacity of the pipe line should be designed to allow the flow within the limits, without much pressure loss and tendency for increased erosion.
9. Carrying capacity of milk (kg/h) should be considered.

Table 2.2. Capacity of Flow through Stainless Steel Pipes

Outside Dia. of Pipe (mm)	*Approx. capacity (kg/h)*
25	0 – 2000
38	800 – 7200
50	7200 – 15000
63	15000 – 22700
75	22700 – 36000

i) Next larger size for heavy liquids, like cream, Ice cream mix.
ii) Similarly, next larger size if the flow is by gravity
iii) Criteria for selection is velocity of flow and cost

The following points may also be considered while designing the sanitary pipe lines

1. Piping lengths should be kept to a minimum.
2. Piping layout should not be a hazard or restriction to operators.
3. All valves and instruments should be located so that they can be handled easily during operation and maintenance.
4. Pipe layout should be conducive to adequate individual support, and group supports wherever possible.

5. Piping system should be designed with proper drainage and avoiding any pockets. If pockets are inevitable then proper drain valves should be provided.
6. Pipe lines with valve clusters (for remote operations) to be kept away from path of operators.

Glass Piping

It is generally used where the visibility of the product is important. Due to its brittleness and fragile nature, it has to be again protected by external piping of suitable metal. It is also used in the level gauges, sight and light glasses for milk storage tanks. In certain situations, however, it can be used as an inner liner for the storage tanks or pipelines. The glass has to be of non-splinter type.

Plastic Tubing

In addition to being used as packing material, some of the pipes and flexible tubing is being made of plastics. They have to be of food grade, and should not impart off flavours. Its smoother surface gives lower resistance to the fluid flow and is light in weight. It has high resistance to detergents. However, it has lower resistance to temperature, and organic solvents. The maximum continuous exposures for various plastics are:

Table 2.3. Temperature Resistance of Various Plastic Material for Tubing

PVC	50°C
Polystyrene	65°C
Polyethene (linear)	60 90°C
Polypropylene	70 - 100°C
Polytetrafluoroethylene	325 - 33°C

Plastic tubing is easily wetted by aqueous solutions and it is more easily wetted by fats. This can cause fat to penetrate into the surface. It can be removed by process of Extraction and if it is not removed, a rancid odour develops on the surface of the plastic material due to fat decomposition. The consequences can be discolouration, crevice formation, irreversible distortions of shape, bacterial contamination in fine crevices, taints in milk due to released plasticizer.

Gaskets

The role of Gaskets is to prevent leakage between the two metal surfaces. In addition to being flexible, the dairy industry demands them to be fat-resistant, resistant to detergents and sanitizers. They will have to resist certain degree of heat too in applications like Pasteurizer gaskets. The various other applications in dairy industry are SS pipe fittings. Storage tank manhole gaskets, milk pumps, cram separators, homogenizers, diaphragm valves *etc.*

The following are the different types and their heat tolerance limits.

Table 2.4. Gasket Materials and their Temperature Resistance

Nitrile rubber (NBR)	130°C
Ethelene propylene (EPDM)	65°C
Silicone rubber	175°C
Viton rubber	180°C

Another commonly used material is Acrylonitrile Butadiene Styrene (ABS), mostly used in electro-pneumatic valves used in automation.

To keep the gaskets in place, various methods are used. Glues, gasket cements and mechanical methods are more commonly used. In case of glues and gasket cements, special procedure is needed to fix the gaskets in place. This some time limits their use in high temperature applications.

The gaskets have to be stored in cool and dry place. It should not be exposed to ozone, like in electric arc welding applications.

Installation, Care and Maintenance of Pipes and Fittings

Pipelines in dairy industry are of two types, *i.e.* those which are coming in contact with the milk and milk products, and those which are mainly of service pipe lines. Though we will be considering here only those which are coming in contact with milk and milk products, some of the basic principles will be same.

Installation

Well-designed piping systems require a plant layout that provides adequate space around equipment to ensure that the piping and associated valves are arranged in an operable and maintainable fashion with minimal intrusion into operators' space. Such conditions should be a part of the building's design from its conception. A systematic approach to layout produces a workable end result. The end result should reflect due consideration for the construction, operation, and maintenance factors for every piece of equipment involved.

The first step in installation is to draw the Piping and Instrument diagram showing the arrangement of the process equipment, piping, instruments, valves and other fittings. It should include:

1. All process equipment, identified by an equipment number. This should be drawn roughly in proportion and the location of the equipment.
2. All pipes, identified by a line number. The pipe size and material should be shown.
3. All valves control and block valves, should be shown with an identification number and size.

4. Ancillary fittings that are part of the piping system, such as inline strainers, steam traps, should be shown with an identification number.
5. Pumps, identified by a suitable code number and type.
6. All control loops and instruments, with an identification number.

The Piping and instrument diagram will resemble the process flow sheet. When the diagram is laid out, it is necessary to show the relative elevation of the process connections to the equipment where they affect the process operation; for example, the Net Positive Suction Head (NPSH) of pumps, barometric legs, siphons *etc.* Full details of pipe layout are usually shown in a different drawing, known as Piping Isometric Drawing. In all the drawings, standard symbols for the valves, instrument lines *etc.* should be used.

Sizing of Pipe Lines

Over sizing of pipelines increases installation costs, and under sizing increases pressure drop. It is important to know the distinction between pipe and tube here. Inside diameters vary according to specified schedule or wall thickness. Pipes are labelled by their nominal bore whereas sanitary tubing is labelled by its outside diameter and it has a smaller bore. For example, a 2" stainless steel pipe has an outside diameter of 2.375", and if made of schedule 10S it has a resultant bore of 2.157". A 2" sanitary tube has an outside diameter of 2.000" with 0.065" wall thickness for a resultant bore of only 1.870". This reduction in available diameter, if overlooked, could lead to errors in flow calculations.

The size of the suction pipe of pump should never be smaller than that of the discharge pipe; usually it is one size larger.

Connections

Commonly used methods for connecting pipes and fittings are flanged connections, welded fittings, and threaded couplings. The important aspect is that these should be not only effective, but also should be easy for cleaning and sterilization. It should not have any cavities that cannot be cleaned. The pockets caused by branches in pipelines should not be more than 6 times the diameter. In modern sanitary pipe lines, preference is being given for the welding so as to reduce the number of pipe unions and minimize potential for crevice areas, which are difficult to clean.

As far as feasible, valves, filters, excessive fittings, horizontal bends, elbows and tee joints should be avoided close to the suction side of the pump.

Valves

Most of the valves installed in plants are used for isolation service, usually referred to as Block Valves to prevent unwanted flow past a given point in a piped system. The remainder are used to control either rate of flow or pressure of the flowing medium, at a particular point in the system, referred to as throttling or

Control Valves. The body shapes of Gate and Globe valves render them uncleanable and hence have a potential source of contamination. The alternative is choice between ball, butterfly, plug or diaphragm valves. The ball valve is frequently favoured because its valve stem is isolated from water, has a quarter-turn operation, and hence suitable for CIP cleaning.

Supports

The tubing should be supported on both sides of valves and to all changes in direction. Supports should be such that sloping of lines should ensure complete draining. Sloping of lines of 0.3 to 0.5 per cent is sufficient for post cleaning and gravity flow of low-viscosity fluids.

Materials used for hanger components and floor stand supports should be of stainless steel to avoid external corrosion and galvanic action between dissimilar materials. Rubber insert pads may be used to isolate tubing from the support to reduce vibration. Rigid anchoring should be replaced by floating supports which avoid distortion of tubing systems due to temperature changes. Distance between supports should be sufficient to avoid sagging or excessive load on joints.

Table 2.5. Spacing for Support of Pipes

Size (mm)	*Distance between Supports (m)*
50	2.4
75	3.4
100	4.0
125	4.5

Defects in Pipe Lines

The pipelines should be clean, free from longitudinal grooving, both inside and outside, should be smooth, free from surface defects, straight, cylindrical in shape when viewed from end, uniform in thickness.

Instruments and Fittings

Instruments should be made from materials considered safe for direct contact between the flowing media and the instrument itself, preferably highly polished 316L stainless steel. Connections of pipe should avoid cavities or pockets likely to encourage accumulations of solids.

Maintenance of Pipes and Fittings

1. Worn-out gaskets in pipeline joints should be replaced periodically, before they become brittle, deformed or losing their shape permanently.
2. During preventive maintenance inspection, distress cracks on walls and footings near the pipe anchors will indicate improper layout with regard to thermal expansion.

3. Check for the supports of the pipelines, whether they are firm or if they are dislodged from their installed positions.
4. Check if vibrations, expansions caused by the process equipment like Homogenizer, Pasteurizer are prevented from being carried on to pipelines connected to them.
5. Check if any modifications done like expansion, shifting of equipment is not affecting the pipeline installations, and properly supported and connected without any stresses.

Chapter 3

Milk Storage Tanks, Silos, Road and Rail Tankers

Introduction

Milk received at the collection centres in villages is commonly stored in storage tanks. The current trend is to store the milk in Bulk milk cooling units. Milk is then transported by Road tankers to the dairy plants where it is transferred to raw milk storage tanks. If the distance for transportation of processed milk from the dairy plant is very long then Rail milk tankers are also used for economy. Further, for processing operations like mixing, ripening, culturing, aging *etc.*, the storage tanks require special features. The requirement of modern quality standards, make the use of open vessels obsolete.

Milk Storage Tanks

They are usually cylindrical and mostly insulated. They may be horizontal or vertical the choice depending on availability of floor space. For large quantity of storage, vertical Silo tanks are provided, outdoors, to avoid construction of tall buildings to accommodate them. In case of horizontal tanks, space may be conserved and construction cost reduced by "through-the-wall" installations. It provides ready access to all the necessary fittings and at the same time conserves the expensive sanitary space in the processing room, as most of the horizontal tank is outside the processing area.

The inner shell of storage tank is made of Stainless Steel, and is covered with suitable insulation material. The outer shell may be made of Stainless Steel or Mild steel (Figure 3.1). The outer mild steel is painted to prevent corrosion. It may be two coats of antirust and two coats of enamel paint of suitable colour (usually cream colour). The inner shell of storage tank should be smooth, with surface finish usually of No 4, or mill finish of 150 grit. There should not be any sharp corners, so that proper cleaning is possible. The edges must have a knuckle radius of at least 25 mm and the welds, if any should be ground to a smooth finish. The shell should have a drain from lowest point, preferably at the centre of the bottom. If this is not possible, the bottom of the vessel should be sloped towards the outlet valve to ensure complete drainage. The slope should be at least 1:12.5 towards the outlet.

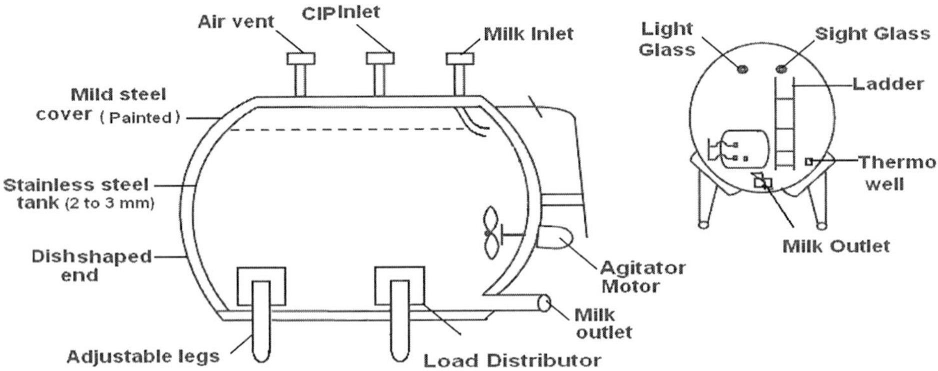

Figure 3.1. Horizontal Storage Tank

The shell has openings for agitator, non-foaming type milk inlet, inlet for cleaning solutions through a spray device, air outlet. It has also fittings for sight class and light glass, for inspection of inside of the tank for level measurement, to ensure proper cleaning *etc.* The glass has to be of non-splinter type. The opposite end of sight class has level indicator markings made with food grade paint, or made by buffing of the surface. The tank has usually 15 per cent excess capacity than the nominal capacity. The technology of using load cells for remote measurement of level is also being introduced in the modern dairies, for automated operation, especially for the silo tanks of large capacities, located outside the processing area.

The inner shell is also be supported by four legs, through load distributing MS angles, to prevent change of shape or damage to inner shell. The load distributors also have a lifting eye at top to lift the entire storage tank with cranes. The thickness of shell depends on the size of the tank, to withstand its weight. Usually the thickness is 2 to 3 mm for horizontal tanks, and 4 to 5 mm for silos. The ends of the tank usually are dish shaped to give strength and it a greater knuckle radius at the inner joint of dish end to the horizontal tank.

Thermocole (expanded foam of polystyrene) is usually used for insulation for lower temperatures, and for higher temperatures, glass wool or mineral wool is preferred. The thickness should be such that, the increase in temperature of milk in the tank should not rise more than 2°C in 18 hours, under a test conditions of 35°C temperature difference between inside and outside of the tank. There should be no hollow space in between the inner SS and outer MS. In some of the designs however, this gap is provided with a breathing outlet at the bottom, so that pressure is not built up in it when hot cleaning solutions are circulated. The outer jacket is effectively sealed and provided with a draining vent hole of vermin-proof design. The number of supporting legs may be 4 or 6, with adjustable ball feet.

The inlet which is at the top of the tank, with bent tube towards the inner wall, will prevent foam formation during filling of the tank. The milk outlet is at the bottom, with a flange type valve fitting. It is usually provided with a two-way valve. It is usually of large size (51 mm or 63.5 mm dia.), to allow rapid unloading. For Silos the inlet is at the bottom, with a non-return valve.

To keep the milk inside the tank properly mixed, an agitator at slow speed is operated. It may be horizontal, vertical, or inclined (for Silo tanks). Most commonly, the agitator is vertical, with one or two sets of mixing blades. It is supported at the bottom by a bush, to prevent lateral oscillations. The agitator is connected to motor through a speed reduction gear. The gear box which contains lubricant has to be perfectly sealed to prevent it from leaking into the tank.

Tanks are also provided with other essential fittings such as manhole, sampling cock, thermo-well, ladder in front to access sight glass, CIP pipe line connection with adequate support at the top. The manhole is of swing-back type, which is oval shaped and can swing both horizontally and vertically, so that it can be positioned in place from inside. There is a tightening arrangement to prevent milk leakage from manhole door. The sampling cock is so positioned that a sample can be obtained even when milk is at 5 per cent of the rated capacity of the tank. It is spring loaded push type of lever, or a screw type of opening. A provision is also made for a Thermo-well, in which the sensing element of temperature indicator can be placed. It is an SS tube, with its lower end sealed, and welded in inclined positioned to the inner shell. It is filled with oil or ceramic powder for improving the contact between the sensing element and the wall of thermo-well. The air outlet from the tank is provided at the top with vermin proof cover to prevent dust and insects entering the tank.

Vertical Storage Tanks/Silos

The vertical storage tanks are mainly used, when the floor space is limited, and the roof of processing hall is high enough (Figure 3.2). Products with higher viscosity like cream and ghee are also stored in this type of tanks.

A manhole is provided at the top, and a ladder is provided for access. The top is made conical to prevent accumulation of water or dust. The sight and light glass are also provided at the top. The agitator is long, and provided with more than two sets of blades. In some designs an internal ladder is provided for manual cleaning of inside of the tank. Other fixtures like CIP cleaning attachment or spray ball, thermowell, sampling cock, lifting lugs or eye at the top, are provided as in horizontal storage tanks. Both the inlet and outlet are at bottom of the tank. A slope of 1:10 to 1:15 towards the outlet is provided for free and complete drainage of liquid. The inner wall is of Stainless steel sheet of 3 mm thickness while outside is 2 mm. The floor is of 5 mm thickness. The tank is provided with four supports with SS ball feet and with provision for height adjustment.

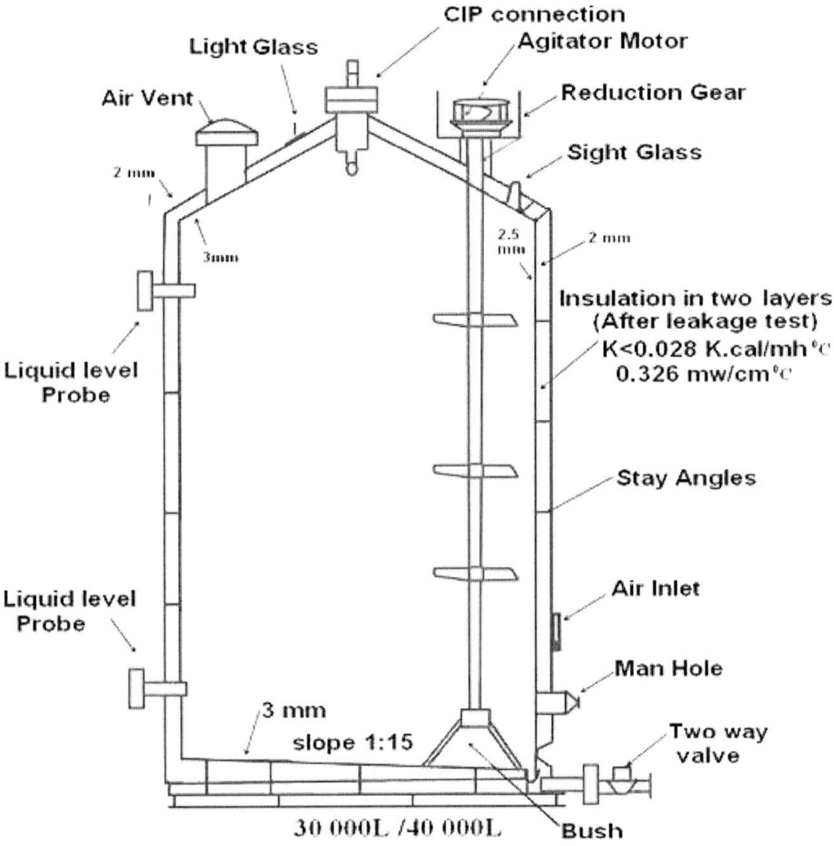

Figure 3.2. Vertical Storage Tank

Silos are vertical storage tanks with capacity ranging from 30,000 to 1 lakh litres. The compressed air is used for agitation (Figure 3.3) though, it is being replaced by suitable mechanical agitators, as the air absorbed into milk is undesirable. The inlet and outlet are controlled from inside of the processing hall through an opening called Alcove. Inlet valve is provided with non-return valve. An overflow line is also provided to indicate if the tank is full. The silos have load cells or similar electronic instrumentation to indicate level of milk inside the tank, even remotely.

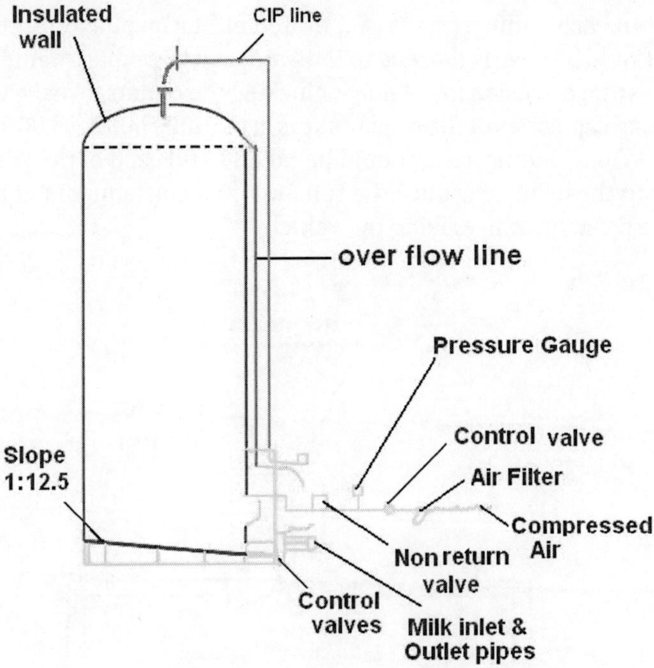

Figure 3.3. Milk Silo Tank

Road Tankers

The milk collected at collection centres is transported to dairy plants by Road tankers. The tankers should withstand shocks, vibrations, jerks, extreme weather conditions and other environmental factors, while transporting milk. Therefore, the essential features include, thicker SS sheet for inner and outer shells, better insulation, weather proof manhole doors covered with a hood, two or three compartments to prevent rolling over of milk during jerks. The tank is mountable on a truck, or an articulated trailer, with all the limitations of manoeuvrability while driving. While ordering the tank, the type of vehicle and model should be specified to suit the chassis. The capacity, length of tank and its mounting should be suitable for village roads and these factors impose a severe limitation. The tanker should be mounted on the vehicle chassis, keeping the centre of gravity low.

The insulation used is a combination of glass wool and Polystyrene (Thermocole) in two or three layers, with bitumen as sealing material to prevent moisture seepage. The barrel of the tankers preferably should be oval shaped to keep centre of gravity low. It is supported by M.S. supports which are affixed to the chassis of the vehicle. The contact points are separated by asbestos pads. The vehicle has a ladder on the rear and also walk way (suitably chequered to avoid slippage) to access the manhole for inspection, sampling and manual cleaning.

The outlet from each compartment is operated either by plug or butterfly valves. The outlets should be clearly marked to indicate which compartment; they belong. There could also be provision for sealing or locking the outlets to make them tamper proof. The usual capacities of the road tankers are 7,000, 9,000, 13,000, 15,000 and 21,000litres. While loading, care should be taken to fill each of the compartments of the tanker to the brim, to reduce the roll over, and consequent churning action, and to reduce difficulties in driving the vehicle.

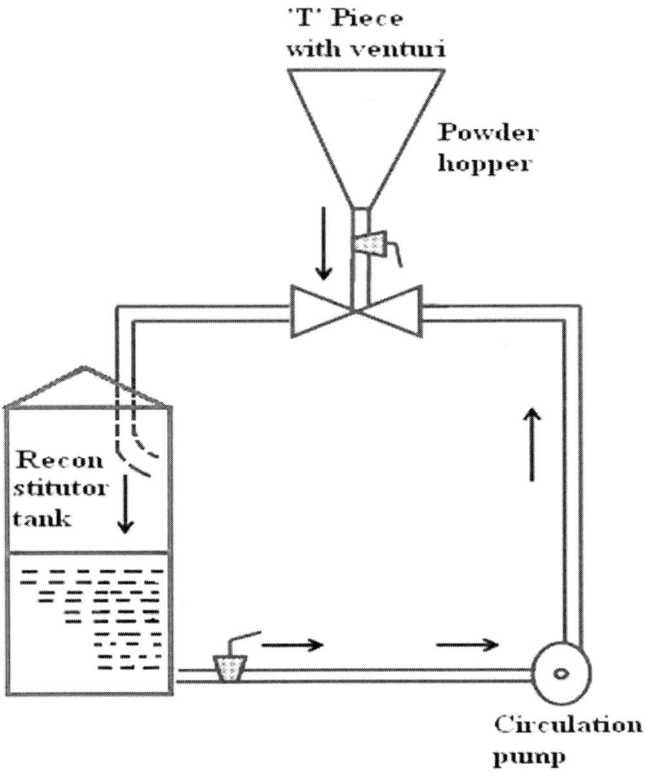

Figure 3.4 Reconstitution Tank.

Rail Tanker

Railway tankers are used for transporting milk over long distances and have capacities 20,000 to 40,000litres. The tank is divided into two or more compartments. The final structural dimensions and design features should meet the requirements of Indian Railways and be suitable for hauling by passenger trains. The top of the tank has a dust cover, pressure relief valve and vacuum relief valve, manhole door. The Manhole door is provided with gasket made of Acrylonitrile butadiene copolymer or polychloroprene. An internal ladder is also provided. A dip stick is provided to measure the quantity in the tank approximately. Insulation provided is 100 mm thick. This insulation ensures that when the tank is full and for temperature differential of 35 °C, and initial temperature of 4 °C, the rise in temperature is only 2 °C in 24 hours,. For quick unloading or loading of the tanker, the outlet pipe size provided is 63.5 or 76.2 mm. and it is located at the centre. Two ladders are also provided on each side of the tanker. The inner vessel should withstand a pressure 34.3 kPa, and should be tested for leaks by Radiographic and dye penetration tests.

Other Types of Tanks

There are other types of tanks designed for specific applications such as cream storage, ghee settling *etc.* These are double jacketed. An SS perforated ring is installed in the jacket to allow a gentle flow of cold/hot/well water to maintain the desired temperature of the product inside the tank. These tanks are usually vertical, with all the essential features provided on the horizontal milk storage tank. Milk reconstitution tank (Figure 3.4) is a special purpose tank with additional accessories of a funnel with venture, a valve at bottom, a circulation pump and a potable water inlet to the tank. The tank is does not require insulation, as it is operated at about 40 °C.

Bulk Milk Coolers

Introduction: Chilling milk at village level to improve the quality of milk and overall shelf life, and also to avoid losses, has been recognized as the technology intervention that is needed. Though electricity-based Bulk Milk Coolers are being operated, they are also being provided with electricity generators using diesel as fuel, so that the power supply is not interrupted, when the raw milk is being collected, and cooled to below 5°C. The present technologies are striving towards BMCs to be operated with solar energy.

Advantages BMCs are that they reduce the dependence on transportation of milk by cans, reduce the microbial contamination as well as proliferation, during the time the raw milk reaches the Dairy plant for further processing. The BMCs are designed to chill milk to below 4oC, in various options, like entire capacity of bulk milk being cooled or half the capacity to be cooled in specified time.

The BMCs are being classified mostly as per ISO 5708 standard, which laydown the capacities and operating conditions. The BMCs categorisation as per safe operating temperatures of ambient is designated as A, B and C corresponding to 43, 38 and 32°C respectively. Also, the rate of cooling is also designated as I, II, III, IV corresponding to the rate of cooling of first milking from 35 to 4 °C in 2.5,3.0,3.5 and manufacturer declared performance, respectively. For example, a BMC performance designated as meeting ISO 5708 – 2 A II, indicates that, the BMC is designed to receive milk, twice a day and can operate with ambient temperature up to 43 °C. Further it can cool the first milking in 3.0 h.

The BMC basically has a storage tank, provided with arrangement to cool milk either directly with Direct expansion type of refrigeration system, or by indirectly cooling with chilled water using Ice Bank Tank type of Refrigeration system. Depending on the requirement in the village clusters, the capacity of BMCs is in the range from 500 L to 5000 L, having various designs like open type, closed type or cylindrical or semi cylindrical. The important controls are the temperature of the milk, the intermittent operation of agitator, CIP cleaning cycle as an option etc. The design features of typical BMCs are shown below.

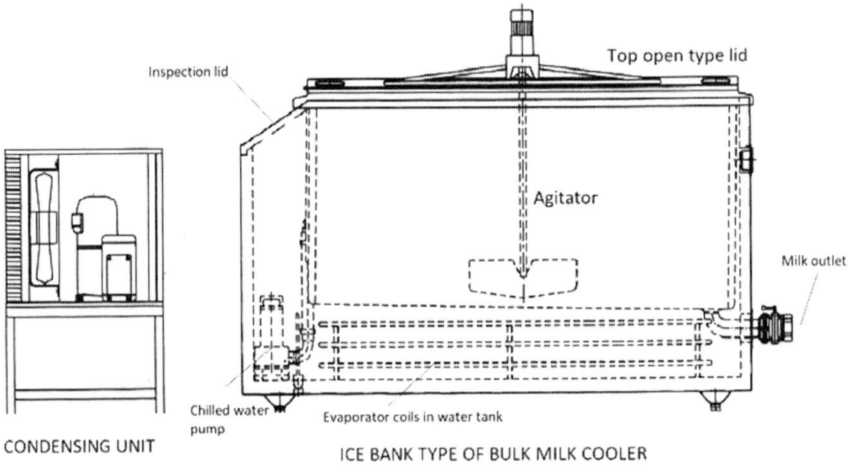

Figure 3.5 Bulk Milk Cooler with Condensing Unit

Figure 3.6 BMC Liked with Solar Panels and Battery Backup

Chapter 4

Can Washing, Bottle Washing and CIP Systems

Description and Working of Can Washers

The type and capacity of can washer used depends on the number of cans to be washed and the speed at which the cleaning operation is required to complete. Further, the size and type of can washer also determine the methods employed for washing of cans. Mainly three types of can washers are in use:

1. Hand washing or can scrubbing
2. Rotary can washer
3. Straight through can washer.

Manual Washing or Can Scrubber

This method is commonly adopted at all milk chilling centres where a smaller number of cans are washed. A single person can operate this unit conveniently. The can scrubber is illustrated in Figure 4.1(b). The can scrubber consists of a tank in which two nylon brushes are mounted on revolving shafts. The tank is made of 10 SWG G.I. sheet or M.S. sheet that is galvanised. It is painted with 'epoxy' paint to avoid corrosion. The extensions of shafts are properly encased in sealed enclosures and provided with suitable bearings and lubrication ports. Gland packing is used to prevent leakage through the shaft. A stationary nylon brush with a bracket is fitted on the inner wall for cleaning outside of the cans. Side stationary brush is replaceable. It can be fitted on either side of the inner wall. Cylindrical nylon brushes revolve at a low speed of about 80 to 100 rpm, in opposite directions by a motor and reduction gear unit. The brushes give a thorough and effective cleaning on either side of the can. Steam and water supply connections are given to the tank for washing operation. At the bottom of the tank a drain valve is fitted for easy draining and cleaning of the tank.

The can scrubber is prepared for can washing operation, first by filling it with water and then adding 0.8 per cent washing soda and 0.2 per cent tri-sodium phosphate. Steam is supplied to raise the temperature of the detergent solution to 55 °C. Each can is rinsed in warm water before it is washed with this equipment. After starting the motors to rotate the brushes, each can is slid into the position manually, and held for about 10 seconds. The can is then taken to the Can steaming block [Figure 4.1(a)], where it is rinsed with hot water at about 60 °C. Finally, steam is injected into the can for sterilization. The can Steaming block has a spout, over which the can to be sterilized is placed in an inverted position. The spout is connected to water and steam supply lines, which are controlled by pedal valves. The appropriate operation of pedals gives either hot water (when both the pedals are pressed) or only steam.

(a) (b)

Figure 4.1. (a) Can steaming block (b) Can washing tub

Operational Precautions: Steam supply should be given through a steam mixing battery to avoid vibrations. If the brushes fail to rotate, check the direction of rotation and correct it or tighten the idler wheel, provided on the chain. It is hard to insert the can on a new brush. In such case the nylon bristles should not be cut to reduce brush diameter to reduce the labour. Cutting of the brushes reduce the scrubbing and make cleaning ineffective. Periodically the lubrication and alignment of gearbox and sprockets should be checked to avoid breakdowns.

Straight through Can Washer

Sequence of Operation (Figure 4.2 a)

Draining

Before passing the can to the can washer milk, cream or any other fluid should be drained by passing the can in an inverted position over a drip pan before it is rinsed.

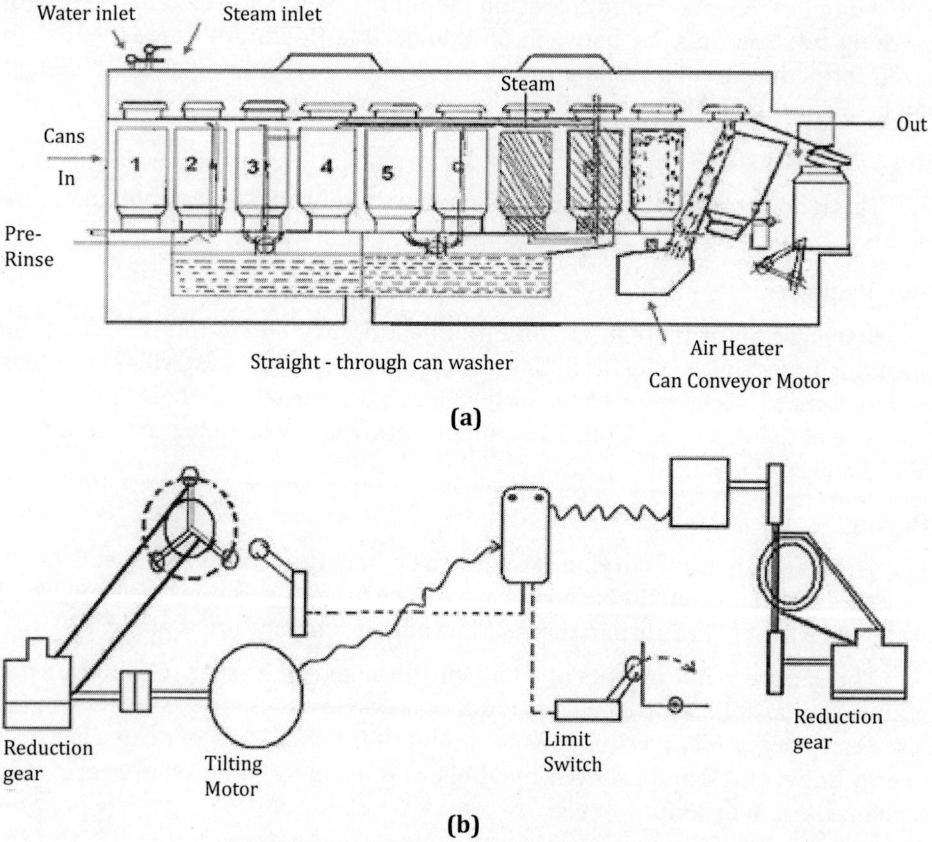

Figure 4.2a-b. Straight through Can Washer and Can Movement Control

Pre-Rinsing

Rinsing of can is done by passing it over a jet of water to remove the milk film remaining in the can. For proper rinsing, water at a pressure of 100 kPa to 200 kPa (1.0 to 2.0 kg/cm^2) is used for 3-6s.

Warm Water Rinsing

The cans are rinsed by clean hot water. The temperature of hot water is maintained at about 40-45 °C.

Caustic Rinsing

Caustic rinsing is done by passing the washing solution through jets at a sufficient pressure (about 200 kPa) to remove all the dried milk and cream film inside and outside of the cans. Cleanliness obtained depends on temperature and strength of washing solution. When using an alkaline solution, alkalinity should be less than 0.15 percent. Caustic soda should not be used as detergent, but sodium carbonate with a corrosion inhibitor is suitable. The temperature of washing solution must be between 65°C and 70°C. Operation time is 9-18s. To avoid formation of milk stone on cans, temperature of washing solution should not be higher than 70°C.

Hot Water Rinsing

This is done to remove the washing solution thoroughly. The water temperature in this section is about 85-90°C and this water is drained after use.

Sterilization

Sterilization of the can is done by injecting dry steam into the can. The steaming proceeds usually for 5-10 seconds. Steaming process sterilizes the cans and increases the can temperature which facilitates the ease and quick drying of the cans in the drying section. The temperature, quality of steam influences the effectiveness of steam.

Drying

The main purpose of drying of cans is to prevent the corrosion of metal due to moisture and to prevent the bacterial growth. The drying operation is accomplished by blowing a jet of hot air into the can. The hot air temperature is about 120 °C.

The service requirements of a straight-through can washer (6 to 8 cans per minute) are as follows: steam consumption is between 0.85 to 1.8 kg per can (at a pressure of 0.4 MPa), water consumption is 4.0 to 4.5 litres per can, electrical energy between 1.0 and 2.4 kWh, and boiler horsepower required to operate the machine is 23.0 to 43.0.

Controls

The sequential operation of can forward movement, and the can tilting back to upright position is achieved by electro-mechanical arrangement as shown in Figure 4.2.(b) The can movement has to be intermittent, so that can is positioned above the jet of hot water, detergent solution, steam and hot air remains above each jet for few seconds. When, the can reaches end of the can washer the can is tilted back to the upright position, during which the motor for can movement mechanism is stopped. When can tilting is completed, a lever operates a limit switch, which disconnects the motor for tilting mechanism and starts the motor for initiating the can movement. As the next can approaches the end of the conveyor, a lever operates another limit switch, which stops the can movement motor and starts the can tilting motor.

The temperatures of hot water and detergent solution are maintained by controls that operate the valves supplying steam to hot water and detergent solution tanks. Water for rinsing is heated by exhaust vapours from the can washer.

Maintenance

Dirty cans or cans with high bacterial count, or wet cans will exit the can washer if the washer working is improper or poorly maintained. The water valves should be unscrewed, cleaned and a water pressure (2.0 to 4.0 kg/cm^2) should be ensured. If the pressure is less it may be due to clogged jets and or defective pump. In such case the clogged jet should be cleaned and the pump should be inspected for packing leaks, clogging and worn parts. The pump strainer, inlet and outlet lines should be inspected for the accumulation of lime and scale. Majority of the mechanical problems can be avoided by proper, lubrication of the moving parts, to reduce wear and replacing the worn-out parts promptly.

The temperature and pressure of the washing solution, hot water, steam and drying air should be maintained at optimum level. The proper strength of the detergent solution is an important factor for proper washing. Heavy scale deposits in can washers are usually the result of inadequate every day cleaning, use of hard water and use of solution at incorrect temperature and strength. The scales can be removed with inhibited acid at a temperature of 50°C.

Bottle Washers and factors affecting Washing Operations

Soaker Type Bottle Washers

In this type of bottle washer, the bottles are loaded on to a continuous conveyor chain, having rows of bottle holders. The conveyor chain passes through various stages of cleaning and sterilizing such as: 1) Pre-soaking 2) Water jetting 3) Detergent soaking 4) Flushing 5) Rinse water jetting 6) Hot water jetting 7) Warm water jetting 8) Cold water jetting. The bottle washer may have loading and unloading of bottles on same end, or on opposite ends. The Figure 4.3(a) shows the bottle washer of soaker type with loading and unloading on same side (Make: Stork, Holland).

All the directional changes of the conveyor chain are guided by chain wheels, which are provided with bearing boxes of plastic material. The shafts are made of stainless steel. All the bearings are self-lubricating. The sections of the bottle washer where heating is required are provided with steam heated coils. The steam condensate from the heating coils is removed through steam traps.

The various sections of the bottle washer are so arranged that there is a definite time – temperature treatment in each section (Figure 4.3 b). The temperature variation never exceeds 25 °C in spite of high temperatures used for cleaning. This will prevent thermal shocks to the bottles, thereby reducing breakages. There are no acids involved in cleaning. The total time taken for complete cleaning of bottle while passing through all sections of the bottle washer is about 15 to 20 minutes. The cleaning time is determined by the speed of the conveyor chain movement and is adjustable.

Description of Operation

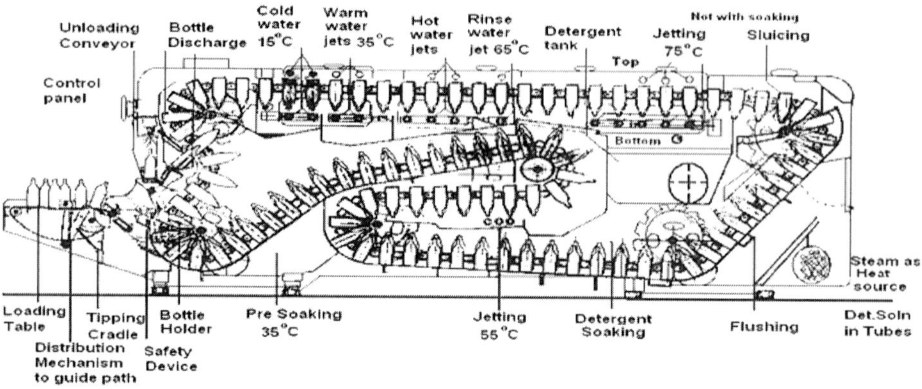

(a)

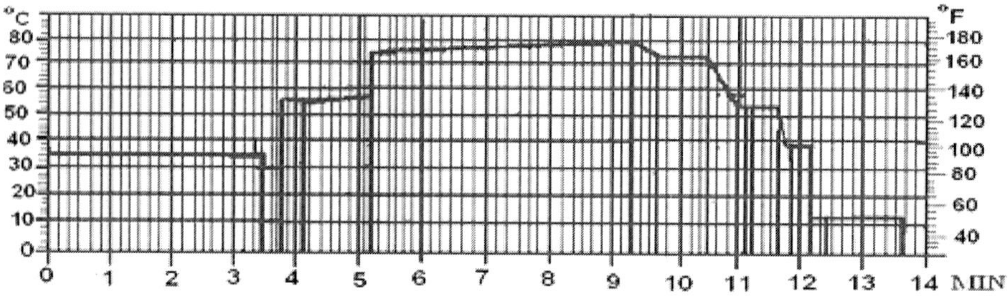

(b)

Figure 4.3. Soaker Type Bottle Washer (a) and (b)

The bottles are loaded on the loading table, which is a flat chain made of stainless steel. Then a distribution mechanism guides the bottles in to individual bottle holders in each row. Each row has 8, 12 or 16 receptacles, depending on the capacity of the bottle washer. The Tipping cradle ensures that the bottles are loaded in inverted position in the receptacles. The Tipping mechanism is also provided with a safety switch, which stops the movement of chain, when a bottle is stuck between the tipping cradle and the conveyor chain. This prevents the breakage of bottles when the conveyor moves to receive next set of bottles.

The conveyor chain moves the bottles first to the Pre-soaking section, where water at 35 °C, wets inside and outside of the bottle and removes residual milk film sticking to it. Next the bottles move to the jetting section, where hot water at 55 °C is sprayed into the bottles to dislodge any milk solids still sticking to the surface. The chain then moves the bottles into detergent soaking section, where detergent at 75 °C sprayed to dissolve the fat and greasy substances sticking to the surface of bottles. The detergent solution is flushed out of the bottles. Then the bottles move to the jetting sections, where the detergent solution under pressure is jetted into the bottles. The bottles then are moved to different water jetting sections, designed to progressively remove the detergent residue and cool the bottles. These include operation of rinse water jet (65 °C), hot water jet and warm water jets (35 °C). The bottles are finally washed with tap water at ambient temperature.

The bottles are discharged in vertical position on to a flat stainless steel belt conveyor and moved to the bottle filling section. A hood is provided on the conveyor to prevent external contamination. An observation point with a magnifying glass is provided at a convenient location along the conveyer for visual inspection of the bottles and removal of those with minor cracks, chipping at the mouths or at bottom. This helps in avoiding further damage and breakage of bottles during filling and subsequent operations of crating, storage and loading.

Maintenance

The proper maintenance of bottle washer is essential. It includes periodical inspection, repair and replacement of worn-out parts. The clogging of jets occurs frequently. Therefore, the jets should be inspected by observing them daily, through the leak proof inspection doors provided on the sides of the enclosure. Safe lighting is provided inside the bottle washer for this purpose. The specified steam and water jet pressures should be maintained. If scaling occurs on the motors or pipelines it should be removed. The spillage of water jets outside of the bottles must be avoided. This problem occurs due to insufficient or excess water pressure. The free and smooth movement of chain conveyor should be ensured by preventing excess friction at various wheels provided for the change of direction. The steam condensate traps must be inspected for proper operation and draining of the condensate. The strainers provided before the jetting rows should be cleaned regularly.

Design and Cleaning of the CIP System

The cleaning and sanitization of dairy equipment are essential. The Cleaning –in-Place (CIP) operation involves circulation or recirculation of water and chemical cleaning solutions through the equipment which in completely assembled state. All the contaminated surfaces are cleaned to high and consistently reproducible standards. The CIP operation involves use of water and cleaning solutions at much higher concentrations and temperatures than those employed in manual cleaning, The filters should be removed during CIP cleaning to prevent damage

For effective CIP cleaning the following are important: 1) the design of plant, materials used and the ancillary equipment should be suitable for CIP cleaning. 2) The cleaning procedure to be adopted is determined by the type of soil and detergent, the concentration and temperature of detergent, contact time, type of the cleaning surface. 3) The CIP cycle must function effectively and automatically with minimum human intervention, 4) some supervision is important although high degree of automation is available.

The favourable effect of increased temperature on cleaning is due to:

1. Loosening of bond between soil and the surface, and improvement in wetting.
2. A greater degree of turbulence due to decreased viscosity
3. Increased solubilisation of constituents of soil with higher solubility *e.g.* lactose, minerals
4. Increased chemical reaction rate which greatly increases the cleaning

It is a common practice to keep the minimum temperature of the detergent solution 3 - 5°C more than the melting point of milk fat. The maximum temperature of the detergent solution is dependent upon the denaturation of the milk proteins.

The turbulence of detergent solution influences the cleaning efficiency. A velocity of 1.5 m/s (5 feet/second) of the detergent solution at the surface is satisfactory. The hardness of water should kept be below 50 – 60 ppm for effective cleaning. Presence of Iron above 3 ppm and Manganese above 0.2 ppm in water causes stain to the surface under cleaning. Hard water employed for cleaning leads to sludge and scale formation, resulting reduced heat transfer rate, clogging up of jets, and wearing of moving parts. Cleaning becomes difficult when hardness of water is above 200 ppm.

Various Types of CIP Cleaning

There are two types of CIP cleaning systems. These are Single use and Re-use systems. Throwing away of the cleaning solutions after use as in Single use system is not in practice. Currently there are combined systems which take advantages of both single use and reuse systems.

Single Use System

It is used for cleaning one process circuit at a time. All the cleaning liquids are drained on completion of cleaning. This method of cleaning is used for units like HTST pasteurizer. The equipment to be cleaned and cleaning system should be close. A Typical cycle includes 1) Three Pre- rinses each of 20 s duration with a time gap of 40 s to remove gross soil contamination.2) Cleaning with a detergent solution at certain temperature for 10 to 12 min. 3) Two rinses with cold water 4) Rinse with an acid solution with at pH about 4.5 to 5.0, for 3 min. 5) Flushing of acid remnants with water for 5 min.

Cleaning cycle for cold milk equipment is slightly different. 1)Alkaline cleaning solution of 0.1-0.2 per cent concentration 2) Surface active and anti – foaming agents are added (up to 0.08 per cent). 3) Liquid chlorine sanitizing agent is also added to obtain a concentration of 50 to 100 ppm. 3) Flushing away of the remnants of detergent solution with water.

Re-Use CIP System

In this system, the cleaning solutions acid, and alkali are reused by circulating them between the equipment and solution storage tanks. The concentration and temperature of the solutions are maintained in these tanks. Due to Pre – Rinses, pollution from detergents is much less.

The soil to be cleaned primarily consists of milk or milk product residues. These residues may be more or less modified by the processing treatment. The soil may be one or more of the following: liquid milk films, air-dried films, heat precipitated films, heat-hardened films, milk-stone and miscellaneous foreign matter. The following cleaning stages are arranged in succession in cleaning applications. The sequencing of various stages is dependent on the nature of the soiling matter:

W →C→W

W→C→W→A→W

W→A→W→C→W

W→A→W→C→W→A→W

C= Caustic rinse; A= acid rinse, W= water rinse

An acid rinse can be omitted if there is no milk stone deposit, such as in can washing and storage tank cleaning. The positioning of the acid rinse in the cleaning sequence depends on the position of the milk stone, *i.e.* whether it is above, below or in the middle of the deposits.

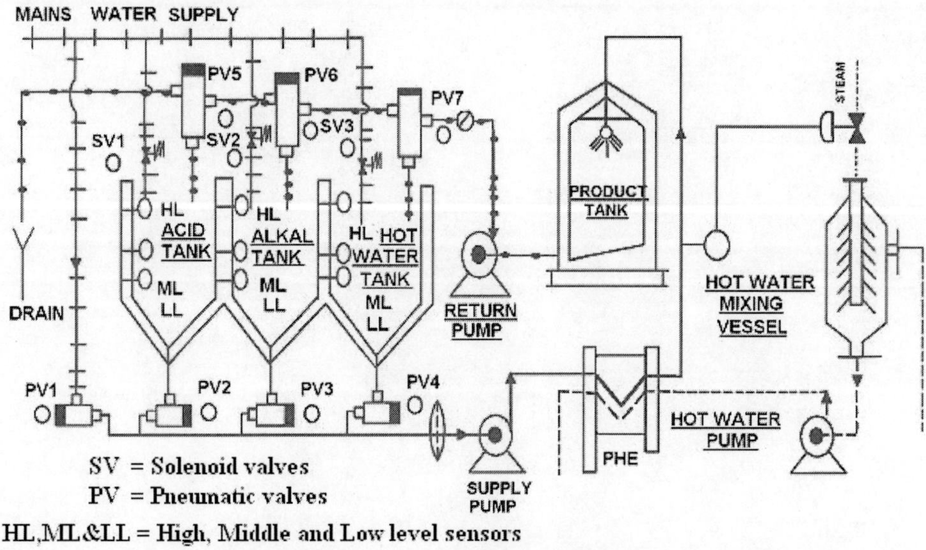

Figure 4.4. Automated CIP System

For cleaning with acid solutions based on Nitric acid, the concentration of acid should not exceed 5 per cent. The solution temperature can be up to 100 °C and contact time is up to 1 hour. However, temperatures above 80 °C should not be used with nitric acid solution to prevent attack on gaskets. At times, discolouration of surface (yellow, brown, dark blue) may occur at higher temperatures with alkaline as well as acid cleaning, but this is not an indication of corrosion.

Automation of CIP Cleaning

As the operation of CIP cleaning is inherently hazardous and not without danger of leakages and potential harm to operating personnel and material, its operation is automated in most of the dairy plants. The arrangement of tanks piping and control valves in a CIP cleaning system are shown in Figure 4.4.

The sequence of operation and timing for pumping of solutions are controlled by a programmable PLC controller. The sequence for circulation of cleaning solutions, and the time duration of circulation can be selected in the program. Such operation is made possible because of electro-pneumatic valves installed both at the inlet and outlet of the tanks. A Plate Heat Exchanger is used in maintaining the temperatures of the cleaning solutions. The levels and concentration of cleaning solutions in each tank are sensed continuously. Filling up the tanks and maintaining the concentration of solutions in the tanks is done automatically.

Chapter 5

Separation Equipment

Mechanical Separation Fundamentals Involved in Separation

Introduction

In the manufacture of various food products, separation of useful components from waste is an essential preliminary step. Efficient separation processes have been instrumental in economical manufacture of useful components. Some of the physical separation processes are Sieving, Filtration, Membrane separation, Gravity separation, Centrifugal Separation *etc.* They may also be classified broadly in to categories of Liquid- Liquid, Liquid – Gas, Solid – Solid, Solid- Gas separations. Each operation would involve specific application for separation.

Some of the common operations in dairy industry are the separation of fat from milk, separation of extraneous matter from milk, separation of dried powder from cooling air, separation of fine powder particles from exhaust air of spray drier, filtering moisture from various dairy products *etc.*

Filtration is the process of passing a fluid containing suspended particles through a porous medium. The filter medium traps the suspended solids producing a clarified filtrate. Filtration is employed when the valuable component of the mixture is the filtrate. The liquid is forced through the porous membrane.

Gas-Solid Separations

Gas-Solid separation involves the application of particle mechanics to the design and application of dust-collection systems. It includes wet collectors, or scrubbers, for particle collection. It could involve equipment for removing entrained liquid mist from gases.

Purpose of Dust Collection

Dust collection is concerned with the removal or collection of solid dispersoids in gases for purposes of:

1. Air-pollution control, as in fly-ash removal from boiler flue gases
2. Equipment-maintenance reduction, as in filtration of engine intake air for air compressors *etc.*
3. Safety- or health-hazard elimination, as in collection of siliceous and metallic dusts, some operations and flour dusts from milling or bagging operations.
4. Product-quality improvement, as in air cleaning in the production of long life products like UHT milk.
5. Powdered-product collection, as in pneumatic conveying; the spray drying of milk, eggs, and soap; and the manufacture of high purity zinc oxide and carbon black.

Measurements of the concentrations and characteristics of dust dispersed in air or other gases may be necessary (1) to determine the need for control measures, (2) to establish compliance with legal requirements, (3) to obtain information for collector design, and (4) to determine collector performance.

The basic operations in dust collection by any device are (1) separation of the gas-borne particles from the gas stream by deposition on a collecting surface; (2) retention of the deposit on the surface; and (3) removal of the deposit from the surface for recovery or disposal. The separation step requires (1) application of a force that produces a differential motion of a particle relative to the gas and (2) a gas retention time sufficient for the particle to migrate to the collecting surface. The principal mechanisms of aerosol deposition that are applied in dust collectors are (1) gravitational deposition, (2) flow-line interception, (3) inertial deposition, (4) diffusional deposition, and (5) electrostatic deposition.

Cyclone Separators

The most widely used type of dust collection equipment is the cyclone separators, in which dust-laden gas enters a cylindrical or conical chamber tangentially at one or more points and leaves through a central opening. The dust particles, by virtue of their inertia, will tend to move toward the separator wall, from which they are led into a receiver. A cyclone is essentially a settling chamber in which gravitational acceleration is replaced by centrifugal acceleration. At operating conditions commonly employed, the centrifugal separating force or acceleration may range from 5 times gravity in very large diameter, low-resistance cyclones, to 2500 times gravity in very small, high-resistance units. The entrance to a cyclone is usually rectangular.

Venturi Scrubbers

The venturi scrubber is one of the most widely used types of particulate scrubbers. The designs have become generally standardized, and units are manufactured by a large number of companies. Venturi scrubbers may be used as either high- or low energy devices but are most commonly employed as high-energy units. The units originally studied and used were designed to the proportions of the classical venturis used for metering, but since it was discovered that these proportions have no special merits, simpler and more practical designs have been adopted. Most "venturi" contactors in current use are in fact not venturis but variable orifices of one form or another. Any of a wide range of devices can be used, including a simple pipe-line contactor. Although the venturi scrubber is not inherently more efficient at a given contacting power than other types of devices, its simplicity and flexibility favor its use. It is also useful as a gas absorber for relatively soluble gases, but because it is a concurrent contactor, it is not well suited to absorption of gases having low solubilities.

Current designs for venturi scrubbers generally use the vertical down flow of gas through the venturi contactor and incorporate three features:

(1) a "wet-approach" or "flooded-wall" entry section, to avoid dust buildup at a wet-dry junction; (2) an adjustable throat for the venture (or orifice), to provide for adjustment of the pressure drop; and (3) a "flooded elbow" located below the venturi and ahead of the entrainment separator, to reduce wear by abrasive particles. The venture throat is sometimes fitted with a refractory lining to resist abrasion by dust particles.

Liquid – Solid Separations

These operations involve, the unit operations like Crystallization, Leaching, Gravity sedimentation, Filtration, Centrifugal separation *etc.* Of these Filtration and Centrifugal separation involve mechanical action in separation.

Sedimentation is the partial separation or concentration of suspended solid particles from a liquid by gravity settling. This field may be divided into the functional operations of thickening and clarification. The primary purpose of thickening is to increase the concentration of suspended solids in a feed stream, while that of clarification is to remove a relatively small quantity of suspended particles and produce a clear effluent. These two functions are similar and occur simultaneously, and the terminology merely makes a distinction between the primary process results desired. Generally, thickener mechanisms are designed for the heavier-duty requirements imposed by a large quantity of relatively concentrated pulp, while clarifiers usually will include features that ensure essentially complete suspended-solids removal, such as greater depth, special provision for coagulation or flocculation of the feed suspension, and greater overflow-weir length.

Clarifiers

Continuous clarifiers generally are employed with dilute suspensions, principally industrial process streams and domestic municipal wastes, and their primary purpose is to produce a relatively clear overflow.

They are basically identical to thickeners in design and layout except that they employ a mechanism of lighter construction and a drive head with a lower torque capability. These differences are permitted in clarification applications because the thickened pulp produced is smaller in volume and appreciably lower in suspended solids concentration, owing in part to the large percentage of relatively fine (smaller than 10 mm) solids.

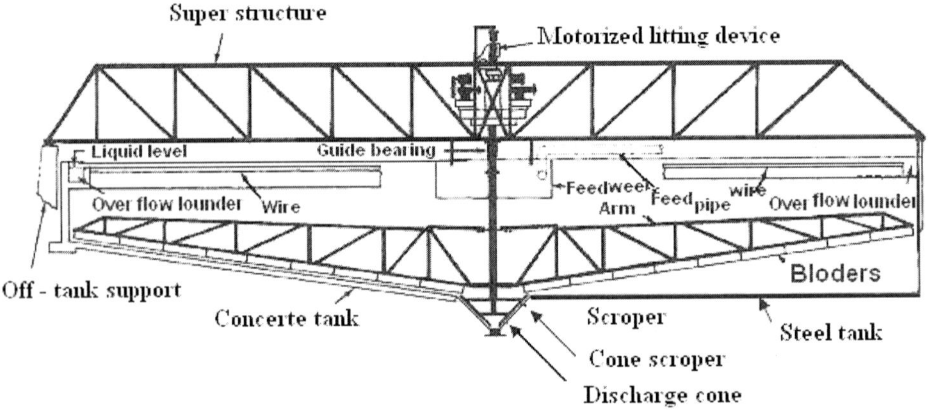

Figure 5.1. Clarifiers

Rectangular Clarifiers

Rectangular clarifiers are employed primarily in municipal water and waste treatment plants, as well as in certain industrial plants, also for waste streams.

Circular Clarifiers

Circular units are available in the same three basic types as single-compartment thickeners: bridge, center-column, and peripheral-traction. Because of economic considerations, the bridge-supported type is limited generally to tanks less than 20 m in diameter. A circular clarifier often is equipped with a surface-skimming device, which includes a rotating skimmer, scum baffle, and scum-box assembly

Lamella or tilted-plate separators have achieved increased use for clarification. They contain a multiplicity of plates inclined at 45 to 60° from the horizontal. Various feed methods are employed so that the influent passes into each inclined channel at about one-third of the vertical height from the bottom. This results in the solids having to settle only a short distance in each channel before sliding down the base to the collection zone beneath the plates. The clarified liquid passes in the opposite direction beneath the ceiling of each channel to the overflow connection.

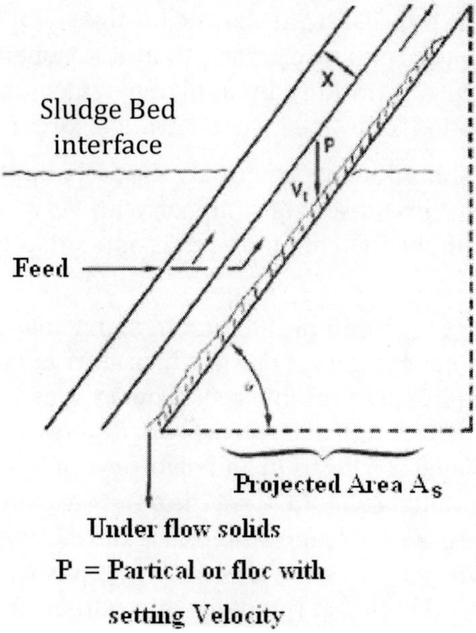

Figure 5.2. Tilted Plate Clarifiers

Principles Involved in Filtration

Introduction

Filtration may be defined as that unit operation in which the insoluble solid component of a solid-liquid suspension is separated from the liquid component by passing the latter through a porous membrane or septum which retains the solid particles on its upstream surface, or within its structure, or both. A broader definition could be that the fluid may be a liquid or a gas; the valuable stream from the filter may be the fluid, or the solids, or both. The solid – liquid suspension is known as the feed slurry or 'prefilt', the liquid component that passes through the membrane is called the filtrate and the membrane itself is referred to as the filter medium. The separated solids are known as the filter cake, once they form a detectable layer covering the upstream surface of the medium. Often the feed is modified in some way by pre-treatment to increase the filtration rate, as by heating, re-crystallizing, or adding a "filter aid" such as cellulose or diatomaceous earth.

Filter, must provide a support for the filter medium, a space for the accumulation of the solids, channels for the introduction of the feed. A means of inducing the flow of filtrate through the filter and medium must also be provided.

Classification of Filtration on basis of Force

The fluid flows through a filter medium by virtue of a pressure differential across the medium. The flow of filtrate may be brought about by means of gravity alone, by the application of a pressure greater than atmospheric pressure upstream of the medium (pressure filtration), by applying a vacuum downstream of the medium (vacuum filtration) or by means of centrifugal force (centrifugal filtration).

Filtration under the influence of gravity alone is limited in application to slurries containing very free-draining solids or with very low solids contents. It has a very limited use in the food industry but is applied to the water and sewage treatment.

The applications of filtration in the food industry may be considered to fall into three categories. First category is one in which slurries containing appreciable amounts of insoluble solids are present. In such operations, a cake is formed on the upstream surface of the medium and the process is known as cake filtration. The second category is termed clarification and involves removing small quantities of insoluble solid from a valuable liquid. Here the object usually is to produce a clear liquid and the solids are generally unwanted. The third category, often referred to as microfiltration, involves the removal of very fine particles, of the order of 1μ or less, and is generally directed at removing microorganisms from liquid foods.

In the Dairy Industry, the application of filtration is during:

1. Filtration of raw milk at dump tank, either at milk collection centre or at chilling centres. They may be open to atmosphere or in-line filters.
2. During processing, while passing through pasteurizer
3. Reconstitution of milk powder
4. Removal of ghee residues from ghee
5. Separation of whey during cheese manufacture
6. Separation of whey during preparation of Channa, an indigenous dairy product.

Rate of Filtration

In a continuous process, the rate of filtration can be expressed in relation to the pressure, area of filtration, properties of the fluid *etc.*

1. The basic filtration equation is

$$\frac{dv}{d\tau}=\frac{(-\Delta P)A}{\mu\left[\sigma\frac{W}{A}+r\right]} \qquad \text{(Eqn.5.1)}$$

W= w. V

where, w= mass of dry cake per unit volume of filtrate, kg/m^3 filtrate

$$\therefore \frac{dv}{\delta\tau} = \frac{(-\Delta P)A^2}{\mu[\sigma.w.V + A..r]} = \frac{(-\Delta P)A^2}{\sigma.\mu.w\left[V + \frac{A.r}{\sigma.w}\right]} \quad \text{(Eqn.5.2)}$$

where,

$(-\Delta P)$ = Pressure – drop across the filter medium, N/m^2 or Pa

V= volume of filtrate collected in time τ, m^3

τ = duration of filtration, *s*

A= area of filtering surface, m^2

μ = dynamic viscosity of the filtrate, Pa.s or kg/m.s

σ = average specific cake resistance

W= mass of cake on dry basis, kg

r= resistance of the filter medium per unit area.

2. Rate of filtration can also be expressed in other forms. Transposing and integrating Eqn. 5.2. We get

$$\mu.\sigma\frac{w}{A}\int_0^v V.dv + r.\mu\int_o^v dV = (-\Delta P).A.\int_o^\tau d\tau \quad \text{(Eqn.5.3)}$$

Or
$$\mu.\sigma.\frac{w}{A}.\frac{V^2}{2} + r.\mu.V = (-\Delta P).A.\tau \quad \text{(Eqn. 5.4)}$$

Or
$$\frac{V}{\tau} = \frac{(-\Delta P)A}{\mu.\frac{\sigma}{2}.\frac{W}{A} + r.\mu} \quad \text{(Eqn. 5.5)}$$

This is also an expression for constant-pressure process of filtration and is valid only for those slurries that form incompressible cakes *i.e.*

σ = constant

If the cloth resistance is neglected, the equation simplifies to

$$\frac{V}{\tau} = \frac{2(-\Delta P)A}{\mu.\sigma.W}.A^2 \quad \text{(Eqn. 5.6)}$$

(3) Constant- Rate Filtration

For this process, the rate of filtration is constant *i.e.*

$$\frac{dv}{d\tau} = \text{constant} = \frac{Vo}{\tau_o} = \frac{V}{\tau} \quad \text{(Eqn. 5.7)}$$

Therefore, integrating the basic equation of filtration, we get for constant-rate process (*i.e.* constant – rate of slurry feed to the filter)

$$\frac{V}{\tau}=\frac{(-\Delta P)A^2}{\mu\left[\sigma.\frac{W}{A}+r\right]}=\frac{(-\Delta P)A^2}{\mu.\sigma.w.V+\mu.r.A} \qquad \text{(Eqn. 5.8)}$$

Or, $$\frac{V}{\tau}\left(\mu.\sigma.\frac{W}{A}+\mu.r\right)=(\Delta P)A$$

Or

$$\frac{V}{\tau}.\mu.\sigma.\frac{W}{A}=(-\Delta P)A-\mu.r.\frac{V}{\tau}=(-\Delta P)A-P_1.A=[(-\Delta P)-P_1]A \qquad \text{(Eqn.5.9)}$$

where, P_1 = Pressure – drop through the filter medium

$$=\frac{\mu.r}{A}.\frac{V}{\tau}$$

$$\therefore\frac{V}{\tau}=\frac{[(-\Delta P)-P_1]}{\mu.\sigma.w}A^2 \qquad \text{(Eqn. 5.10)}$$

This is the expression for constant-rate filtration process.

If the resistance of the filter medium is neglected, then equation simplify to

$$\frac{V}{\tau}=\frac{(-\Delta P)}{\mu.a.W}.A^2=\frac{(-\Delta P)}{\mu.\sigma.w.V}.A^2 \qquad \text{(Eqn. 5.11)}$$

Vertical Plate Pressure Filters (Filter Presses)

A vertical drainage plate supporting a filter medium is the basic filtering element in a vertical plate press. A commonly used design in the plate-and -frame press is shown in Figure 5.3. In this type, the grooved plates, covered on both sides with filter medium, alternate with frames in a rack. The assembly of plates and frames can be squeezed tightly together by a screw, hydraulic or pneumatic mechanism to form a liquid-tight unit. The filter medium also acts as a gasket, preventing leakage between the plates and frames. Both plates and frames are provided with openings at one corner and when the press is closed these openings form a channel through which the feed slurry is introduced. In addition, the hollow centre of each frame is connected by an auxiliary channel to this feed channel.

The feed slurry enters the frames and the cake builds up in the hollow centre of the frames. The filtrate passes via an outlet channel in each plate. Filtration is

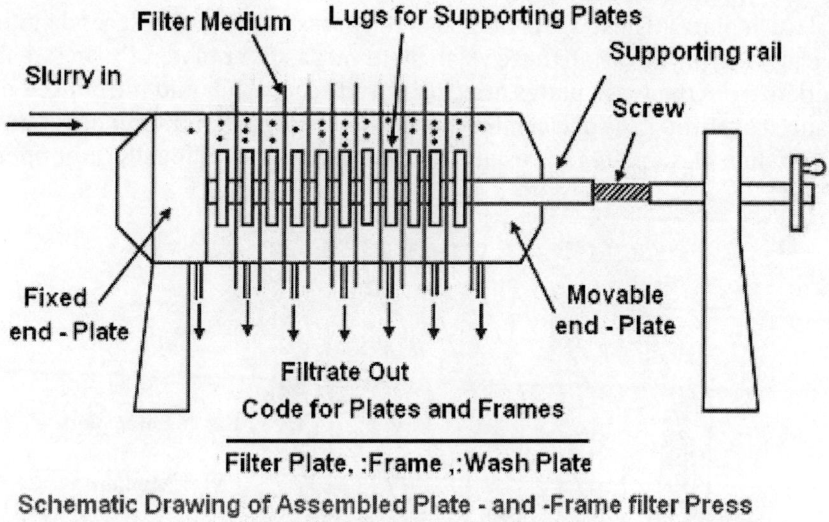

Figure 5.3. Plate and Frame Filter

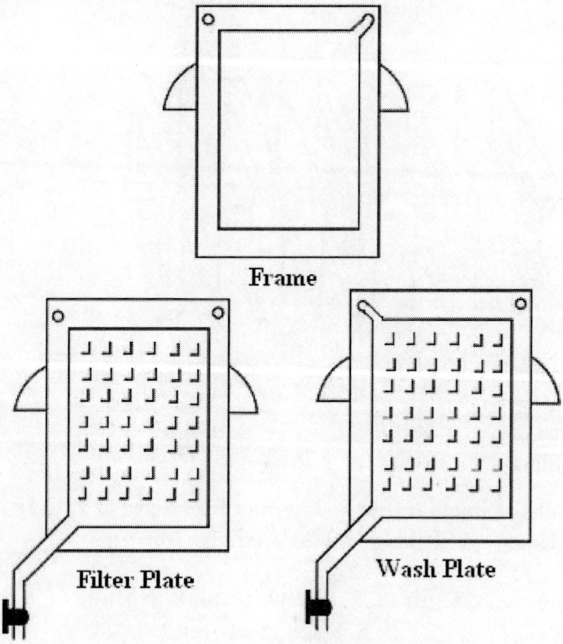

Figure 5.4. Filter Plate and Wash Plate

continued until the flow of filtrate drops below a practical level or the pressure reaches an unacceptably high level, due to the cake packing tightly in the frames. After filtration, washing of the cake may be carried out by replacing the flow of feed slurry with wash liquid. However, more effective washing is obtained by the use of special wash plates. These are arranged in the press so that every second plate is a wash plate. During filtration these wash plates act as filter plates. During washing the outlets from the wash plates are closed and the wash liquid introduced on to their surfaces through a special inlet channel. The flow path for both filtration and washing when wash plates are used. The cake is removed manually after opening the press.

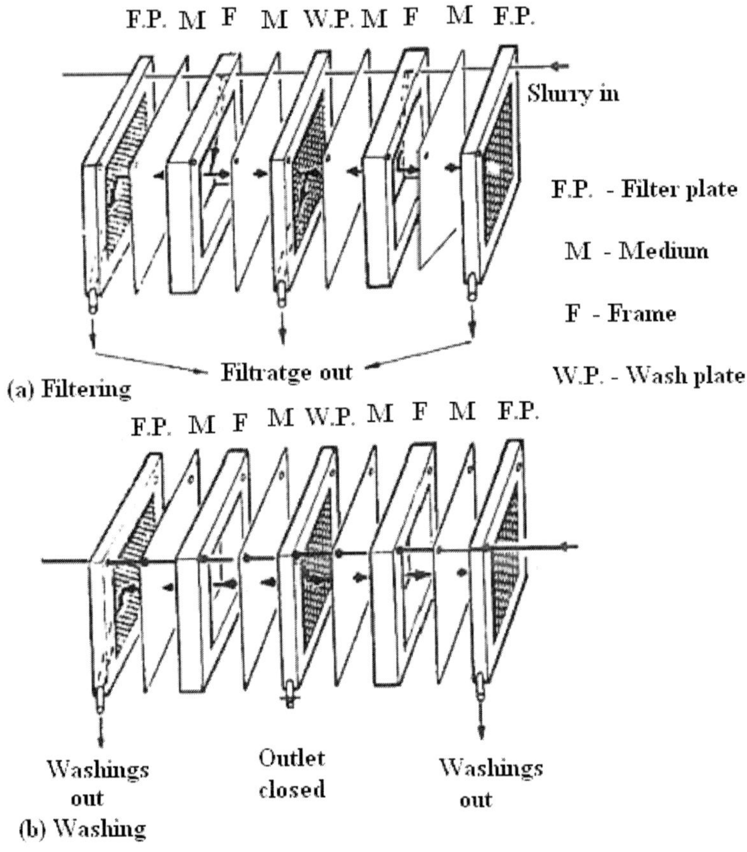

Flow paths of liquid through plate-and-farme press, fitted with wash plates,during (a) filtering and (b) washing.

Figure 5.5. Filtering and Washing

Another common design of filter press is known as the recessed-plates press. In this filter the cake accumulates within recesses in the plates and no frames are used. The feed is usually introduced centrally and removed via an outlet on the corners of each plate. For washing, the feed slurry is replaced by wash liquid entering through the same inlets. Many other types of vertical plate filters are available. Larger presses have facilities for lifting or moving plates and frames mechanically.

The vertical plate filter has found very wide application in industry. It is simple in design and operation, compact, flexible and can be used to handle a wide variety of types of slurry. It is relatively cheap initially. On the other hand labour costs and filter cloth consumption are high and washing of the cake is not always efficient.

Separation of immiscible liquids by Gravity Settling

Separation of components of immiscible liquids by use of density difference is a well-established method. In the dairy industry, the application of centrifugal force is done particularly in separating fat from rest of the milk. It is also being used in separating bacteria from milk.

Different components of milk have the following specific gravities:

Fat: 0.93 gm/cc

Skim milk: 1.036 gm/cc

Whole milk: 1.028 to 1.032 depending on composition

Stoke's Law

The rapidity of separation depends on various factors, and can be estimated by use of Stoke's law for gravity separation (Eqn. 5.14) modified to include the centrifugal force, instead of gravitational force. If the density of the particle is higher than that of the surrounding medium it sinks and if it is lower it rises. The particle remains in suspension if densities are equal.

Considering the particle in the Figure 5.6, the particle is acted upon by two forces, *viz.* Gravitational force and Viscous force. The Gravitational force pulls the particle down, and the viscous force is resisting the movement of the particle. The particle will come down, if its density is greater than the fluid, and it will move up, if its density is less than that of the fluid surrounding it. The expression for these forces are given below.

The Gravitational Force is Expressed as:

$$\frac{\pi d^3}{6}.(\rho_s - \rho_f).g \qquad \text{(Eqn. 5.12)}$$

where, d= Diameter of the sphere

ρs = Density of serum mass

ρ*f* = Density of sphere

G = Gravitational constant

The Viscous Force is expressed as:

Co – efficient of resistance

$$\frac{\pi d^2}{4}.\left(\frac{\rho_s}{2}\right)V^2 \qquad \text{(Eqn.5.13)}$$

where,

V = Velocity of separation of sphere

Since the movement of particle is generally down, the co-efficient of resistance can be

$$=\frac{24}{R_e}=\frac{24.\mu}{\rho dv}$$

$$\frac{\pi}{6}d^{3^2}(\rho_s-\rho_f).g=\frac{\pi d^2}{4}.\frac{\rho_f}{2}.V^2.\frac{24^{12^3}}{V.d\rho_f}$$

$$\frac{d^2(\rho_s-\rho_f)}{6\times 3}.g=V.\mu \qquad \text{(Eqn. 5.14)}$$

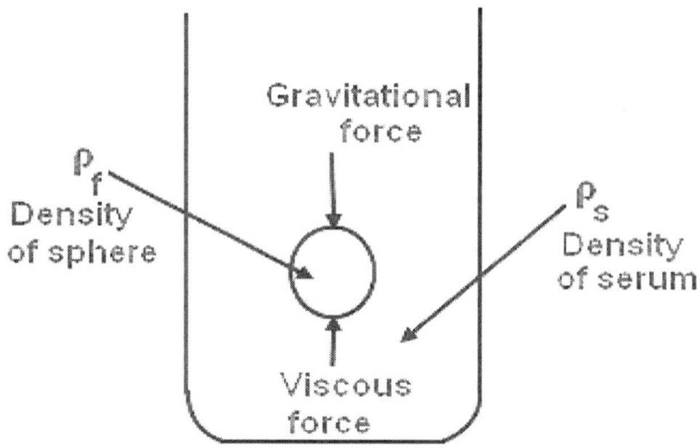

Figure 5.6. Viscous Force

Velocity of fat globule rise under gravitational force,

$$V = d^2 g(\rho_s - \rho_f)/18\mu \qquad \text{(Eqn. 5.15)}$$

where,

V is in cm/s

d = diameter of fat globule (cm)

g = gravitational constant, 981 (cm/s^2)

ρ_s = density of serum or skim milk, 1.037 (gm/cm^3)

ρ_f = density of fat globule, 0.92 (gm/cm^3)

μ = co – efficient of absolute viscosity poise (gm/cm.s)

For Milk it is 2.12 Cp at 21°C or 1.04 x 10^{-3} kg/m.s

Or 1.79 x 10^{-3} Pa/s

REFERENCE

Perry's Chemical Engineer's Handbook, (1999), 7th Edn.,MacGraw Hill Inc, New Delhi

Chapter 6

Centrifugal Separation Equipment

The solid-liquid separation under gravitational force is a slow process, and this method is not suitable to separate fat from milk. To hasten the process, application of centrifugal force is necessary.

The Stoke's law employed for gravity separation can be applied for use with centrifugal force, by replacing the term 'g', from the Eqn. 5.14 with the term representing centrifugal force.

Noting that the angular acceleration= $\omega^2 R = , \left(\frac{2\pi n}{60}\right)^2 R$ the Eq. 5.14 modifies to

$$v = \frac{c.d^2 n_s^2 R(\gamma_s - \gamma_f)}{\mu} \quad \text{(Eqn.6.1)}$$

In the above equation,

d= diameter of the smallest fat globule to be separated,cm

n = speed of rotation,1/s

γ_s and γ_f = the densities of serum and fat, gm/c.c

μ= viscosity of milk, centipoise

R = Radius of bowl,cm

C = a constant

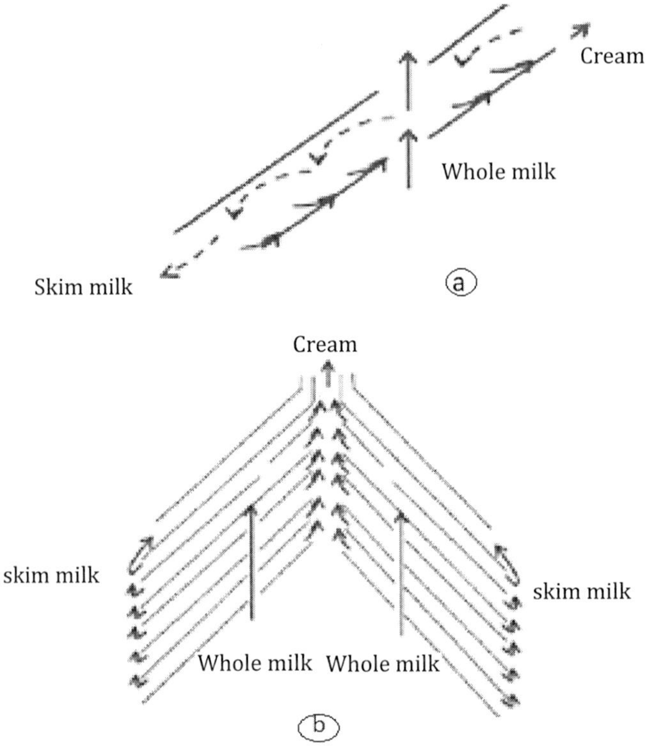

Figure 6.1a-b. Principle of Centrifugal Cream Separation

Energy required for the Cream Separator

The approximate power required to the cream separator assembly can be determined by assuming the stack of discs with the bowl as a hollow cylinder of uniform density as shown in Figure 6.2. The centrifugal force from the bowl, W can be written as,

$$W = \frac{1}{2}\omega^2 \sum (mR^2) \qquad \text{(Eqn. 6.2)}$$

In the above,

$\Sigma(mR^2)$ = moment of inertia

where,

m=mass of a particle

R= distance of particle from axis

$\omega = 2\pi n_s$ for bowl

$$W_B = \pi p_B n_s^{\ 2}(R_0^4 - R_i^4)h$$

where,

h = height of bowl

R_i, R_o = Inner and outer radii

n_s =Rounds per second

ρ_B = density of bowl

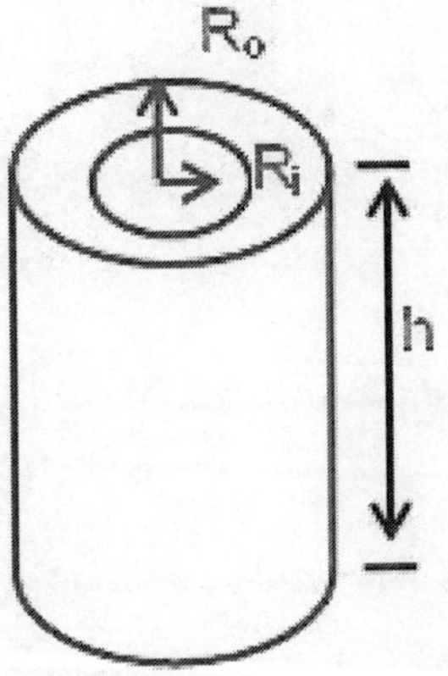

Figure 6.2. Hollow Cylinder with Uniform Density.

The power required to rotate this cylinder of uniform density can be expressed by dividing W_B with the term "t",

$$\text{Power} = \frac{W_B}{t} \qquad \text{(Eqn. 6.3)}$$

Capacity of Centrifugal Separator

The capacity of the centrifugal separator depends on the number and size of discs, speed of rotation, the viscosity of the fluid *etc.* An expression for the throughput of a centrifugal cream separator can be obtained. It is possible to obtain effective cream separation if one succeeds in transporting fat globules separated by centrifugal force to the upper surface of one of the discs while the milk flows through the spaces between the discs. The skim milk travels outwards towards the underside of the truncated cone-shaped discs. The fat globules move to the upper surface of the disc and form a film which is driven inwards towards the axis of rotation. The flow paths of the cream and skim milk, as explained here are illustrated in 6.1 (a) and 6.1 (b).

The Figure 6.3 shows an instance during cream separation when a fat globule in milk enters at the radius R_n, in the space between the discs and it must be separated

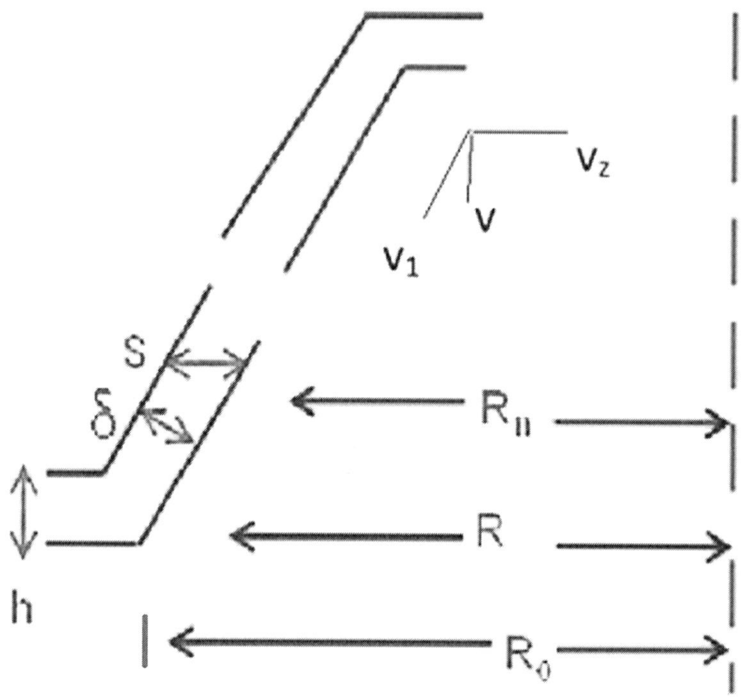

Figure 6.3. Capacity of the Centrifugal Separator.

from milk, before it reaches R_0. Two forces are acting on the fat globule: The drag force V_1 due to Volumetric Flow Q (M^3/Sec)

$$V_1 = Q/(2\pi R \delta z) \qquad \text{(Eqn. 6.4)}$$

Velocity of flow due to centrifugal force is given by,

$$Vz = d^2 \frac{(\rho_s - \rho_f)}{18\mu}\left(\frac{2\pi n}{60}\right)^2 R \qquad \text{(Eqn. 6.5)}$$

where,

n = rounds per minute.

The actual velocity V and the direction of the particle can be derived from the velocity vectors V_1 and V_z (Figure 6.3). In the figure, S is the largest distance between the discs the fat globule to be separated from the milk, has to travel horizontally across the stream of milk.

Let in time dt, the volumetric flow rate be 'Q' flows.

$$dt.\ Q = 2\pi R\, dR\, h\, Z \text{ or } dt = 2\pi R\, dR\, h\, Z/Q \qquad \text{(Eqn. 6.6)}$$

Z = No. of gaps or one less than the number of discs including top disc.

During the same time let the fat globule travel a distance 'ds'.

$$ds = Vz\,dt = \frac{d^2 \Delta\rho}{18\mu}\left(\frac{2\pi n}{60}\right)^2 R.\frac{2\pi R\,dR\,h\,z}{Q}$$

(Eqn. 6.7)

Choosing 'S' as the upper limit for a particular fat globule to travel longest horizontal distance and integrating both sides.

$$\int_o^s ds = \frac{d^2 \Delta\rho}{18\mu}\left(\frac{2\pi n}{60}\right)^2 \frac{2\pi\,h\,z}{Q}\int_{R_n}^{R_o} R^2.dR$$

(Eqn. 6.8)

$$S = \frac{d^2 \Delta\rho}{18\mu}\left(\frac{2\pi n}{60}\right)^2 \frac{2\pi\,h\,z}{Q}\left(\frac{R_0^3 - R_n^3}{3}\right)$$

(Eqn. 6.9)

$$Q = C\frac{(2\pi)^3}{3}\ z.\tan\alpha\ (R_0^3 - R_n^3)\ n_s^2\ \frac{d^2 \Delta\rho}{18\mu}$$

(Eqn. 6.10)

In Eq. 6.10, $\tan\alpha = \frac{h}{s}$

n = speed of rotation.

The value of C, varies between 0.5 to 0.7

α =usually varies from 45 to 60°

The constant 'C' is added to account for the deviations caused by,

1. Velocity profile is not laminar
2. Reduced cross sectional area due to back flow of lighter phase
3. Non- ideal distribution over the feed inlet
4. Geometric inaccuracies in the stacks of discs.

The position of neutral zone or the location where the milk enters the gaps between adjacent discs will influence the ease of separation of the denser and lighter phases of the fluid. In applications such as Clarifier, where the component to be separated is the particles of dust, extraneous matter which are denser, the neutral zone should locate towards periphery. In cream separation, the component that is to be separated is the fat, which is a lighter phase, the neutral zone should be towards centre of the disc, in order to reduce the distance of travel for the separation. The location of neutral zone measured from centre can be obtained as a function of the density differences between the two phases (Figure 6.4).

The radius at which the milk enters the gaps beween the discs, r_i and r_l the at which the heavier and lighter liquids leave the separators are related by their density ratios, ρh and ρl.

$$r_i = \sqrt{\frac{r^2{}_l - \left(\frac{\rho h}{\rho l}\right) r^2{}_h}{1 - \left(\frac{\rho h}{\rho h}\right)}} \qquad \text{(Eqn. 6.11)}$$

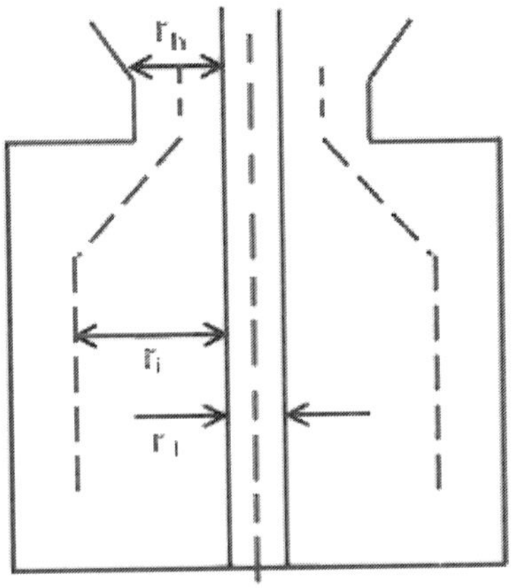

Figure 6.4. Neutral Zone

Different types of Centrifuges and their application in the Dairy Industry

Use of mechanical separation by centrifugal force in dairy industry can be classified in broad terms as shown below, depending on the dried milk physical state of the components being separated.

1. Liquid from liquid — *e.g.* cream from milk by cream separator
2. Solid from liquid — *e.g.* Ghee residue from ghee by Clarifier
3. Solids from gas — *e.g.* powder from air by Cyclone separator
4. Vapours from liquid — *e.g.* vapours from Concentrated milk in liquid vapour separator

The Centrifuges used in dairy industry can be classified in Figure 6.5.

Open type of Cream Separators

These types of cream separators are fed with milk open to atmospheric pressure, and flow in separator discs is by gravity from a vessel open to atmosphere (Figure 6.6). These separators are of smaller capacity, maximum of 1000 lph. They are suitable for sour milk separation and are driven by belt and pulley, powered by small motor. The feed from the vessel flows by gravity through a spout. A float positioned on the top of the inlet prevents entry of air, and allows only milk to enter into the separator bowl.

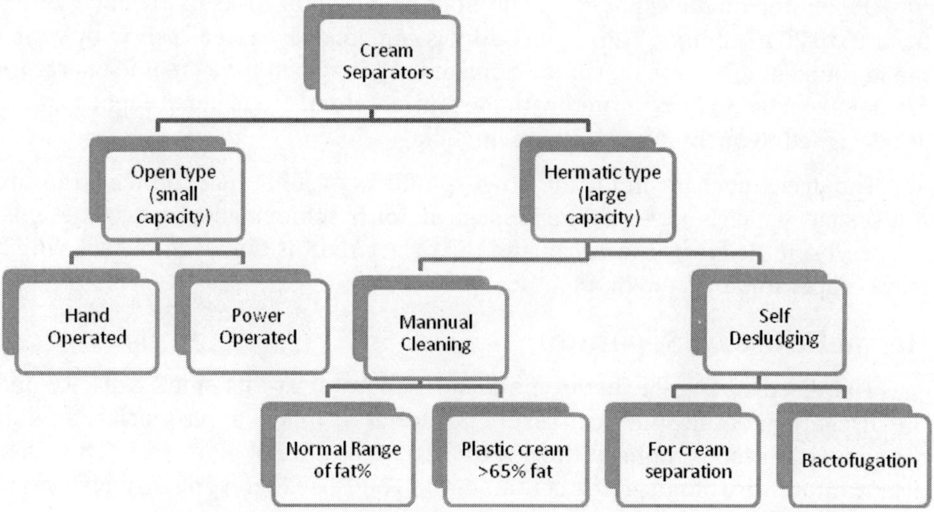

Figure 6.5. Classification of Centrifugal Separators

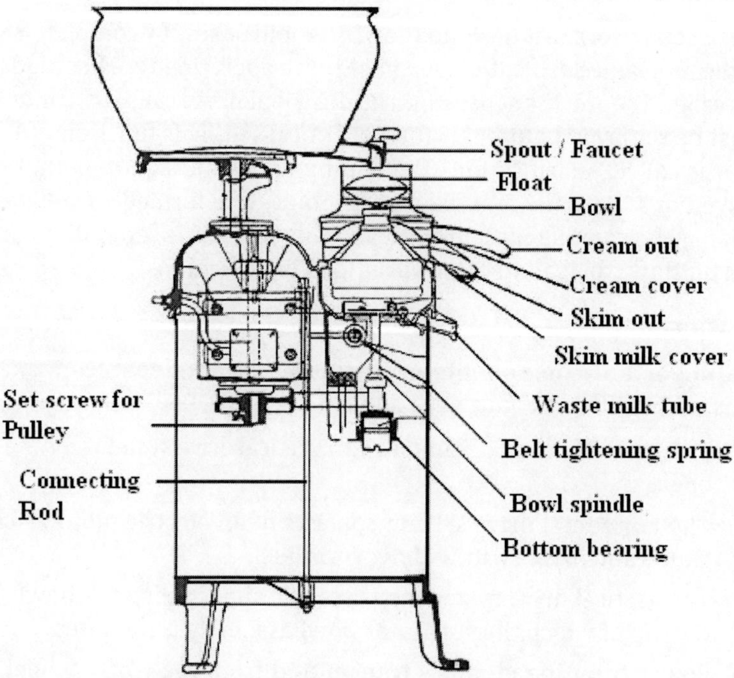

Figure 6.6. Open Type Cream Separator

Depending on the capacity, the number of separator disks in the bowl varies from 16 to 25 in number. The cream outlet is controlled by a small screw, by which the fat percentage in cream can be adjusted. As the open type cream separators are not expected to be operated with pasteurizers, and are of small capacity, they are designed to operate with lukewarm milk.

The drive mechanism for the bowl spindle is by a flat belt connected to the driver shaft, which is driven by single speed motor. In some smaller capacity designs, the bowl spindle forms the worm and the driver shaft is the worm wheel, which may be operated by hand or a motor.

Hermetic Cream Separator

The disadvantage of Open type cream separator are its small capacity and the discharge at atmospheric pressure. It cannot be linked to pasteurizer so that the separation can be done at the optimum temperature of 40 to 45°C. The other disadvantages are atmospheric contamination and also ingress of air, which make the product foamy.

The above problem is solved with Hermetic cream separator, which also has a larger capacity. Its milk inlet and outlet are connected to piping, making it possible to operate under completely air tight condition.

These separators are designed for three purposes, to complete separation, standardization and clarification. As these three operations are handled in a single unit, it is called Tri-process separator. The dust, leukocyte cells *etc.*, contained in the milk must be removed before it is further processed. This function of Tri-process separator is called "Clarification". Depending upon the requirement, the fat may be removed only partially to a fixed percentage. It is termed as standardization. The milk may be completely skimmed, and this process is separation. The control of cream outflow can lead to these variations in operations.

Description

The important components of Hermetic Cream Separator and its drive mechanism identified in Figures 6.7 and 6.8 are described as below:

1. The separator has horizontal and a vertical drive systems for the separator bowl.
2. The horizontal drive system consists of an electric motor, a centrifugal clutch and drive shaft with worm wheel.
3. The vertical drive system consists of a shaft, called the bowl spindle on the top of which the separator bowl assembly is mounted.
4. Power from the motor is transmitted from the worm wheel on to the horizontal drive shaft to a worm gear of the bowl spindle. Thus the bowl spindle rotates at a higher speed (speed of 4 to 5 times) than the horizontal drive shaft.

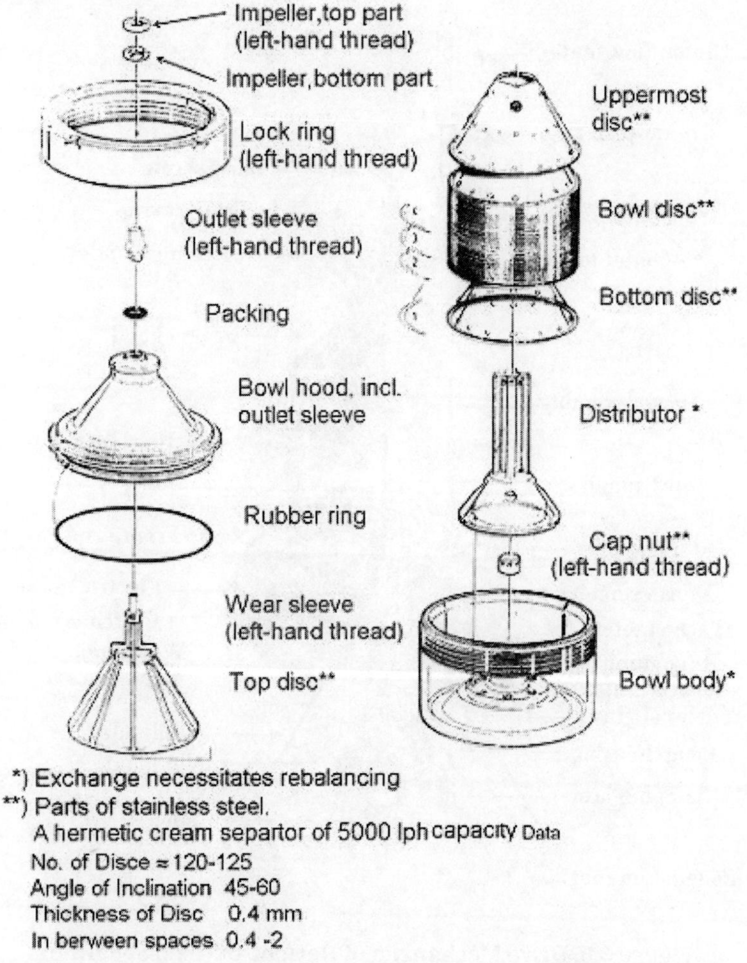

Figure 6.7. Internal Parts of Separator Bowl

5. Precision ball bearings support the drive components to eliminate vibration.
6. Also mounted on frame is a hand brake acting on the clutch disc, a tachometer and revolution counter button for speed indication, an oil filter plug,a drain plug, an oil level gauge, *etc.*
7. The bowl consists of two main sections: the body and the hood. These are held together by a threaded lock ring.
8. The disc stack is clamped between the hood and the bell shaped distributor at the axis of the bowl.
9. The gap between discs is about 0.5 mm. The disc wall thickness is about 0.4 mm and angle of inclination to the horizontal is 45 to 60 degrees.

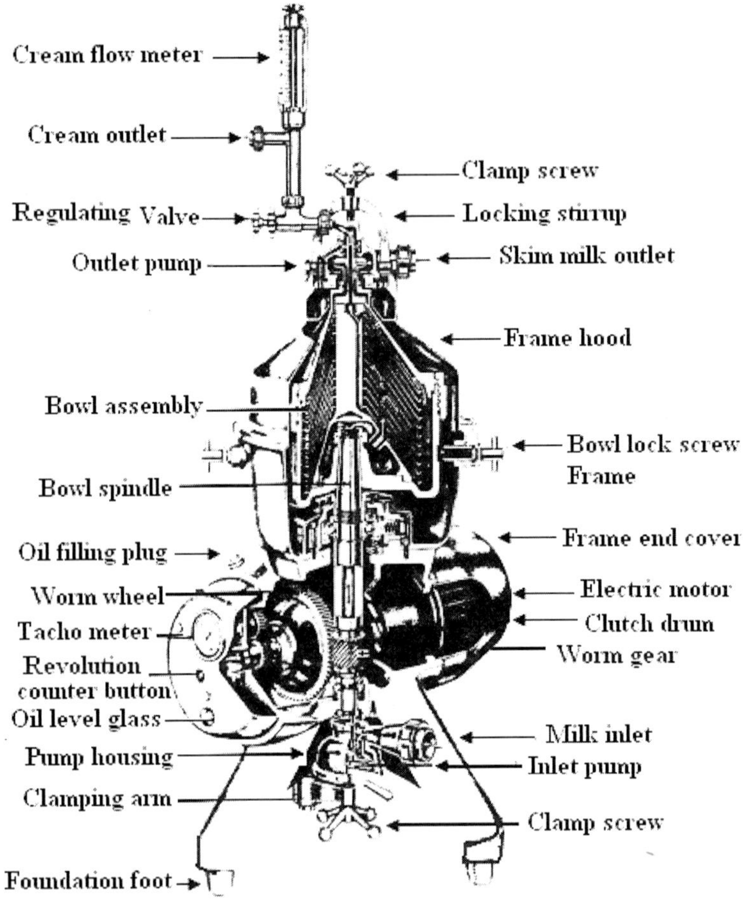

Figure 6.8. Drive Mechanism of Hermetic Cream Separator

10. The bowl speed is 5500 to 6000 rpm, for 5,000lph and 7500 for 10,000 lph capacity separators
11. The bearings in the frame are splash lubricated by the worm wheel, which runs partially, immersed in an oil bath in the base of the frame.
12. The Separators remove about 98.5 to 99 per cent of fat in milk. The skim milk contains remaining 0.02 to 0.05 per cent fat.
13. Cream outflow from the separator normally represents about 9 to 10 per cent of the total milk inflow.
14. The feed supply to the separator is by means of a pump, in case the supply tank is far away. Otherwise, a centripetal pump leads the milk suction line into the spindle, from where the milk enters the bottom of the distributor.
15. For standardization, the cream outlet is kept at a higher pressure than that of skim milk outlet. The flow rate of cream is measured by a rotameter.

Self-Desludging Centrifuge

In the operation of Hermetic cream separator, the process of separation and also clarification, have to be stopped, and the unit must be dismantled for cleaning after few hours, especially when the quality of the milk is poor. This causes interruption to HTST pasteurization operation also. To avoid this process interruption, a method of ejecting the sediment separated from clarification or cream separation was developed. This required certain changes to the design, especially to the lower part of the bowl which houses the separator discs.

Self-Desludging Operation

The milk sludge, formed during the separation process collects in the sludge space of the bowl, which is designed as double taper (Figure 6.9). The bottom disk

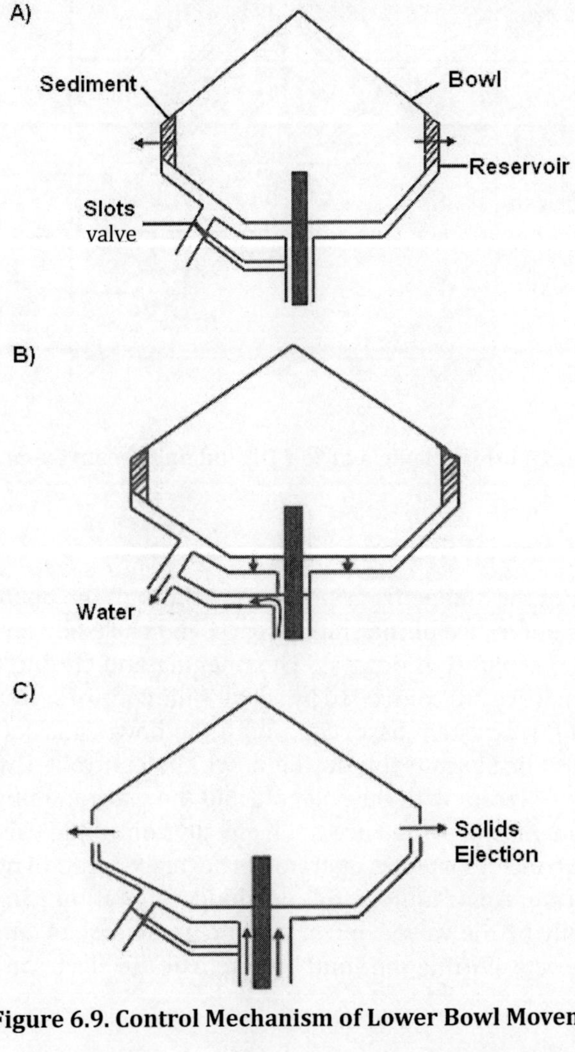

Figure 6.9. Control Mechanism of Lower Bowl Movement

of the sludge space is controlled hydraulically to move downwards, so that the sludge can exit the bowl through the double conical portion, (where the holes at the edge get uncovered). The sludge ejected from the bowl is collected in the upper transmission part made of stainless steel, and fed through the sludge collector to the sterilizing vessel. The sludge is sterilized with caustic soda, at a high temperature. After the sterilizing process is terminated, the vessel must be emptied for fresh sludge ejections. The hydraulically controlled bowl opening mechanism can be operated both manually and automatically.

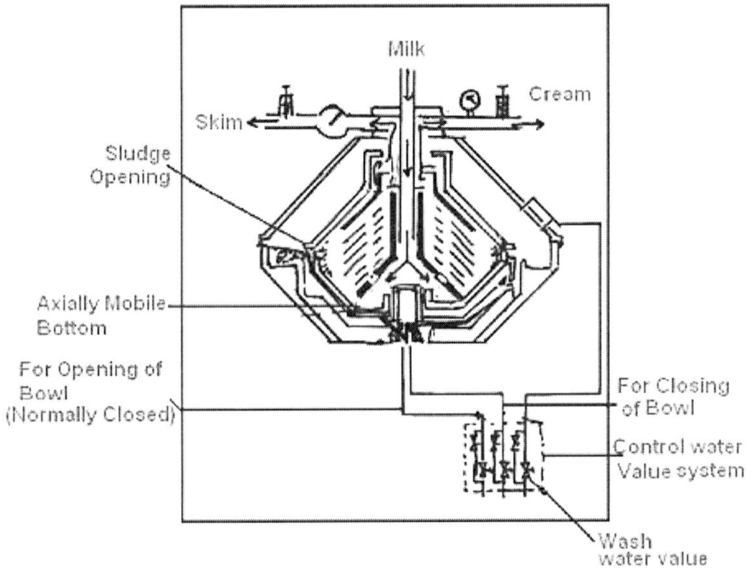

Figure 6.10 Internal parts of Self-Desludging Cream Separator

Operation of Control System

The removal of the sludge from the bowl is through the openings arranged at the widest circumference of the sludge space, and closed by an axially mobile bottom during the separation process. The opening and closing movements of the mobile bottom valve are controlled by a hydraulic control system (Figure 6.9). The automatic control system associated with the bowl consists of the mobile bottom disk hydraulically movable in the bowl, the controlled water feed, the control water valve system with the solenoid and the programming provision for the control system. Also provided are the manually controlled valves for manual control. The bowl control system is operated in normal operation by a preselected desludging program. Any required full de-sludge operation can be controlled semi- automatically by the valve control system. In the case of current failure or other possible defects, putting the control system or the electrical control valves out of operation, the separation process can be brought to an end by operating the manually-operated valves.

The water used for hydraulic operation must be clean, soft fresh water. The control water pressure also must be sufficient, usually about 245 kPa (2.5 kg/cm^2)

Controlling Fat Content of Cream in Hermetic Separator

Exiting a the fat content of cream hermetic cream separator can be controlled by two methods:

1. By using cream flow meter and a throttling valve at the exit of the cream outlet.
2. By in line automatic control of fat using Diaphragm type constant pressure control valve in the skim milk outlet pipe and Density transmitter in the cream pipe line.

Fat content of Cream

The whole milk supplied to the separator is discharged as, skim milk and cream. The cream normally represents about 10 per cent of the total throughput. The proportion of milk input discharged as cream determines the fat content of the cream. If the whole milk contains 4 per cent fat and the throughput is 6000 litres/hour lts/h, the total amount of fat passing through the separator will be 4 x 6000/100 = 240 litres/hour. Let that cream with a fat content of 40 per cent is required. This quantity of fat must be diluted with a certain amount of skim milk. The total quantity of cream discharged with 40 per cent fat will then amount to 240 x100/40 = 600 litres/hour. This gives 240 litres/hour of pure fat, and 360 litres/hour of skim milk. Installation of throttling valves at the cream and skim milk outlets makes it possible to adjust the relative volumes of the two streams in order to obtain the required content of fat in the cream.

a) Controlling of Fat content using Cream Flowmeter and the Throttling Valve

Controlling of fat content in cream using cream flow meter is shown in Figure 6.11. The volume of cream discharged is controlled by a throttling valve (2) and the cream flow rate measured by a built-in cream flowmeter (3). On opening the valve gradually increasing amounts of cream is discharged resulting in lower fat content in the cream.

A given rate of cream discharge consequently corresponds to a given fat content in the cream. If the fat content of the whole milk is 4 per cent and cream with 40 per cent fat is required, then the discharge from the cream outlet must be adjusted to 600 litres/hour. The pressure at the skim milk outlet, (Figure 6.11) adjusted and set by a regulating valve at a certain value according the separator and the throughput. Then the throttling valve (2) at the cream outlet is adjusted to give the flow rate corresponding to the required fat content. Any change in the cream discharge will be matched by an equal, and opposite, alteration in the skim milk discharge. An adjustable constant pressure nut is fitted in the skim milk outlet to maintain the back pressure at the outlet constant, regardless of changes in the rate of cream flow.

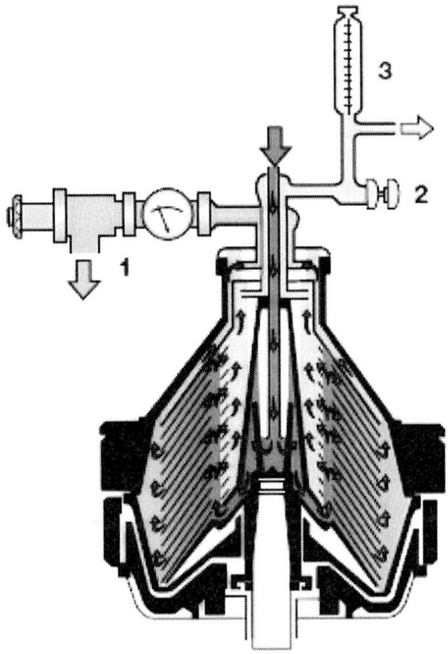

1. Skim milk outlet with pressure regulating valve
2. Cream Throttling Valve
3. Cream Flow meter

Figure 6.11. Hermetic Cream Separator with Cream Throttling Valve and Cream Flowmeter

(b) In-line Automatic Control of Fat Using a Diaphragm Type Constant Pressure Control Valve in the Skim Milk Line and Density Transmitter in the Cream Line

As seen, fat content of the cream from the separator is determined by the cream flow rate. The fat content in cream is inversely proportional to its flow rate. Some standardization systems employ flow meters to control the fat content in the cream. This is a quick method and, as long as the temperature and fat content in the whole milk before separation do not change, results from this method remain consistent. Several types of instruments can be used for continuous measurement of the fat content in cream. The output signal from the instrument adjusts the cream flow so that the fat content in the cream is correctly maintained. This method is accurate and also sensitive to variations in temperature and fat content of the milk being separated. However, the control response is slow and it takes a long time for the system to adjust the correct fat content when a disturbance occurs.

As can be seen in Figure 6.12 there are two transmitters (2) one each for measuring the flow of standardized cream and skim milk. With the data from these two flow meters, the control system (4) calculates the flow rate of whole milk into the separator. A density transmitter (1) measures the cream density and converts into fat content. Combining fat content and flow rate data, the control system actuates the modulating valve (3) to obtain the required fat content in the cream.

An automatic control valve fitted at the skim milk outlet for controlling the fat percentage in cream for a hermetic separator is shown in Figure 6.12. The valve shown is a diaphragm valve. The required product pressure is adjusted by means of compressed air pressure acting above the diaphragm. During separation process the diaphragm is acted upon by the constant air pressure on its top and the product (skim milk) pressure on its bottom.

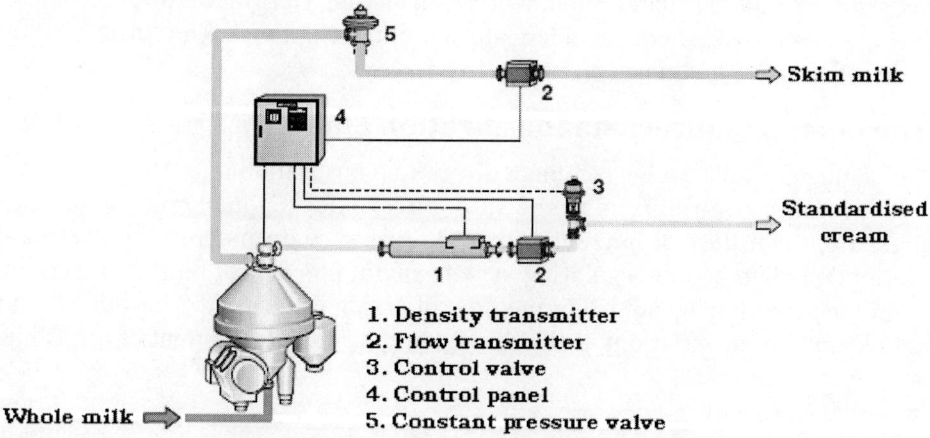

Figure 6.12. Control System for In-line Standardisation of Fat content in Cream

The air pressure forces the diaphragm down if the pressure of the skim milk drops. The valve plug, fixed to the diaphragm, then moves downwards and reduces the passage area. This throttling increases the skim milk outlet pressure. The opposite action takes place when there is an increase in the skim milk pressure, and the pre-set pressure is again restored. By controlling the skim milk outlet pressure the fat percentage in the cream can be controlled.

Direct In-line Standardization of Milk

In modern milk processing plants direct inline standardization is usually combined with milk separation. Previously the standardization was done manually but with large quantity of milk under process, a need for rapid standardization methods, independent of variations in raw milk fat content, has become necessary. Control valves, flow and density meters and a computerized control loop are employed to adjust the fat content of milk and cream to the desired values.

The pressure at the skim milk outlet must be kept constant for ensuring accurate standardization. This pressure must be maintained regardless of variations in flow or pressure drop caused by the equipment on downstream side. This is done with a constant-pressure valve located close to the skim milk outlet.

For maintaining precision in the process it is necessary to measure parameters such as 1) Fluctuations in the fat content of the inlet milk 2) fluctuations in throughput 3) fluctuations in preheating temperature.

Most of the variables are independent; any deviation in one stage of the process often results in deviation in all stages.

Most commonly the whole milk is preheated to 55–65°C in the pasteurizer before separation though in earlier time it was around 40–45°C. Following the process separation the cream is standardized at pre-set fat content and subsequently, the calculated amount of cream intended for standardization of milk is routed and remixed with an adequate amount of skim milk. The surplus cream is directed to the cream pasteurizer.

The Complete Direct Standardization Line

Figure 6.13 illustrates complete direct standardisation line. The diaphragm type pressure control valve at the skim milk outlet (5) maintains a constant pressure, regardless of fluctuations in the pressure drop across downstream equipment. The cream regulating system maintains fat content in the cream from the separator by adjusting the flow of cream discharged. This adjustment is independent of variations in the throughput or in the fat content of the inflow

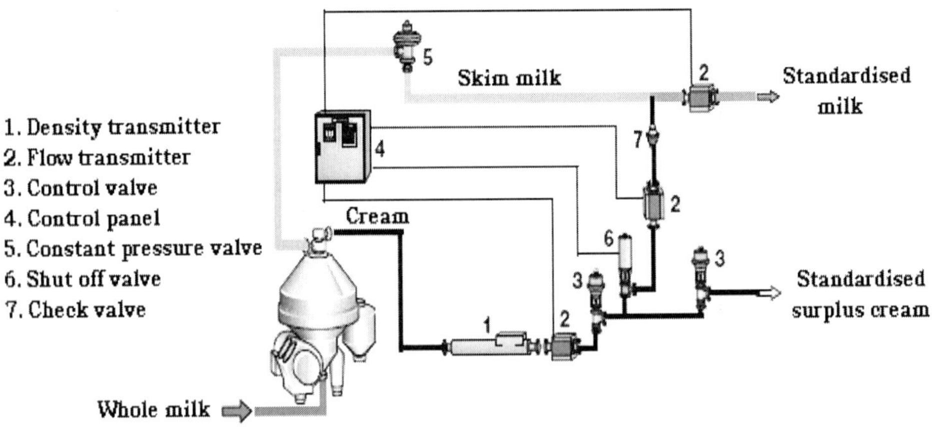

Figure 6.13. The Complete Process of Automatic In-line Standardization of Milk and Cream

milk. Finally, the ratio controller mixes the cream of constant fat content with skim milk in necessary proportions to give standardized milk of a specified fat content.

Bactofugation

Bactofugation is the process of separating the microorganisms from milk by application of centrifugal force.

Application of Bactofugation

1. It is useful for UHT processing of milk
2. In the manufacture of Low heat dried milk powder, both the aerobic and anaerobic spore formers are considerably reduced. The low heat powder is useful in preparation of recombined milk.
3. In the manufacture of special cheeses such as Mozzarella and Camembert using unpasteurized milk.
4. Most of the Spores can be removed by bactofugation, and only a portion around 2.5 to 3.0 per cent of the milk can be given the UHT treatment, thereby saving energy and promoting nutritional quality.
5. Useful before separation of Whey Protein Concentrates (WPC)

Densities of Milk Constituents

As the separation of fat from milk by centrifugal force depends on their differences, it is useful to know the densities of various components in milk,

Table 6.1. Densities of Major Milk Constituents

	Density(g/cc)
Milk	1.028 to 1.032
Skim milk	1.036
Fat	0.93
Bacteria	1.07 to 1.13

As we observe from the above values, the bacteria is denser than most of the components of milk, and hence, it can be separated by centrifugal force. Care must be exercised to safely dispose off the sludge. A self-desludging type of separator is most desirable.

Effect of Bactofugation

From milk at 50 to 55 °C the bactofugation removes about 89 to 93 per cent of the bacteria. Bactofugation can still be more effective, if the milk is first clarified, At higher temperatures the protein loss can occur, but at lower temperatures, the efficiency of separation is poor.

Table 6.2. The Composition of Slime

	Clarifier Slime	*Bactofugal Slime*
Total solids	18.2 per cent	16.0 per cent
Fat	0.03 per cent	0 per cent
Protein	12.8 per cent	11.3 per cent

Bactotherm

In this process, the design of the centrifuge is such that the heavier portion of liquid, which has higher concentration of bacteria, is sterilized, and mixed with rest of the milk. This method which is illustrated in Figure 6.14 is especially useful in the manufacture of UHT milk, where the final bacterial load depends on the initial bacterial load. Therefore the initial bacterial load must be minimum.

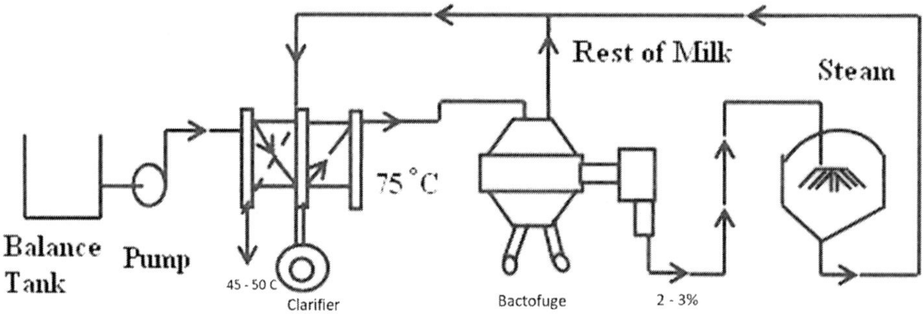

Figure 6.14 Bactofuge.

Maintenance of Separators and Clarifiers

Centrifugal separators are operated at high speed with clarifier bowl having discs held vertically by support of bearings.

General Maintenance Practices

1. Rubber gaskets, 'O' rings must be replaced before they wear out
2. Lubrication oil in the oil sump must be of required quality and quantity.
3. Bearings holding the spindle must be replaced regularly, to prevent bending of the shaft and rapid wearing out of 'o' rings, which cause drop in efficiency of separation and capacity.
4. Discs should be cleaned with softer material and not with wire mesh, and should be placed, stored properly without causing damage.
5. Assembling of discs and other components must be properly done.
6. Cream separator should not be run dry.

7. No adjustments, assembling, disassembling should be done while the cream separator is in operation.
8. Worn out worm or worm wheel should be replaced before damage is caused to other parts due to jerks, and loss of speed of rotation.
9. Centrifugal clutch shoe if work out should be replaced to prevent any loss of power transmitted to worm wheel. It should be roughened from time to time to provide grip.
10. Bearings to the shaft of worm wheel should be replaced when worn out.
11. Under no circumstances, the direction of rotation of spindle should be changed. This is likely to happen, when motor for the drive is repaired and reconnected. The direction of rotation must be checked before while connecting to the three phase power supply rotation.
12. Milk being fed to the separator must meet the recommended feed pressure.
13. While assembling and disassembling, the nature of the threads whether right hand threads or left hand threads should be observed and accordingly tightened.
14. Under no circumstances, the weight of the connecting pipe lines should cause load on the separator.
15. In self-desludging cream separators the lubricating oil is further cooled by cooling water. It is important that the cooling water used should be soft, sufficient flow and at required pressure. Under no circumstances, the cooling water by pass valve from the control water line should be closed or disturbed once the requirement is set.
16. The cream separation should not be done at temperatures higher than recommended, as it will reduce the life of the rubber collars, packing *etc.*

Problem 1: Problem on Gravitational separation

Calculate the rate of rise of fat globule of diameter 3 microns at 20 °C with a density of solids – not – fat as 1030 kg/m^3 and that of fat globule as 930 kg/m^3 and the viscosity of milk as 1.79 x 10^{-3} Pa. S

Solution

$V = d^2. (\rho_s - \rho_f). g/(18. \mu)$

$= 3 \times 3 \times 10^{-12} \times (1030 - 930) \times 9.81/(18 \times 0.00179)$

= 2.74022E-07 m/sec

= 0.009864804 m/h

Problem 2: If the problem 1 is solved with application of Centrifugal force.

For the above problem if,

radius of bowl, R, is	0.14	m
RPM is	5500	rpm
Velocity of separation of fat, V		
	0.09330817	m/sec
	335.9094114	m/h

Problem 3: Find the capacity of Hermetic Cream separator with the following data:

No. of discs	105	
Disc Angle	55	o
Outer Radius, Ro	0.13	m
Inner Radius, Rn	0.075	m
limiting diameter of fat globule, d	1.6	micron
Number of revolutions, n	6000	rpm 100 rps
Density difference of milk and fat globule, Δρ	116	kg/m^3
Correction factor, C	0.6	
Absolute Viscosity, η	1.04x10(-3)	0.00104 Pa.s
Q=C (2π) 3 z Tanα (Ro3 - Rn3) n x		
d2 10(-12) Δρ/18 μ	7530	l/h

Problem 4: Cream separation and Adjustment of fat content of m1 and m2 are the two partial form before mixing

(m1+m2) then mass balance equating

m1f1+m2f2= (m1+m2) fm

or after transformation

m1/m2 =fm-f2/f1-fm

Ex if f1=3.5 per cent, f2=30 per cent fm=10 per cent

F1	3.5
F2	30
Fm	10
m1/m2=	3.076923

Chapter 7

Homogenization

Introduction

Homogenization is used to prevent or the formation of a cream layer in full cream milk. It is accomplished by reducing the diameter of the fat globules in milk.

In the past, pasteurized milk used to be without being homogenized, although the flavour of milk fully develops by homogenization. A certain amount of cream was permitted to form to show the consumer the full cream character of the milk. The Sterilized milk, evaporated or condensed milk and sterilized cream are generally homogenized. The Ice cream mixes, milk in yoghurt production and milk for milk powder production are also homogenized.

The purpose for which the following products are homogenized is:

Product	*Purpose of Homogenization*
Milk, cream, condensed milk	Prevention of cream separation,
Coffee cream	Improvement in flavour, increased
	Whitening power, increase in
	Viscosity (some time)
Yoghurt	A more stable gel
Ice cream mix	Less fat separation during freezing
Full cream milk powder	Less separation of free fat

Mode of Operation

In milk, a larger part of the fat consists of globules with diameters ranging from 2 to 6 µm. A few fat globules may exist with diameter of 10 µm or more. Milk fat contains also a large number of small fat globules with diameters down to 0.1 µm, but these do not influence the total volume of the fat. The largest fat globules in milk intended for only a few days' storage need not be smaller than 1 - 2 µm. If sterilized milk is to be made suitable for several weeks' storage, the range of diameter of the fat globules should be between 0.2 and 0.7µm. Research studies showed that the diameter of 0.7 µm is critical for clumping of fat. With a diameter less than 0.7 µm the dispersion of fat globule-milk serum is stable, since aggregates of fat globules are separated again by Brownian movement (Figure 7.1). This is not so for formation of clusters at higher fat contents (> 20 per cent) where the sub-units of the casein micelles keep the fat globules together by bridge formation.

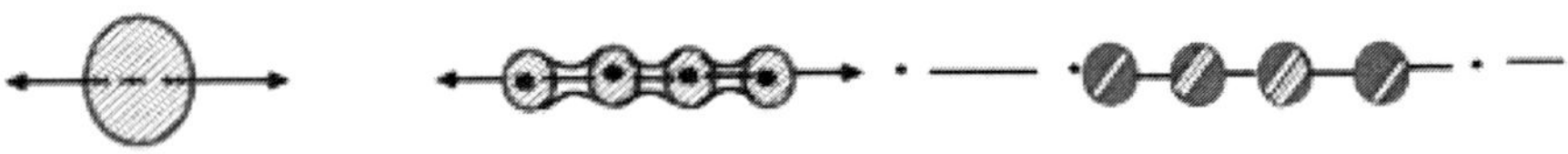

Figure 7.1. Mechanism in Breaking of Fat Globules

Homogenization breaks the fat globules into smaller droplets with diameters less than 1 µm. The droplet size depends on the homogenization pressure. The reduction in fat globule size is accomplished by forcing the milk at high pressures through a narrow opening, of size slightly larger than the diameter of the fat globules. The velocity of milk through the slit is in the range 100 to 200 m/s. This high velocity causes high shearing stresses, cavitation and micro-turbulence. The globules are deformed, and become wavy, under this condition of flow and finally break up.

Extensive research in recent years elucidated the actual mechanism of homogenization. The research studies established that there is a definite relationship between the Laplace number La, a dimensionless homogenizing pressure, and the degree of particle size reduction, irrespective of whether flow is laminar or turbulent. The Laplace Number is given by

$$La = \frac{\Delta p . d_{mo}}{\sigma} \qquad \text{(Eqn. 7.1)}$$

where,

Δp=pressure difference (N/m^2)

d_{mo}= mean initial particle size (m)

σ =Interfacial tension (N/m)

When cavitations was suppressed, the degree of homogenization was the same at the given Laplace number, and was independent of the type of flow. This shows that turbulence does not influence homogenization. However, the effect of homogenizing can be improved if cavitations take place.

According to the theories of fluid mechanics, particle size reduction is possible only if the viscosity ratio μ_{Oil}/μ_{water} is less than 4. For homogenization of milk this ratio is found to be 2 to 4 times larger. Nevertheless, the shearing effect at the homogenizer slit cannot be excluded since the high shear gradients as well as very high velocity gradients at the slit can have accelerating as well as a retarding action on the flow. As soon as a liquid thread is produced by deformation only forces of smaller magnitude are needed to break it up. Research studies established the following proportionality relationship for the mean globule diameter d_m:

$$d_m \sim h._{\sigma.} (1/Re^{1/3})/V_o\, \rho.\, \upsilon \qquad \text{(Eqn. 7.2)}$$

where,

h = width of the slit

v_o = velocity at the slit aperture

σ = interfacial tension

ρ = density of the continuous phase (milk)

υ = kinematic viscosity of the continuous phase

Re =mean Reynolds number of the flow through the slit

The cavitation occurs when the kinetic energy ($m.v^2/2$) increases during flow through the slit and when, according to Bernoulli's equation, disregarding friction and deformation losses, the potential energy ($P.\upsilon$) decreases to such an extent that the static pressure becomes as low as the vapour pressure of the liquid. Shortly after the liquid enters the homogenizer slit, the initial homogenizing pressure decreases sharply due to the sudden increase in velocity. Depending on the value of the back pressure behind the slit, the pressure can drop to as low as the saturated vapour pressure of the liquid. A critical back-pressure is must for the cavitation to occur.

Implosion, of cavitation bubbles is one of the main indications of good homogenization. Similarly, sudden pressure changes are also known to occur in the valves which are of type that widen continuously. Compression shock occurs when a uni-phase flow accelerated up to supersonic velocity abruptly changes into flow at sonic velocity.

The requirements for a compression shock can be fulfilled in homogenization, depending on the shape of the slit and the pressure conditions. The intensity of the compression shock depends on the geometry of the slit. It can be influenced as by the amount of the applied back pressure. As in the valve with varying gap, the compression shock can also reach a maximum value within the homogenizer slit. These are conditions for optimum homogenizing results.

It is not clearly understood how implosions of bubbles disintegrate the fat globules. The following hypotheses however, have been proposed: Extremely high local pressure gradients form during spreading of the shock waves. The overlapping of several shock waves produces cavitation noises. The implosion of a single bubble cannot contribute to the disintegration of a fat globule. According to one of the hypotheses, only the overlapping of shock waves, which cause cavitation noises, stimulates the fat globules to resonate near their own frequencies causing disintegration to occur when the critical amplitude is exceeded.

With regard to shear stress, fat globule disintegration occurs mainly in flows which have high shear gradients. Flows which cause cavitation are responsible for pressure fluctuations.

Types of Homogenizer Valves and their Profiles

Figure 7.2 shows several types of homogenizer valves. The valve plugs are pressed with an adjustable force F onto the corresponding valve seat. In this manner the homogenizing slit is formed.

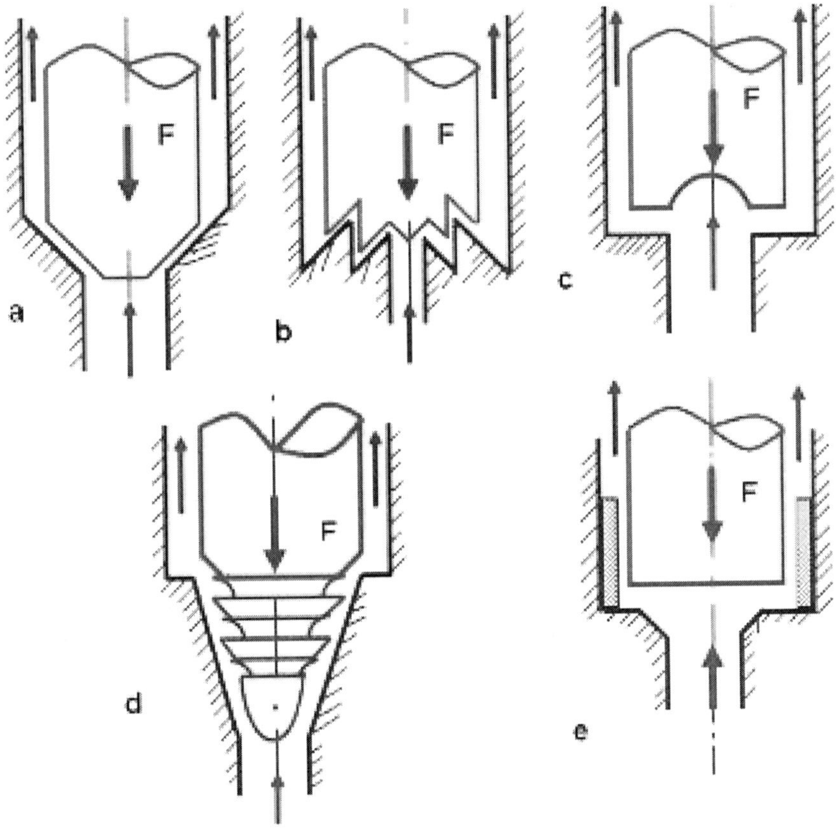

Figure 7.2. Homogenizer Valve and their Profiles

A conical shape, "b" causes changes in direction of flow due to its profile, "c" shows a simple plate valve and "d" a conical shape but with a grooved valve face which creates alternating stresses on the liquid to be homogenized, "e" shows a valve with a breaker ring which has a flat valve face and a conical seat. Changes to the cross section can be brought about by means of conical shape. By varying the 'internal diameter of the breaker ring, the back pressure and therefore the position of the zone of cavitation can be adjusted.

For two stage homogenization, two valves are arranged in series for homogenization of the liquid before the liquid comes out of the homogenizer. The milk fat globule has a native membrane, picked up at the time of secretion. This membrane lowers the interfacial tension resulting in a more stable emulsion. During homogenization, the fat globules are subdivided into smaller globules and there is increased surface area of the newly homogenized fat globules. The original milk fat globule membrane material is not sufficient to cover this. Proteins, particularly casein micelles migrate from serum phase to form new membrane material with the existing milk fat globule membrane. This may result in sharing of the casein micelles with fat globules and therefore some aggregation of fat globules could take place thereby defeating the purpose of homogenization. A first stage pressure may be about 175 kg/cm^2, depending on the product to be handled homogenization therefore, becomes essential at reduced pressure at almost 20 per cent of the first stage pressure or upto 35 kg/cm^2. This enables aggregated fat globules to be disrupted for formation of stable emulsion of finely dispersed fat globules. While most of the fat globule reduction takes place in the first stage, there is a tendency for clumping or clustering of the disrupted fat globules after first stage homogenization. The second stage valve permits the separation of those clusters into individual fat globules. The principal advantage of the two stage method is that it improves dispersion of fat globules and is useful in controlling viscosity of cream and ice cream mix.

Milk is pumped into the homogenizer with help of a Triplex type Piston pump. Milk passes at high pressure through a manifold and enters into the centre of the valve seat. The internal diameter of the valve seat is smaller than the manifold. The high pressures created by the piston pump causes milk to push the valve which is held in position by the heavy duty spring and form a narrow gap between the valve seat and the valve. The gap is maintained against the feed pressure by a counter force exerted by the adjustable heavy duty spring. As milk first enters the valve, liquid velocity is about 4 to 6 m/s. It then moves into the gap between the valve and the valve seat when its velocity is increased to 200 to 300 m/s in about 0.2 millisecond. When milk moves across the face of the valve seat shearing of fat globules takes place due to the velocity variation between the fat globules caused due to the friction between the surface of the homogenizing valve and milk. Velocity variation also causes grinding action between the fat globules resulting in disintegration of fat globules. High velocity of the liquid in the valve also causes turbulence resulting eddy currents within the flow. Milk then moves across the face of the valve seat (the land) and exits in about 50 micro seconds. When milk exits the narrow passage, the pressure suddenly drops to below saturation vapour pressure. This permits microscopic bubbles to form due to cavitation for a few microseconds before collapsing. Milk then strike on the perpendicular breaker ring with this velocity. These effects on milk contribute to the disruption of fat globules and completion of the first stage of homogenization.

Milk after the first stage of homogenization now passes through a passage for the second stage homogenization. For the second stage homogenization milk passes under a pressure of 35 kg/cm^2 and strike on the valve and push the valve against the action of the spring and forms a narrow passage for the milk to pass through. Here as the clustered fat globules pass through the narrow passage, they are separated and later when it strikes the breaker ring, separation of all the clustered fat globules will be completed and milk forms a stable emulsion. After the second stage homogenization milk comes out through the outlet. Homogenization always takes place in the first stage. The second stage basically serves two purposes (i) supplying a constant and controlled back-pressure to the first stage (ii) giving best possible conditions for homogenization by breaking up clusters formed directly after homogenization.

Figure 7.3 shows a 3 cylinder piston pump with two homogenizing steps arranged in series. A pump having 3-piston with 120° phase shift draws in the liquid the suction valves open and discharge valves closed. The pistons finally force the liquid through the homogenizer value when the suction and discharge valve positions are reversed. The discharge pressure can be regulated from outside by -manipulating the pressure adjustment springs while the machine is in operation. Beside the triple stage piston pumps, shown in Figure 7.3, five- stage pumps are often used for homogenizing because of their uniform flow characteristics.

Two-stage homogenizing is not always necessary. By a suitable construction of the valve, both pressure stages can be combined. A homogenization procedure of two or several stages has only a slight effect on the mean particle size.

The throughput through a homogenizer can be regulated by either adjusting the piston stroke of a piston pump or by employing infinitely variable speed for the pump motor.

Efficiency of Homogenization

The method recommended by the United States Public Health Service has been the most widely used for checking the homogenization efficiency. It is performed by subjecting a specified volume (one quart) of milk to storage for 48 hours without any disturbance and then testing the fat in upper 100 ml and the remainder of milk. For properly homogenized milk, the percent difference in both the top 100 ml and the remainder milk should not be more than 10 per cent.

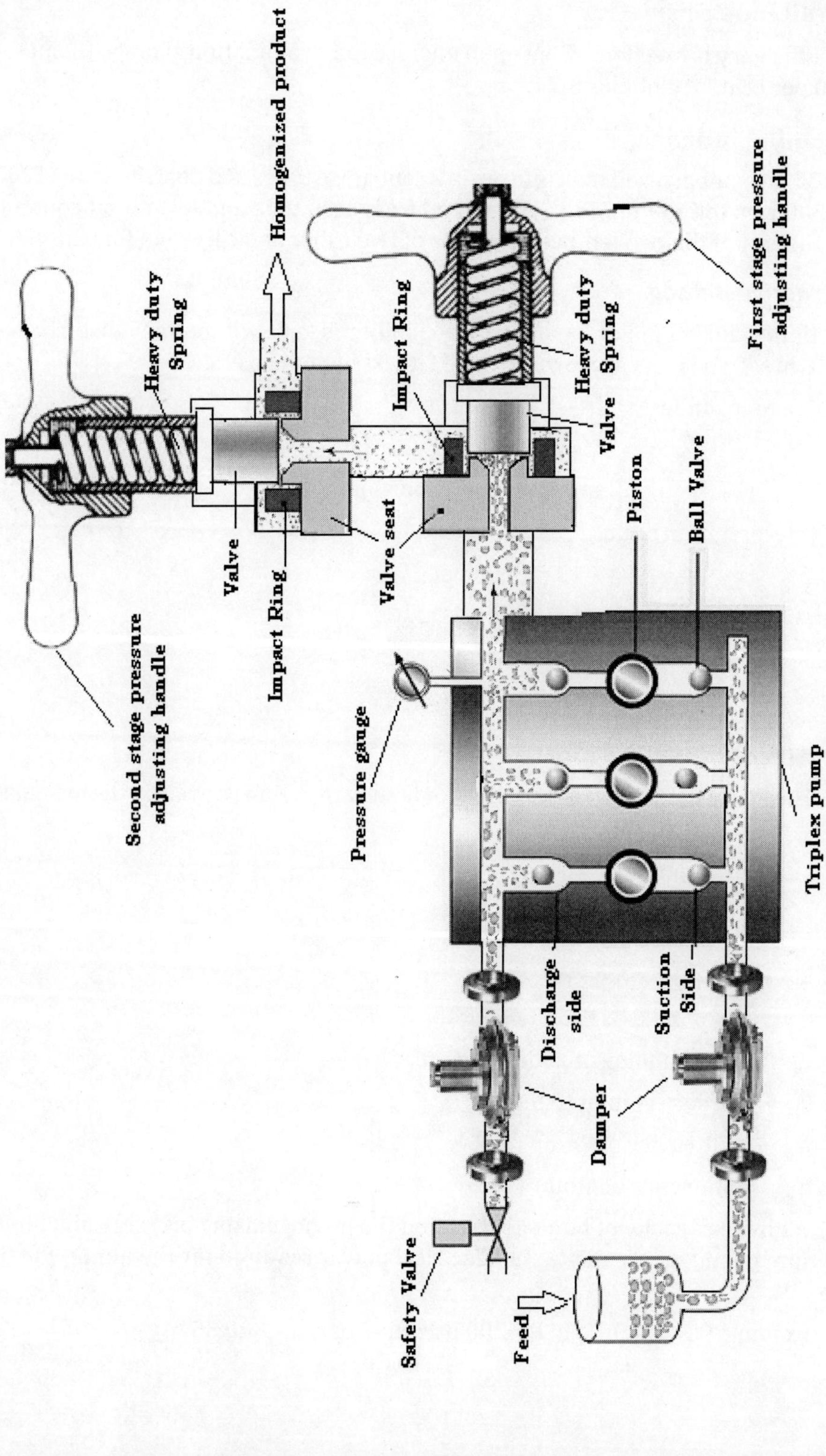

Figure 7.3. Homogenizer Pump and Valves

Farrall Index

Efficiency is excellent when upon microscopic examination average Diameter of 90 per cent fat globules ≤ 2 µ.

Creaming Index

50 ml homogenized milk is taken in 2 centrifuge tubes and Centrifuged at 1200 to 1500 rpm for 15 min. Take upper 5 ml from each tube and test for fat content. Mix the rest of the two left over portions of two tubes and carry out fat test

Alternative Method

Homogenized milk in a measuring cylinder is kept without any disturbance for 48 hrs. Two layers are formed. Take fat test of both layers.

Creaming index = $[(f_1 - f_2)/f_1] \times 100$

f_1 = Fat in upper portion

f_2 = Fat in lower portion

Interpretation scale:

≤ 10 = Excellent

10-20 = Very good

20-30 = Good

30 = Fair

Electrical Power of Motor

The Electrical power (E) required for homogenization is given by the formulae

$$E = \frac{Q_{in} \times (P_1 - P_{in})}{36000 \times \eta_{pump} \times \eta_{el.motor}} kW$$

where,

E = Electrical power, kW

Q = Flow of milk, lts/h

P_1 = 1st Stage homogenization pressure, bar

P_{in} = Pressure at inlet to pump

η_{pump} = Efficiency of Pump (0.85),

η_{Motor} = Efficiency of Motor (0.95)

With the capacity of homogenizer and the homogenizing pressure and input pressure of milk given above, the Electrical power required for the motor will be 123 kW.

Example PQ=18000 I/h; P_1=200 bar; P_2=2 bar; η_{pump}=0.85; η_{Motor} =0.95

Sometimes the electrical power is calculated by the simple formulae without considering the efficiency of Electric motor and Homogenizer Pump:

$E = P_1 \times Q$

where,

E = Electrical power, kW

Q = Flow of milk, m^3/s

P_1 = 1st Stage homogenization pressure, Pa

$E = 200 \times 10^5 \times 18/3600 = 100000$ watts = 100 kW

Temperature rise of Milk during Homogenization

In the process of homogenization, mechanical energy is converted into heat. The rise in Temperature of milk is given by a simple formula:

$E = P_1 \times Q$

$= \rho \times C_p \times \Delta T \times Q$

where,

r = Density of milk kg/m^3

C_p = Specific heat of milk, kJ/kg °C

ΔT = Temperature rise of milk, °C

Q = Flow of milk, m^3/s

P_1 = 1st Stage homogenization pressure, Pa

In the above example, the Temperature rise is obtained as:

$\Delta T = (100 \times 3600)/(1030 \times 3.93 \times 18) = 4.94°C$

The Temperature rise of milk can also be calculated by the formulae:

$$T_{out} = \frac{P_1 - P_{out}}{40} + T_{in}$$

where,

T_{out} = Outlet Temperature of Milk, °C

P_1 = 1st Stage homogenization pressure, bar

P_{out} = Pressure at exit of homogenizer, bar

T_{in} - Initial Temperature of Milk, °C

For every 4 MPa (40 bar) pressure drop the Temperature rises by 1°C

In the above example, the rise in temperature of milk will be 5°C and the outlet temperature of milk will be 70°C.

Aseptic Homogenizers

Homogenizers are used in UHT processing of milk. Homogenization may be carried out before or after UHT heating. If the latter arrangement is adopted, homogenization will cost more, since the process had to be carried out aseptically. There is, however an advantage in homogenizing after UHT heating, since it prevents or reverses protein- protein and fat globule-protein aggregation. It also retards the formation of sediment of heat precipitated proteins. Also it is stated that it is advantageous to homogenize the product after sterilization, since this order provides added heat stability to the protein system. Whether it is homogenized before or after the heating section, the milk is homogenized in the same temperature range, usually 70 - 80°C.

In either location the homogenizer is sterilized with the rest of the system and, therefore, experiences high temperatures during sterilization and cleaning. Hence, the homogenizer components must be compatible with high temperatures regardless of its location. The pump has to be designed with a double-packing plunger design, providing a sterile barrier to avoid contaminating the product by exposure of the working area of the plunger to the environment. The fundamental characteristic of a machine in aseptic execution is that it is necessary to avoid possible contamination of the already sterilized product by the contact with the exterior. In order to achieve this, reciprocating pistons and all the seals are protected in a chamber covered with steam and sterile condensate. The pistons move through an atmosphere of steam. The steam is admitted under pressure at one side of the chamber and discharged as condensate from an opposite side of the sterilizing chamber so that the steam flows along a circular path of travel to contact all parts of the plunger which enter and leave the sterilizing chamber. Hence, a desirable flushing and cleaning action is also realized. Besides, the machines in aseptic execution have to be designed taking into consideration the necessity to withstand temperatures reaching 160°C during the sterilization cycles. The main disadvantage of using the sterile condensate is the high cost of managing the steam production and its further re-condensation.

Care and Maintenance of the Homogenizer

Introduction

Homogenizer operates at high pressures. Hence, the care and maintenance is essential for making its operation safe and useful life long.

The operating efficiency of Homogenizer can be known from the extent of breaking of fat globules. The Homogenizing efficiency is found either direct microscopic method or by cream layer method.

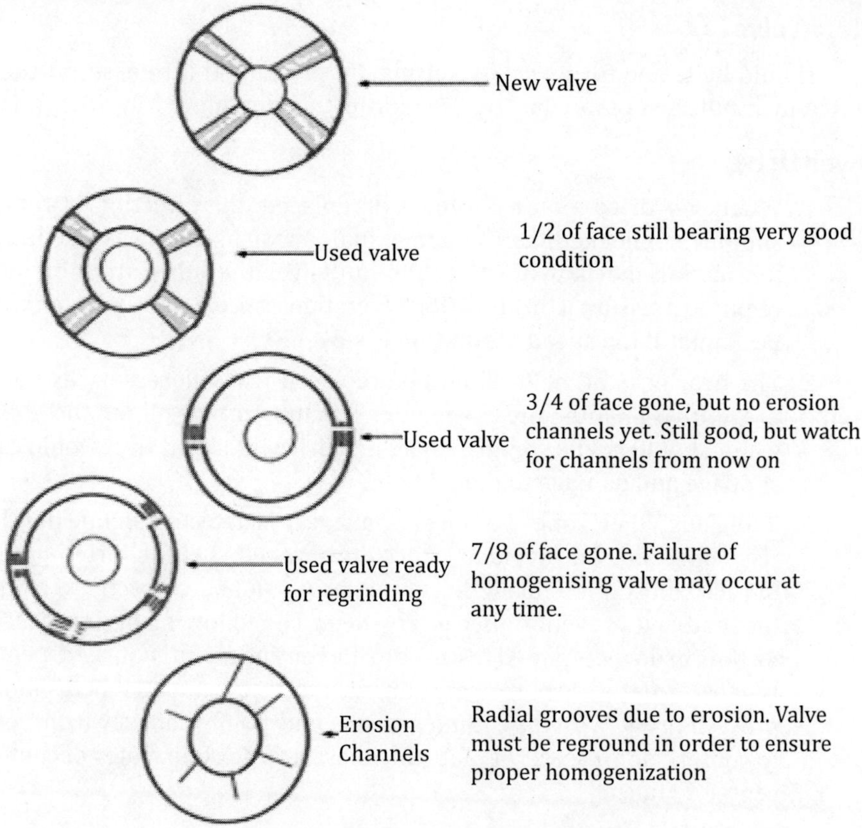

Figure 7.4. Wear of Homogenizer Valve

Valves

The homogenizer valves over a period of use, will lose its proper shape, and may form grooves. It needs to be reground or replaced. If the seals of the valves are leaking, they have to be replaced immediately. The faces of the valves and valve seat must be kept in good serviceable condition to retain an efficient homogenization of the product. This can be recognized by the amount of wear that initiates at the bore of the valve and radiates out across the face (Figure 7.4). If the wear is up to two-thirds of the surface area, re-lapping of the valve can be rectified with carborundum paste applied between the faces. But further wear will allow erosion grooves to appear. This is recognized by the appearance of straight lines radiating outward, and when this occurs loss of homogenization will rapidly follow.

Pressure Gauges

The pressure gauges must be made from stainless steel. The mounting of the gauges must be robust and leak proof. The gauge flange must be leak-proof at the highest pressures. The pressure gauge casing must be filled with glycerol for long life of operation and to damp the vibrations.

Safety Valve

It should be tested for its safety setting, by checking if it releases at the set point. A hydraulic test pump may be used for testing the safety valve.

Power Drive

1. Whenever drive motor is repaired/replaced, the direction of rotation should be checked, as the same shaft most probably is driving the lubrication mechanism, and fails to deliver the lubrication oil at the required pressure if the direction of rotation is incorrect. The result would be similar if the speed of rotation is slow.
2. The bearings and belts should be replaced when necessary as per the preventive maintenance schedule. The lubrication oil, or the cooling water should not be spilling where belt drive is located, as it would cause slippage and damage to the belt.
3. Replacing V-belts one at a time (when wear is detected on one only) is a risky operation and avoided - since the new belt - being shorter - is likely to carry most of the load. Belts must be tensioned correctly to transfer the load and prevent unnecessary wear. The following are suggestions on how to do a simplified testing during routine maintenance inspection. When the motor is not in operation, a correctly tensioned V-belt should, if pressed firmly with the thumb near the mid-point (half way from center to center), depresses 3/4 of its thickness for each one meter of center to center distance.

During assembling and disassembling, the dismantled parts must be kept on a table covered with rubber mat to prevent damage to the parts as well as any contamination.

A daily inspection should be made of the oil level visible through the sight glass. Lubrication oil should be replaced regularly, as per manufacturer's recommendations. The oil pressure must be correctly maintained while the equipment is running. After draining off any condensate from the oil sump through the pet cock, oil should be added if required, before restarting the homogenizer. The oil should be completely drained and replaced, for every 6 months.

The levelling of homogenizer should also be maintained, as it is likely to be disturbed over period of operation due to vibrations. The connecting pipe lines to the inlet and outlet of the Homogenizer should be, well supported, so that they do no exert weight on to the homogeniser, or transfer vibrations to other equipment.

Cooling water should be of the quality and at flow rate as recommended by the manufacturer. The water flow rate to the piston rods should be sufficient, if water is also cooling the oil.

Homogenizer Process Line Setup

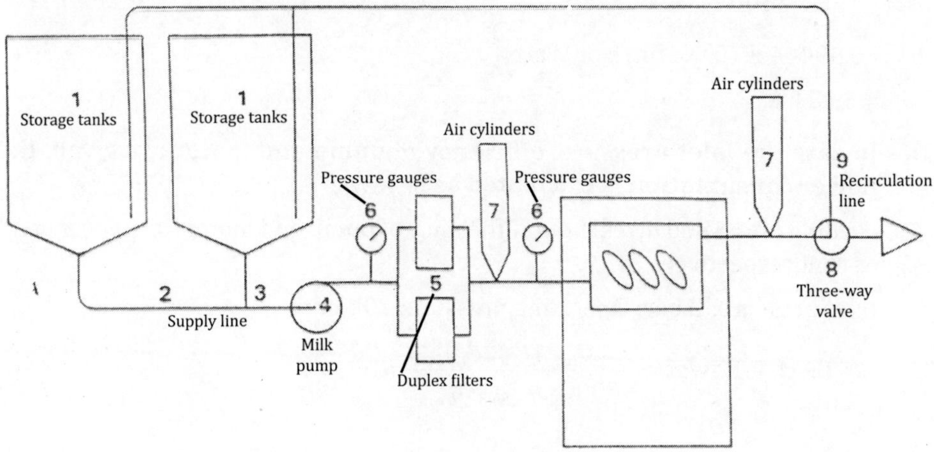

Figure 7.5. Homogenizer Setup.

The installation of the Homogenizer should be completed as per the manufacturer's recommendations, to avoid recurring maintenance problems. The most important precaution to be taken is that the vibrations are not transferred to and from the homogenizer connections. The pipe line supports must prevent this problem. Also, there should not be any leaks from the equipment. The discharge and suction sides of the homogenizer must have air cylinders to reduce the pulsations of the fluid under flow. The discharge side should have a three-way valve, to prevent the closing or blocking of the discharge line. Similarly, the homogenizer should never be starved of milk supply. Hence a quick change over of supply tanks should be ensured. Foam formation or intake of air should be avoided in the storage tanks at low level of milk. The incoming milk flow should also be at a positive pressure.

Solving Numerical

I. Calculate the power required to operate a homogenizer, with the following data

Capacity: 10,000 lph; Maximum operating pressure: 250 bar

Power Consumption = Pressure x Volumetric rate

V = 10,000 litres/hour = (10,000/1000) x (1/3600) =10/3600(m^3/s)

1m^3 = 1000L

P = 250 bar = 250 x 10^5 (N/m^2)

$$E = \frac{10}{3600} \times 250 \times 10^5$$

$=0.69444 \times 10^5$ N – m/s or Watts

= 69.5 kW

II. In case the inlet pressure, efficiency of pump and motor are given, the power consumption is calculated as follows:

Feeding rate: 1800 litres/hour; Efficiency of pump and motor: 85 per cent and 95 per cent respectively

Inlet pressure: 2 bar: Operating pressure: 200 bar

$$\text{Electric Power} = \frac{Q_{in} \times (P_1 - P_{in})}{36000 \times \eta_{pump} \times \eta_{motor}} KW$$

where,

Q_m = feeding rate l/h

P_h = Homogenisation pressure (bar)

P_{in} = Pressure at which milk entering Homogenizer (bar)

n_{pump} =Efficiency of pump

$$= \frac{1800 \times (200-2)}{3600 \times 0.85 \times 0.95}$$

= 122.6 kW

III. Increase in temperature of homogenized liquid is proportional to operating pressure

E = P.V =V. ρ.C. Δ θ

Or, Δθ = P/ρ.C

where,

P = Pressure (N/m^2); C= Heat capacity of milk (j /kg K);

ρ = Density of milk (kg/m^3)

Given: C milk = 3900J/kg K ρ milk=1030 kg/m^3 P=250 Bar = 250×10^5 (N/m^2)

Δθ = P/C. ρ

$$\Delta\theta = \frac{250 \times 10^5}{3900 \times 1030} = 6.2^\circ C$$

IV. Another assumption is that for every 40 bar pressure increase, the temperature of milk rises by 1°C,

$$T_{out} = \frac{P_h - P_{out}}{40} + T_{in}$$

T_{in} = 55 °C

P_h = 175 bar

P_{out} = 2 bar

$$T_{out} = \frac{175-2}{40} + 55$$

= 59. 325 °C

Chapter 8

Pasteurizers

Introduction

The process of Pasteurization involves heating milk to a certain temperature, holding at that temperature for certain period of time and then immediately cooling to below 4°C. This is done most commonly in two methods *i.e.* 1) Low Temperature and Longer Time 2) High Temperature Short Time. The former method is used in most of the Batch processes and involves use of a multipurpose vat. The later method is possible with a Plate type of heat exchanger.

Batch Process

Batch process involves heating of milk to 63°C and holding at that temperature for 30 minutes. For this process, multipurpose vat is most suitable, and its construction is shown in the Figure 8.1.

The Multipurpose vat is a jacketed and insulated tank and is cylindrical in shape. The tank is heated by circulating hot water or steam in the outer insulated jacket. Slow speed agitation provides uniform heating through the wall. The milk is slowly heated and after reaching the required temperature, is held at that temperature for 30 minutes. Then it is cooled to below 4 °C. The bringing up time and cooling time are not accounted for the pasteurization time.

The time taken for heating and cooling is given as:

$$T_h = \frac{m.c}{A.U} l_n \frac{T_s - T_o}{T_s - T_m} \qquad \text{(Eqn. 8.1)}$$

$$T_c = \frac{m.c}{A.U} l_n \frac{T_o - T_{cw}}{T_m - T_{cw}} \qquad \text{(Eqn. 8.2)}$$

where,

t_h, t_c = Time taken for heating or cooling respectively

m = mass of milk

c = Specific heat of milk

A = Area of heat transfer

U = Overall heat transfer co-efficient

T_o = Initial temperature of milk

T_m = Final temperature of milk

T_s = Hot fluid temperature

T_{CW} = Chilled water temperature.

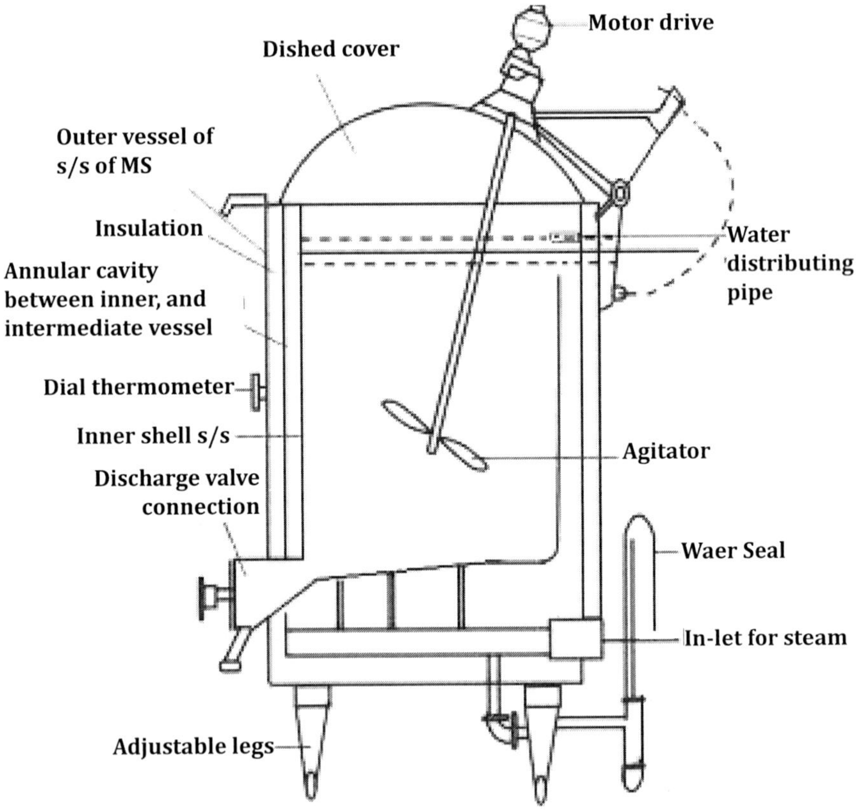

Figure 8.1. Holder Pasteurizer

HTST Pasteurizer

High Temperature Short Time Pasteurizers (HTST), have proved to be the workhorses of processing in Dairy Industry. For milk, the time temperature combination used is 71.5°C for 16 seconds, and then immediately cooled to below 4°C. The heat exchanger that can achieve this duty is the Plate Heat Exchanger.

The Plate heat exchanger (PHE) for HTST pasteurizer is made up of groups of plates that do the duty of heating the milk to the required high temperature, holding at that temperature for the required period(in some of the designs) and cooling it. The schematic diagram of the pasteurizer is shown below (Figure 8.2).

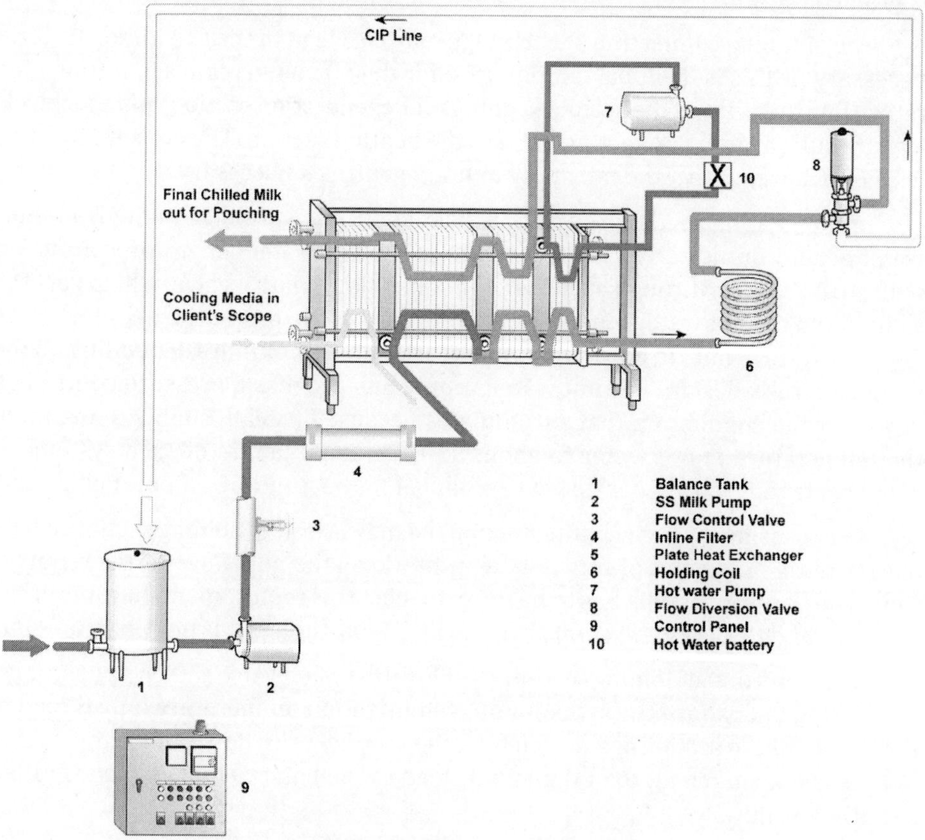

Figure 8.2. Schematic Diagram of HTST Pasteurizer (Courtesy JMD Sonic Engg.)

The milk from storage tank enters the HTST pasteurizer through a Float Balance Tank, which has a float to control the inflow of milk and maintains a constant level of milk in it. The milk is then pumped from here through a flow control valve, which may be ball valve type. Milk enters the Regeneration section where the incoming raw milk is partly heated by the returning hot milk from holding section. As the heat exchange is beneficial by reducing the load on steam and chilled water requirement, the number of plates is maximum in this section (Figure 8.5).

The Regeneration section, may be divided in two segments, so that after the first segment, when the milk is about 40°C, it is lead to duplex filters, which has two sets of vessels with filters, to remove any extraneous material in the milk. The filters are provided with air vents, to remove air in the beginning of the process. A recent development is the cylindrical vessel, which separates extraneous material with the fluid entering tangentially to the cylindrical vessel. The rejected material is drained.

The outgoing connection at this stage may also lead to the Cream separator if necessary, as this is the ideal temperature for the efficient cream separation. The return line then enters the second segment of Regeneration section, and the milk is lead through again a connecting plate to the heating section. The connecting plate here is meant to direct the milk flow to homogenizer if necessary.

In the heating section, the milk is heated with hot water, coming from hot-water generator, aided by a hot water pump. The hot-water generator is provided with steam supply, through an automatic steam flow control valve, which gets the signal from the hot water temperature controller of the pasteurizer controls. The flow rate ratio of milk to hot water is between 1:2 to 1:3, or more depending on the manufacturer's design. The hot water supply line also has a pressure controlled, outlet for the condensate that accumulates because of steam supply to maintain the temperature of hot water to about 85°C. To avoid failure of such systems of condensate removal a separate heat exchanger for heating hot water is being used.

After passing through heating section the milk is lead to holding section, which has stainless steel pipes of sufficient length to keep the milk flow for the required time. The flow rate of milk has to be fixed, to meet this requirement. In some of the pasteurizer designs, this section also is of PHE type, though it is not very common.

At the end of the Holding section, the Flow Diversion Valve (FDV) is positioned, so that, the milk above the set point of pasteurization temperature only is lead to the Regeneration section again, which is called the Forward flow. If the temperature of milk is not sufficient, then the milk is lead to the balance tank, which is called the diverted flow.

The hot milk after FDV in forward flow will enter the Regeneration section II and then I. Here, Milk will give off its heat to the incoming raw milk, and then enters the Cooling section. The milk is cooled to below 4°C by chilled water at 1°C, or at times brine (glycol + water) when viscous products have to be cooled. Here also, the flow rate ratio of Milk: Chilled water is 1:2 or 1:3 or higher. The flow rate difference has to take into consideration the pressure drop on the service fluid side, ie chilled water or hot water. The difference in flow channels and passes are shown in the Figure 8.5.

After chilling of milk to below 4°C, it is sent to Pasteurized milk storage tanks for further utilisation as per the market demand.

Types of Plates and Plate Arrangement

The types of PHE, and their gasket arrangement is shown in Figures 8.3 a and b. The Plate Heat Exchangers of each section consists of heat exchange plates, frame for hanging the same. End plates along with tightening bolts are provided to keep these plates pressured together, with gaskets in between. The groups of plates are provided in between with connecting plates which make it possible to introduce or exit milk after each of the processes, like Regeneration, heating, holding, and chilling.

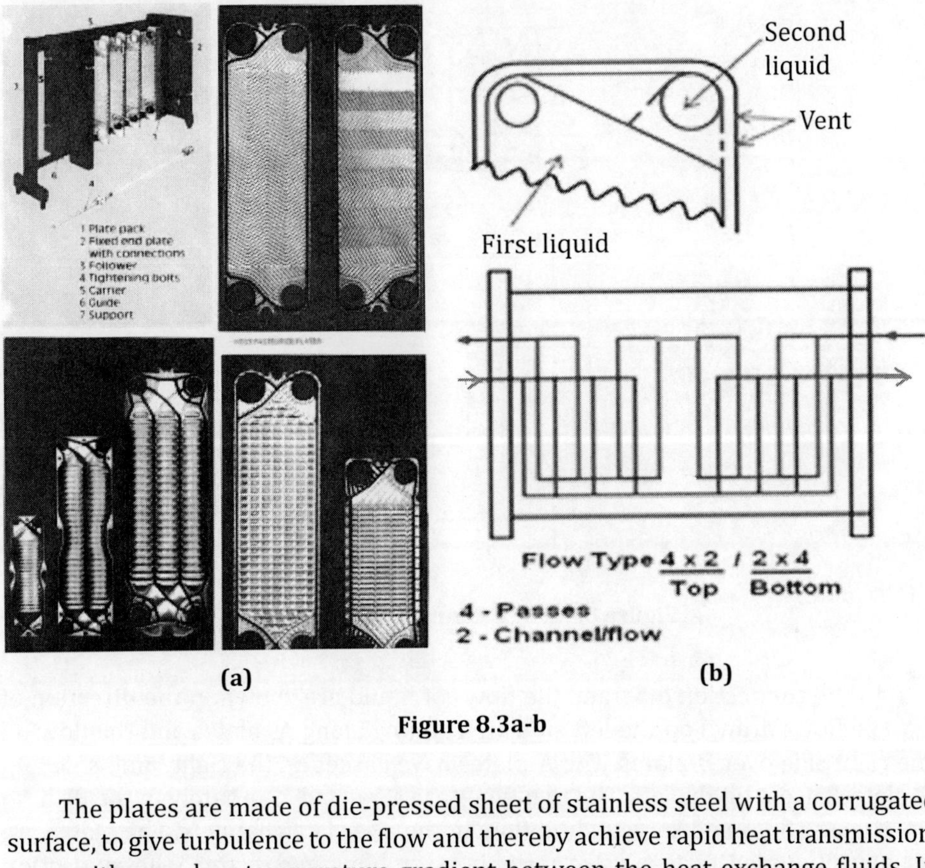

(a) (b)

Figure 8.3a-b

The plates are made of die-pressed sheet of stainless steel with a corrugated surface, to give turbulence to the flow and thereby achieve rapid heat transmission, even with very low temperature gradient between the heat exchange fluids. In the PHE the process fluid or service fluid can be passed in many ways through channels connected in parallel or in series, in the same direction or in opposite directions to achieve maximum heat transfer. The thickness of plates is about 0.5 to 1.25mm. The gap between plates is about 3 to 6 mm wide, depending on the design and products handled.

The arrangement of gaskets makes the flow in a particular direction for the milk and hot or cold water. The plates are of different sizes and configuration of corrugations, and may cater to the needs of wide variety of fluid processing. The manufacturer supplies the plate connection diagram to meet the specific heat exchange duty, which shows the sequence of the plates, arrangement of gaskets, and location of the connections. The nomenclature for the arrangement is explained below (Figure 8.4):

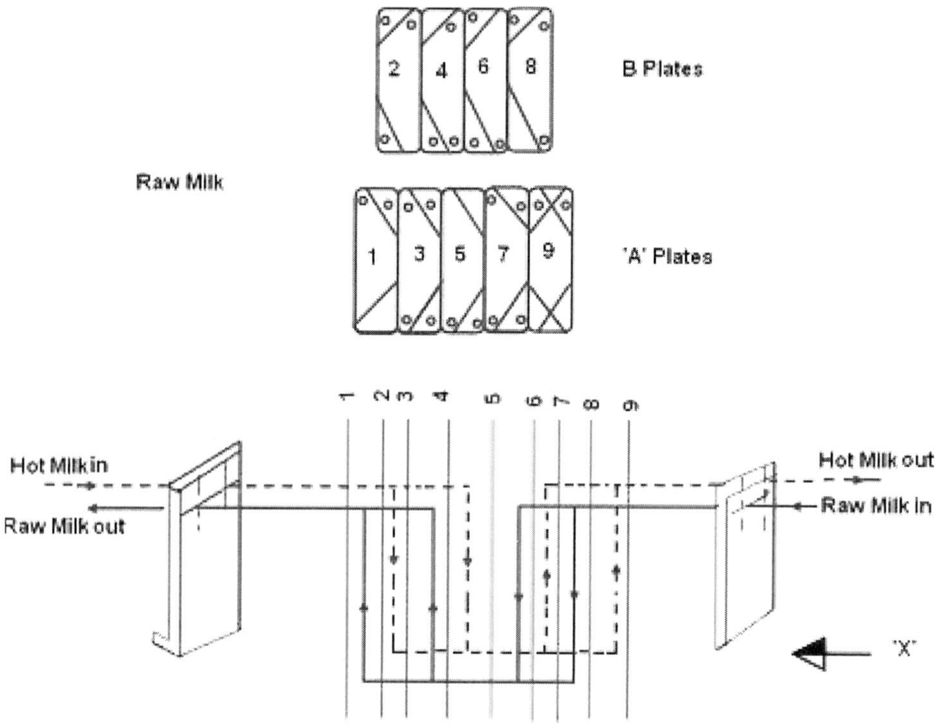

Figure 8.4. Plate Connection Diagram

In the connection diagram, the flows of liquid are shown in the direction of 'x'. The flows drawn on the left side are running along A- plates and the flow on the right side over B-plates. The A-plates have gasket on the right hand side and B-plates have it on the left. The two types are arranged alternately in the PHE, so the process fluid and service fluid flow in the alternate channels. The plates are usually numbered, so as to make easier in assembling, and relate to the connection diagram. Any damaged plate can be replaced by an identical spare plate.

Pasteurizer Controls and Flow Diversion Valve

Continuous operation of HTST pasteurizer depends on the monitoring of temperatures, and thereby controlling the flow of milk in diverted or forward flow condition. Further, it is legal requirement to record the temperature and time of operation of pasteurizer in many countries. Most of the instrumentation are now based on digital form and are much more compact as compared to earlier versions of Pneumatic controls.

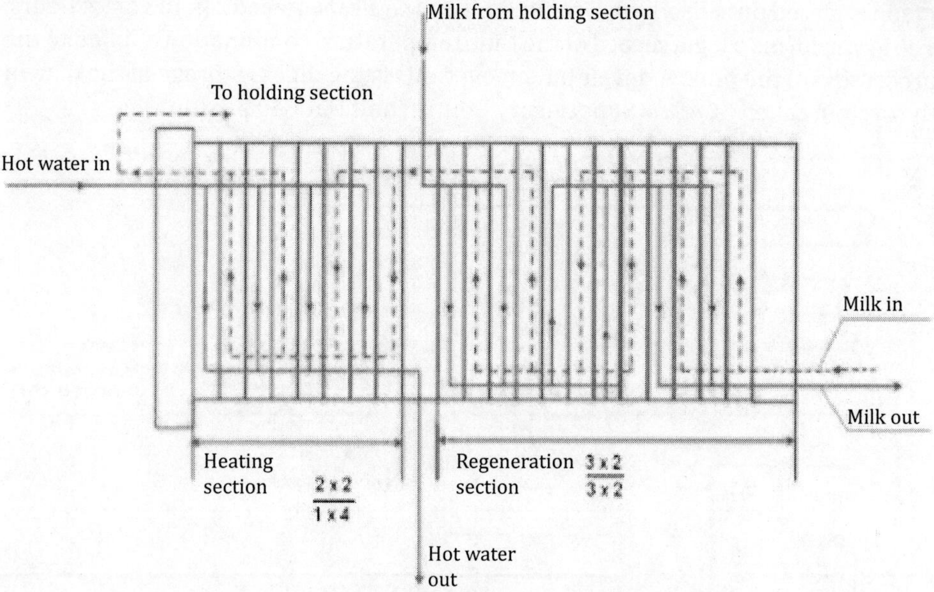

Figure 8.5. Heating and Regeneration Section

Commonly used equipment for control are:

1. Hot milk temperature indicator and controller
2. Cold milk temperature indicator
3. Hot milk and cold milk temperature recorder
4. Flow Diversion Valve
5. Visual and audio alarms

Diagram of Pasteurizer Controls

The temperature of hot milk or cold milk are sensed by suitable sensors. Common among them is the Resistance Temperature Device (RTD), which sends an electrical signal, proportional to the temperature sensed. This signal is received and compared with the Set Point (SP)of temperature by the Temperature Controller (TC). Depending on the deviation of the temperature of milk from the SP, an error signal is sent, which will range from 4 to 20 mA. The error signal is converted into a pressure signal 20 to 103 kPa (3 to 15 psi) (of compressed air), by Electro-pneumatic Converter. This signal will operate the final element, in the form of Diaphragm type of Steam control valve.

In addition to the control of steam supply to hot water, the signals of sensors are also received by Temperature transmitter, which in turn sends a pneumatic signal to pen drive of Thermograph. The Thermograph is a circular graph, with temperature indicated on concentric circles; the radial lines indicate the time. The graph is placed on a disc, which rotates one full circle in either 12 hours or 24 hours. The graph forms a legal record of the time temperature combination applied to the processing of the milk. A development over this is the digital storage of the data in the system called SCADA (Supervisory Control and Data Acquisition).

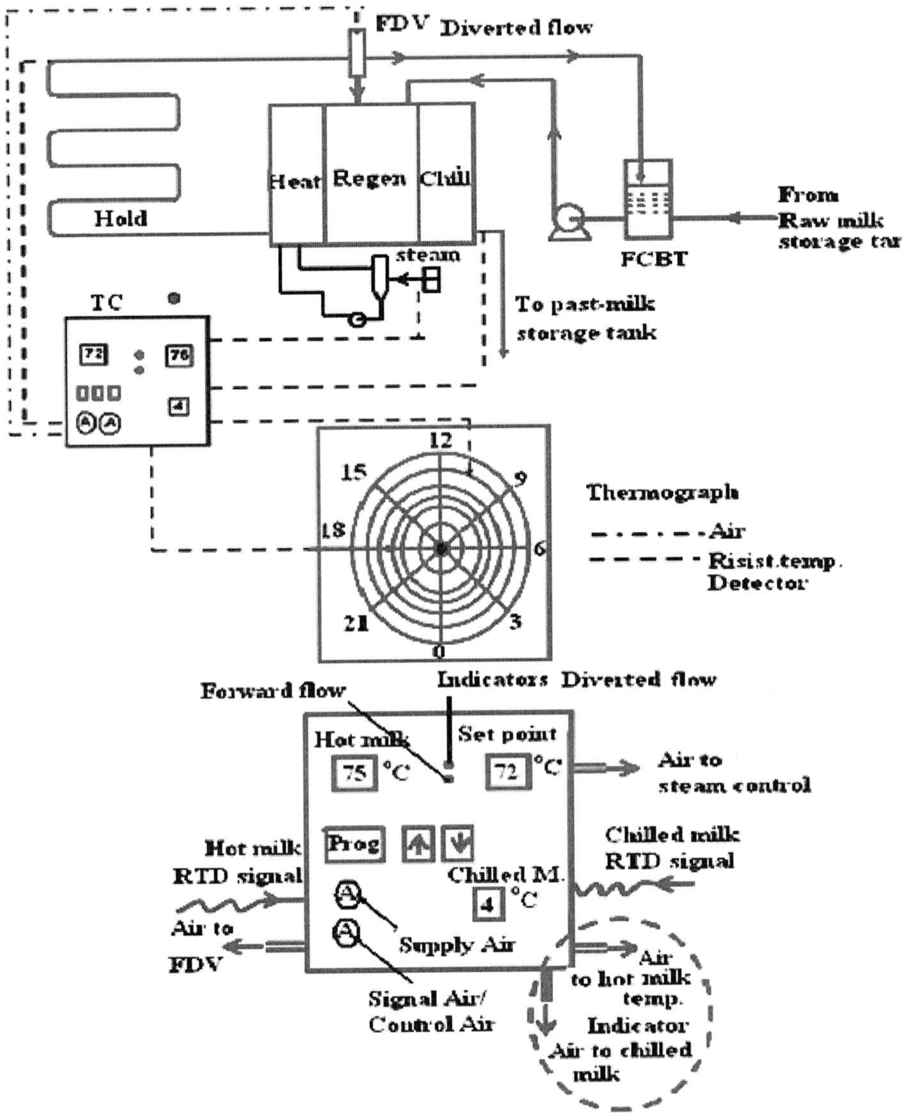

Figure 8.6. HTST Pasteurizer Controls

The controller components are very much standard, and uniform, so that faulty ones can be replaced, if their specifications are known. The motive supply air must be 6 bar and which is then reduced to the required pressure depending on the size of the valves to be operated. The air must pass through filter to avoid moisture and oil.

Flow Diversion Valve(FDV)

The temperature sensor of hot milk also conveys signal to the Temperature controller, where, the signal is compared to the Set Point. If the temperature of milk is less than the Set Point, the electrical signal goes to the Solenoid valve, which stops the compressed air supply to FDV. The compressed air, which will be acting against the spring pressure, is now absent. The FDV valve has three openings, one for supply of milk, and two outgoing. When the compressed air is acting, the FDV will be in Forward flow position, while in the absence of air pressure, will push the valve plunger in the FDV to Diverted flow position. It must be noted here that, whenever electric power interruption is there for the plant, the FDV will be in Diverted flow position.

The signal that triggers the Diverted flow condition of FDV is also fed to the visual signal as well as to a hooter, to alert the plant personnel.

Care and maintenance of Pasteurizer

The performance of the pasteurizer is an important parameter for the dairy plant as a whole, as this is the heart of the dairy plant operations. Its efficiency in pasteurizing the milk and with minimum of services requirement reduces the cost of operation of the plant. One of the important parameters to judge the pasteurizer efficiency is the Regeneration Efficiency, the step in which, the hot milk and cold raw milk exchange heat.

Regeneration in HTST Pasteurizer

Thermal regeneration is a technique devised to reduce service loads and thus improve energy usage. Percentage regeneration may be defined in several ways and due to these different methods, heat exchangers having identical thermal performance may have, seemingly, different regeneration efficiencies.

One definition of Regeneration efficiency is "The temperature range through which the product is heated by regeneration, expressed as a percentage of the total heating range".

A simple heat exchanger is considered where a product is to be heated from 5°C to 80°C first by 80 per cent regeneration and then finally in a heating section. Having heated the product to 80°C it is held in holding tubes and returned to the regeneration section at 80°C.

The total temperature heating range is 80 -5 = 75°C

The temperature range through which the product is heating by regeneration must be 80 per cent of 75°C = 60°C.

Thus the incoming product will be raised in temperature from 5°C to 65 °C by regeneration and will be heated from 65°C to 80°C in the heating section. The heated product is then returned to the regeneration section at 80°C, where it is cooled regeneratively by the cold incoming product at 5 °C. This can be depicted as follows:

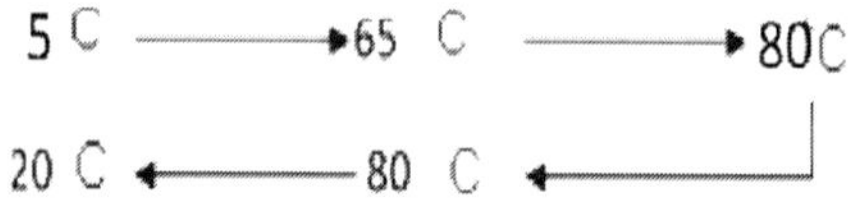

Preventive Maintenance of HTST Pasteurizer

The plant should be inspected externally for leakage of all kinds, and be monitored continuously throughout the process and any abnormalities should be investigated and rectified. The flow rate of milk and media should be checked by measuring the time to process a known quantity of milk.

Plates

1. The plates have to be cleaned manually, time to time, as required in addition to the regular CIP cleaning after day's operations. It should be done with nylon or fibre brush or coir.
2. The proper tightening and dismantling procedures have to be followed. Care should be taken not to tighten beyond the mark, provided on the hanging bar by the manufacturer. The tightening has to be done such a way that plates move parallel to each other and the measurement of length at the top edges of the plate pack should be same as at the bottom.

Gaskets

Gaskets must be of nitrile rubber which can with stand temperatures up to 120°C. Use of aggressive acid solutions during CIP cleaning might damage the gaskets. They should be replaced as a set instead of one at a time, to prevent misalignment of plates. New gaskets should be tightened to the minimum. As gaskets get older, tightening should be increased just to avoid leakage. While replacing the old gaskets with new gaskets, the remnants of old gaskets should be carefully removed from the gasket groove. Old adhesive, grease, dirt *etc.* should be removed by acetone and the plate groove thoroughly cleaned and dried. The new gaskets for replacement should be slightly sand papered at the back and should be cleaned and dried. A thin coat of adhesive should be applied to the gasket groove and back of the gasket with a small brush for even distribution. While the adhesive is still sticky, the gasket should be fitted in to the gasket groove. A slight pressure on plate for some minutes will help in the fixing of the gaskets. Excess of adhesive should be removed if any still present.

Frame

The frame having threads and spindle should be lightly coated with grease, to help easy movement of the plates. The threads should be covered with plastic cover, to prevent accumulating dirt.

Filter

Filter cloth should be checked from time to time for any tear or worn out patches. It should be washed after dismantling. Cloth should be replaced if any damage is observed. So is the case with wire gauge filter screen protecting the filter cloth.

Instruments

All thermometers should be checked periodically for accuracy. All air operated instruments should be supplied with clean dry air. The air filters should be inspected periodically, cleaned or changed when required. The components of controlling instruments should be replaced whenever damaged or functioning inaccurately. Leakages in the control instruments of steam supply should be checked.

Pumps and Motors

The pumps and motors should be thoroughly overhauled once in a year. The shaft seals if leaking should be replaced immediately. Motors if noisy, should be checked for damaged bearings. Misalignment is one of the most common sources of pump troubles. Scales in hot water pump should be removed.

Problem No. 1

Required

A milk pasteurizer, with the following data:

Capacity	12.000 1tr/h
Inlet temperature,	15°C
Pasteurizing temperature	85°C
Outlet temperature	4°C
Holding time & Quantity	15 sec. and 50 ltr.
Regeneration	80 per cent

Available

24.000 l t r/h hot water at 90°C

18.000 l t r/h well water at 11°C

24.000 l t r/h chilled water at 1°C

Branch to centrifuge at about 45 °C. Calculate the number of plates and pressure drop. Make suitable assumptions.

Solution

Assumption: Neglecting the reduction in volume.

When calculating for pasteurizers, we normally do not take into account the separation of cream in the centrifuge, which is about 10 per cent.

Calculation

A regeneration of 80 per cent means that 80 per cent of the heat is regained in the regenerating section so the loss will be 20 per cent.

In this case $\frac{80}{100}(85 - 15) = 56\,^{\circ}\text{C}$ rise in temperature

So the complete temperature course will be:

15 – 71 – 85 – 29 – 15 - 4°C.

Heating Section

Heating by 24.000 l t r/h water of 90 °C

$$12\left(\frac{71-85}{83-90}\right)5 \qquad t = 8\,^{\circ}\text{C}$$

$$N = \frac{12.000 \times 14}{0.375 \times 2320 \times 8} = 24$$

Arrangement 3 × 5/2 × 8 = 32 plates.

Regenerating Section

$$4\left(\frac{15-71}{29-85}\right)14 \qquad t = 14\,^{\circ}\text{C}$$

$$N = \frac{12.000 \times 56}{0.375 \times 2290 \times 14} = 56$$

Arrangement 6 × 5/6 × 5 = 61

However, in the entering range 15 – 71 a branch to the centrifuge has to be provided as close as possible to 45 °C.

Each group means a rise in temperature of $\frac{71-15}{6} = 9.3\,^{\circ}\text{C}$.

If we make a branch after 3 groups the temperature there will be approximately 43°C.

Arrangement reg. section I: 3 × 5/3 × 5 = 31 plates

Reg. section II: 3 × 5/3 × 5 = 31 plates

Cooling

By means of 18.000 l t r/h well water of 11 °C.

$$9\begin{pmatrix} 29-15 \\ 20-11 \end{pmatrix} 4 \quad t = 6,1\ ^{\circ}C$$

$$n = \frac{12.000 \times 14}{0,375 \times 2100 \times 6,1} = 35$$

Arrangement 3 × 6/3 × 6 = 37

Deep Cooling

By means of 24.000 l t r/h chilled water of 1°C.

$$8,5\begin{pmatrix} 15-4 \\ 6,5-1 \end{pmatrix} 3 \quad t = 5,25\ ^{\circ}C$$

$$N = \frac{12.000 \times 11}{0,375 \times 1800 \times 5,25} = 36$$

Arrangement 3 × 6/3 × 6 = 37

Pressure Drops

Milk	(15 × 1, 12) + (6 × 0.78) + (28 × 0,1) + 3 (holding section)	= 27,3 mwc
Hot water	(2 × 1, 75) + (3 × 0.19)	= 4, 1 mwc
Well water	(3 × 1, 75) + (4 × 0, 11)	= 5, 7 mwc
Chilled water	(3 × 3, 00) + (4 × 0, 19)	= 9, 8 mwc

Problem No. 2

	Milk	*Hot water*	*Chilled water*
1. Flow Rate	10,000lph	30,000 lph	30,000 lph
2. Inlet temperature	4°C	85°C	1°C
3. Outlet temp.	4°C		
4. Overall heat Transfer coefficient 'U' k.Cal/m² hr °C	2320	2290	1800

Area of each plate 0.375 m^2; Regeneration Efficiency: 80 per cent; Past. Temp. 75°C

Calculate the number of heat exchanger plates required in each section.

Regeneration Section

Regeneration $\frac{80}{100} = \frac{(x-4)}{75-4}$

X=60.8

Heat balance between raw milk and heated milk, assuming no heat loss to the surrounding

10,000 x 0.93 x (60.8-4) =10,000 x 0.93 x (75-x)

X=18.2 °C

Remember that rate of heat is different than heat balance

Heating Section

Heat gained by milk = Heat lost by hot water

10,000 x 0.93 x (75-60.8) =30,000 x1 x (85-*x*)

$\frac{14.2}{3}$ = 85-xor x = 80.3 °C

Chilling Section

10,000 x 0.93 x (18.2-4) = 30,000 x 1x (*x*-1)

X=5.7 °C

Heat gained by raw milk in Regeneration section is rate of transfer in that section

10,000 x 0.93 x (60.8-4) = $UA\Delta T_{LMTD}$

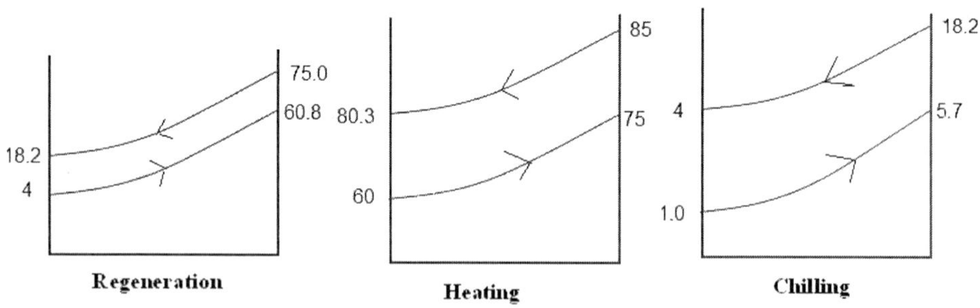

Temperature Profiles in Various Section of Pasteurizer.

Regeneration	Heating	Chilling
$\Delta T_{LMTD} = \dfrac{(18.2-4)-(75.0-60.8)}{\ln\dfrac{(18.2-4)}{(75-60.8)}}$	$= \dfrac{(80.6-60)-(85-75)}{\ln\dfrac{(80.6-60)}{(85-75)}}$	$= \dfrac{(4-1)-(18.2-5.7)}{\ln\dfrac{(3)}{(18.2-5.7)}}$
	$= \dfrac{(19.4-10)}{\ln\dfrac{(19.4)}{(10)}}$	$= \dfrac{(12.5-3)}{\ln\dfrac{12.5-3}{\dfrac{(12.5)}{(3)}}}$
	$= \dfrac{9.4}{\ln(1.94)}$	$= \dfrac{9.5}{\ln(4.16)}$
	$= \dfrac{9.8}{1.42}$	$= \dfrac{9.8}{1.42}$
=14.2	14.24	6.75

Regeneration Section $= \dfrac{10{,}000 \times 0.93 \times (60.8-4)}{2290 \times 0.375 \times 14.2}$

= 43.3 or 44 plates

Heating section $= \dfrac{10{,}000 \times 0.93 \times (75-60)}{2320 \times 0.375 \times 14.24}$

$= 11.26 \approx 12$

Chilling section $= \dfrac{10{,}000 \times 0.93 \times (18.2-4)}{1800 \times 0.375 \times 6.75} = 29$

Answer

The number of plates required in Regeneration, heating and chilling sections are 44, 12 and 29 respectively.

Role of Correction Factor

The above calculations using LMTD, may not be the true representative of counter current flow, as the milk and service fluid are differing in the flow rates. To control the pressures on both sides to be equal, the flow in the channels is made unequal. When the milk and service fluid flow is in the ratio of the range of 0.66 to 1.5, it is possible to have equal number of passes on both sides of the exchangers making the flow to be counter current. When it becomes necessary to go for unequal passes due to large differences in flow rates, it becomes necessary to apply significant correction factor to take into account the non-counter current flow situation, which is similar to cross flow.

The extent of LMTD correction depends upon several factors, including the number of NTU/pass and number of channels per pass. Correction factor is required also because, in the end zones of plates, heat is transferred only on one side, and with even number of plates, one fluid has an extra stream or passage over the other.

ΔT LMTD cross flow = F X ΔT LMTD

$$F = \frac{log_e\left(\frac{1-P}{1-P.R}\right)}{(R-1)logy_e\left(\frac{(R)}{R+by_e(1-P.R.)}\right)}$$

when

$$R = \frac{T_{hi} - T_{ho}}{T_{co} - T_{Ci}}$$

$$P = \frac{T_{co} - T_{ci}}{T_{hi} - T_{ci}}$$

Problem 3

For calculating the time taken for bringing up to pasteurization temperature in a batch pasteurizer the following data is given.

Data

Pasteurization Temperature Tp	75	C	Conductivity of vat wall, k	13
Initial temperature of milk To	4	C	Thickness of wall, x	0.003
Temperature of Hot fluid	85	C	Co-eff of heat transfer, steam	5000
Volume of Milk	3	m^3	Co-eff of heat transfer, milk	860
Density of milk	1030	kg/m^3		
Specific heat of milk	0.93	k.cal kg C		
Area of heat transfer	9.5	m^2		
Overal Heat transfer co-eff	627.5258	$k.cal/m^2h$ C		
Time taken for heating, t =	$\frac{m.c}{A.U}.l_n\left[\frac{T_s - T_o}{T_s - T_m}\right]$	1.008369 h or 61 min	min	

Chapter 9

Sterilization Equipment

Introduction

Sterilization is the term used for the process to destroy all the microorganisms present, both vegetative forms and spores, or at least make them incapable of growth in the product, so that a long keeping quality is obtained without refrigerated storage.

The term "sterilized" here does not imply that the product is sterile in the strict microbiological sense. It has been clear for many years that absolute sterility cannot be obtained by any heat treatment processes. If the treated product is incubated at appropriate temperature, spoilage will be noticed at some parts. This spoilage can be reduced by increasing the time-temperature of the heat treatment process. However, this will cause undesirable chemical changes, color, flavour and or nutritional changes.

Typical time – temperature combinations for in-container sterilization processes are 105-120 °C with effective processing times of 10-30 minutes, the higher temperatures being associated with the shorter times. This sterilization effect can also be obtained by using much higher processing temperatures like 135-150 °C, with shorter holding times of about few seconds. These processes cause much fewer chemical changes as a result of heat treatment than that of conventional sterilization processes. The higher temperature processes are frequently considered as different processes, which have been given the description "Ultra High Temperature "(UHT)" in-container processes, with limited heat transfer rates, do not allow sufficiently rapid change of product temperature. Therefore, UHT processing involves the treatment of product in continuous flow in a heat exchanger.

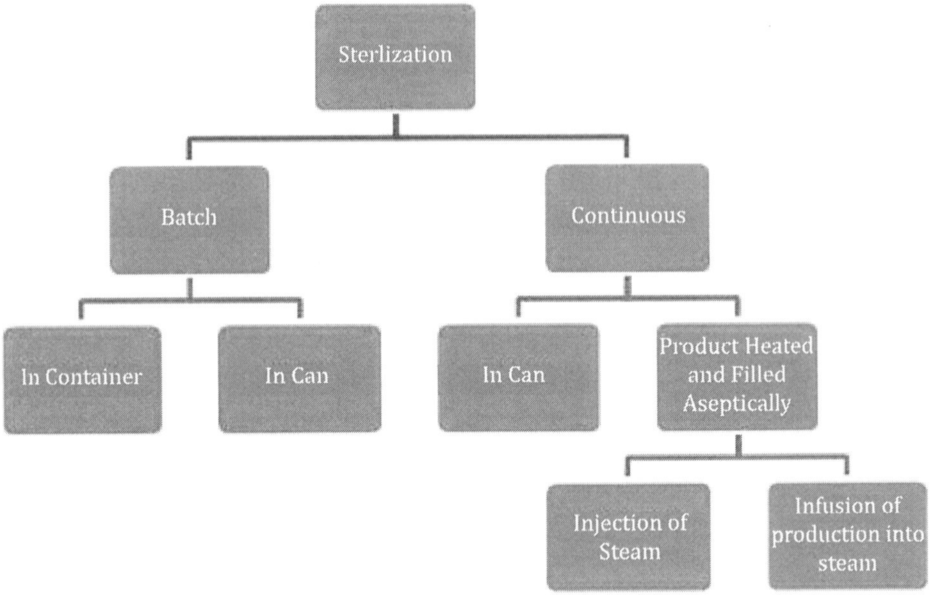

Figure 9.1. Classification of Sterilization Process

Packaging of UHT milk has to be done in aseptic condition. In comparison to In-can sterilized milk, UHT milk cannot completely inactivate the more heat-resistant, proteolytic enzymes, particularly when initially present at high concentrations.

Batch Sterilization Equipment

The Batch Sterilization equipment, (Figure 9.2) usually called Retorts, consists of a cylindrical tank, under pressure, into which the cans containing the food products, placed in cages are lowered. Steam flows into the retort from the top and air is sucked out of the bottom. After heating, cooled water is sprayed into the retort. To prevent the pressure in the cans from becoming higher than that in the retort itself, compressed air is blown into the retort to compensate for the pressure drop, and when the temperature is reduced to below 100 °C the retort may be opened.

The advantage of this type of sterilization equipment is the flexibility with regard to package size and the relative ease to maintain constant temperature. However, the operation of loading is laborious, and the heating is slow because of no agitation of the inside liquid.

Continuous in-Can Sterilization Equipment

These are more suitable to larger capacity of handling. It has basically three sections. In the first section, the cans are heated, in the second section the cans are cooled under high pressure and in the third section they are cooled at atmospheric

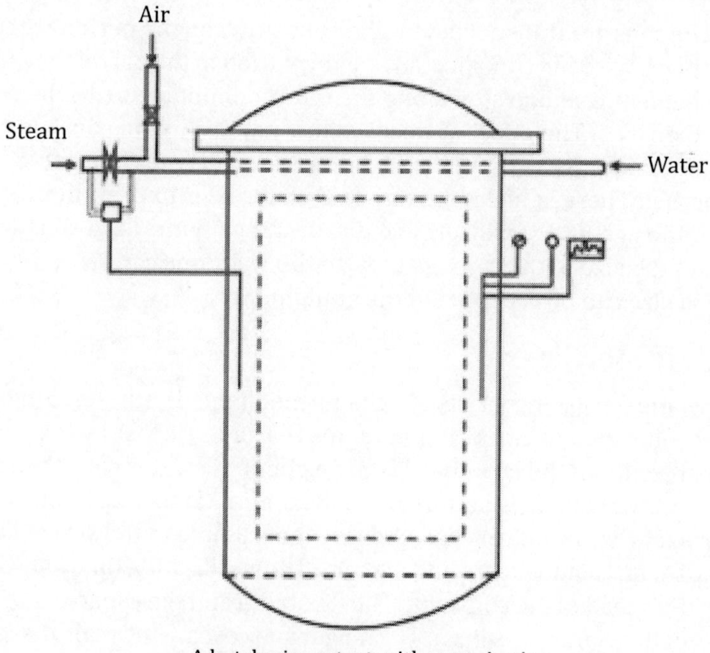

A batchwise retort without agitation

Figure 9.2. Batch Sterilizer

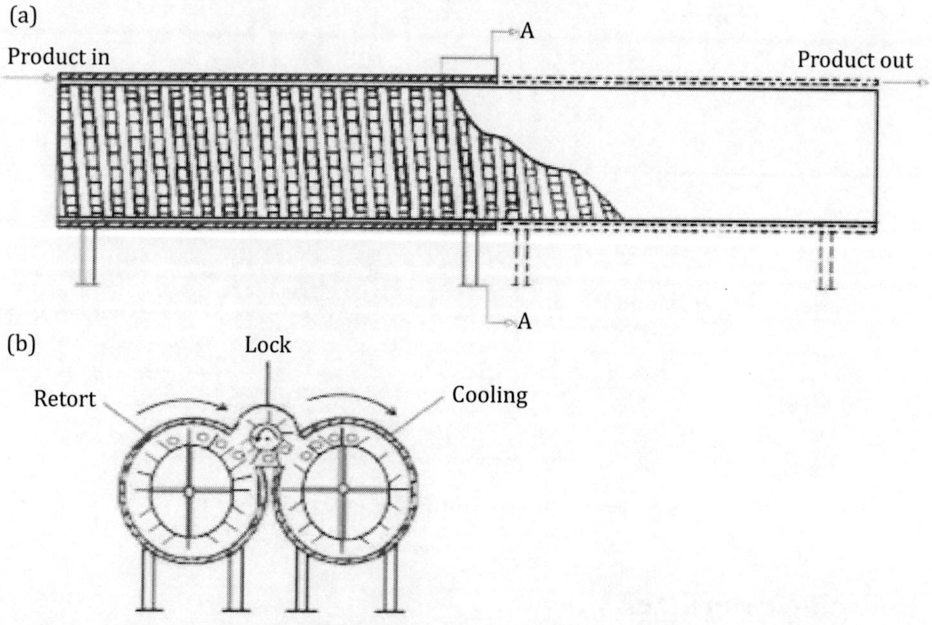

Figure 9.3. Continuous In-Can Sterilizer

pressure. The cans are transported in a helix position on the periphery of a barrel, and when they have been transported one revolution of the barrel they have at the same time been moved one step along the retort. Simultaneously, the can rotates around its own axis. The effect of this is a flow pattern in the package enhancing heat transfer in liquids. As shown in Figure 9.3 the cans are transported from one zone to another. There is high pressure in the sterilizer to prevent differences in pressure in the can and sterilizer. The disadvantage with these sterilizers is the limitation on the size of the cans, as they are built for one can size and only small difference in size can be accepted in the equipment.

Hydrostatic Sterilizer

Hydrostatic sterilizer consists of a chamber equipped with steam injection. The chamber is connected to two water columns (barometric lock) which are used to adjust the pressure in the chamber. If the height of the water columns is changed, the steam pressure changes and thus the maximum temperature obtainable too. These sterilizers are often very tall. A conveyor, suitable to the size of can, travels through the steam chamber carrying the cans. The sterilization time may be changed by varying the speed of the conveyor. The flexibility and the capacity are the major advantages of this type of sterilizer. The height and cost are the main disadvantages.

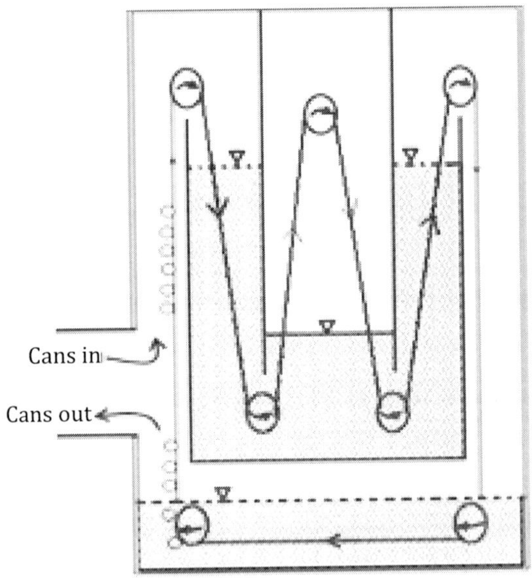

Figure 9.4. Hydrostatic In-can Sterilizer

Hydrolock Sterilizer

This equipment consists of a continuous sterilizer that has a rotating pressure lock, called hydro-lock, which is partly submerged under water (Figure 9.5). The cans pass the hydro lock into a preheating chamber filed with water. They are then transported into the steam chamber. The bottom part of this chamber contains hot water thus further heating the cans. The cans are then transported to the upper part of the chamber containing steam and here sterilization takes place. The product then passes through a cooling section. Hydro-lock sterilizers are flexible but they are rather large and costly. Leakage in the rotating valve of the hydro-lock is a common problem.

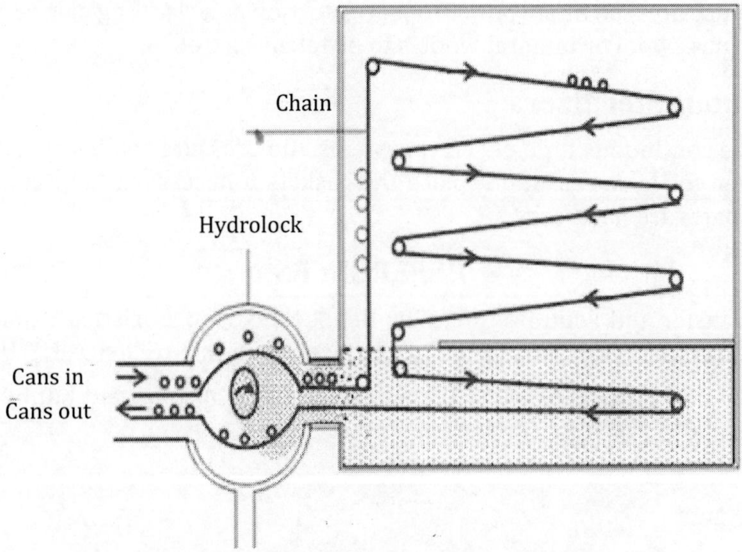

Figure 9.5. Hydro-Lock In-Can Sterilizer

Care and Maintenance of Sterilizers

The components that are liable to be damaged over the use and need replacement or recalibration are a) Main door gaskets b) Steam supply valves c) Steam trap d) Water spray nozzles e) Pressure gauges f) Pressure safety valves.

The gaskets are usually of asbestos tape or rope and will get damaged over the period. The correct dimensions of the gasket are important, as well as seating in the groove provided to the closing door. The hinges of the door will need lubrication, to facilitate smooth opening and closing. So are the bolts and levers that ensure tightness of the closure of door. If the sterilizers are of rotating type, then lubrication the support systems is essential.

Steam supply valves should be leak proof, and if necessary, the gland packing has to be replaced. They should be replaced whenever leakage is noticed, to avoid loss of steam and increased cost of operation, and accidents to the operators. The proper operation of steam trap is to be checked from time to time, and there should not be damage due to corrosion or scale formation.

Steam pressure gauges must be tested for accuracy, as well as the steam safety valve. Steam safety valves also are susceptible to corrosion and not opening at the required set point. It has to be checked whether any local laws are to be satisfied in their installation or operation. Water spray nozzles should be checked for scale formation and chocking.

As the sterilizer operates at high temperature, the insulation and its maintenance are also important. Proper protection and cladding of the insulation whether glass wool or mineral wool is to be taken care of.

Continuous Sterilizers

In the continuous type of retort systems, the pressure sealing system has to be checked for leakages, and replace the gaskets if necessary. Lubrication of all rotating parts are necessary.

REFERENCES

" Heat Transfer and Food Products" by Bengt Hallstrom, Christina Skjoldebrand and Christian Tragardh, (1988), Pub: Elsevier Applied Science, NY, USA.

"Thermal Food Processing: New Technologies and quality issues", Edt by Da-Wen Sun (2006) Pub: CRC, Taylor and Francis Group.

Chapter 10

UHT Processing Equipment

Introduction

Sterilization at higher temperatures are usually referred to as UHT sterilization. The heating to this high temperature for only few seconds, demands that the heating and subsequent cooling has to be as quick as possible to reduce heat damage. The other factors like cost, the technical expertise in handling and maintaining sterility are also important.

Indirect Heating

This type of heat exchanger can consist of an assembly of plates, or tubular or a combination of both. If the product is very viscous, use of Scraped-surface heat exchanger is used. As the product in the plant must not boil at the highest pressure, a back pressure must exist which is more than the vapour pressure of the product at this maximum temperature. At a temperature of 135 °C, a back pressure of about 2 bar is needed. Further, to avoid bubbling of any dissolved air, at least 1 bar over and above the pressure determined by the sterilization temperature is needed. Another factor is the hydrodynamic pressure drop resulting from the product being pumped through the heat exchanger. Hence, the highest internal pressure in the heat exchanger may reach 6 - 8 bar.

Plate Heat Exchanger

Generally, plate heat exchangers are limited to lower temperatures, but, here the special design considerations are required as the operative pressures have to be higher.

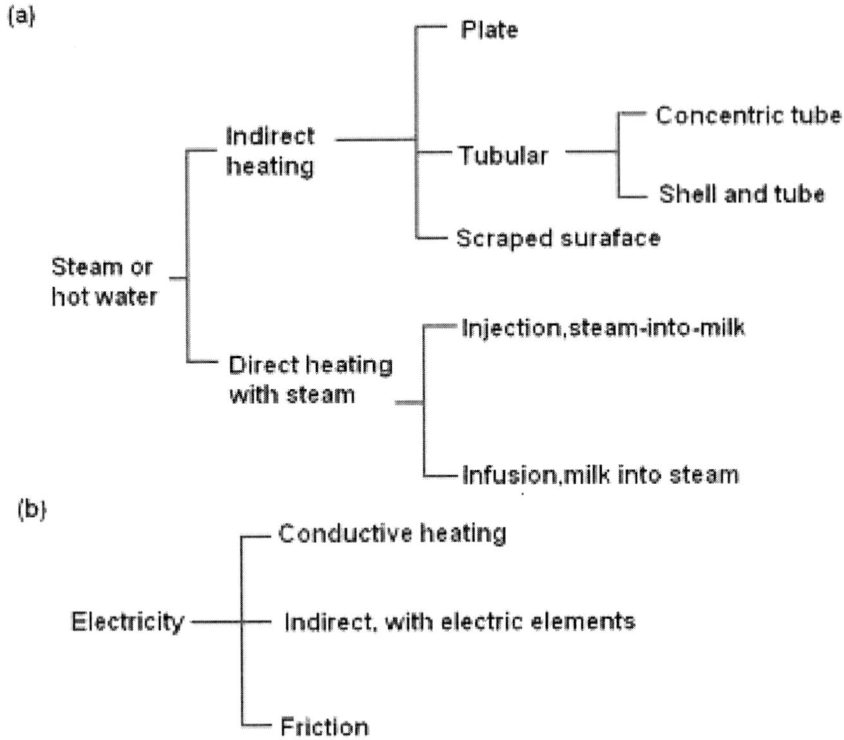

Figure 10.1. UHT Plant and Heating Systems

To withstand the severe conditions of temperature and pressure, the gasket materials used with plates for UHT plant must be more sophisticated. Nitrile rubber is suitable for temperatures up to about 138°C. For higher temperatures, up to 160°C, resin-cured butyl rubber or EPDM (Ethylene Propylene Diene Methylene) are suitable. Gasket performance is also limited by the adhesion of the gasket to the stainless steel plate. A safer option will be with plate and gasket designs where gasket is not bonded to the plate, but is held in place mechanically and can be replaced easily when necessary. High internal pressures, and varying high pressures within the heat exchanger, may cause flexing and distortion of the stainless steel plates. Hence the designs are such that multiple contact points between adjacent plates are provided, to give mutual mechanical support and increase the rigidity of the whole plate pack.

In the sequence of operation of the UHT plant, the heat exchangers can be of regeneration type between milk to be heated and the milk already heated, as in normal pasteurizer. A more efficient one is where an entire water circuit takes care of the heat exchanges progressively (Figure 10.2).

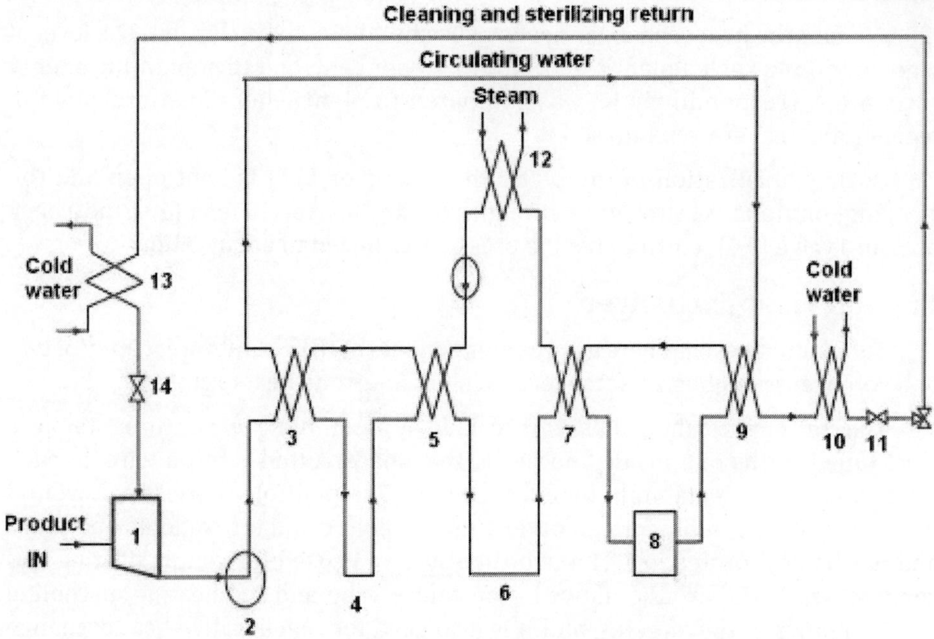

Figure 10.2. Developed Plate-Type Indirect UHT Plant

The untreated product is pumped from balance (1) tank by centrifugal pump (2) to the first plate heat exchange section (3). The end of this heating section, at a temperature of 110-115°C, there may be a holding tube (4) to provide a few seconds holding to reduce fouling of heating surfaces later in the plant. The next heating section (5) brings the product to the final sterilizing temperature of about 140°C, where it is held for the required holding time in the holding tube(6). It is then followed by a cooling section (7) which cools the product to homogenization temperature of about 55°C to 75°C.

The homogenizer (8) being positioned after the sterilization process will have to be of the type of Aseptic type of homogenizer. It is generally considered that if milk has to be processed, the homogenization is before the sterilization process, while if it is products like cream, then the homogenization is preferred after sterilization.

After aseptic homogenization, the product passes through a further cooling heat exchanger section (9), and finally through cold water cooling stage(10) to bring the product to the required outlet temperature. A restrictor (11) is placed after this to provide back pressure required in the plant.

The heat is transferred between product and water circulating in the counter-current flow in a closed circuit. The water in this circuit is heated to a little above the required sterilization temperature by steam under pressure in a heat exchanger section (12). The low temperature differential of not more than 3°C, reduces the amount of fouling on the heat exchange surfaces, and allow longer operating times before the heat exchanger needs to be shut down to be cleaned. The level of regeneration in such a plant can be above 90 per cent, and the operating energy costs are correspondingly low, as compared to plant where heat exchange in regeneration is between milks.

During sterilization of the plant, the restrictor (11) is kept open and the cleaning solutions are circulated through a cooling heat exchanger (13) and a back pressure valve (14) to bring the line pressure to be fed to balance tank.

Tubular Heat Exchanger

Tubular heat exchangers used as components of UHT plants are mainly of two types, concentric tubes, or some form of shell-and-tube heat exchanger.

The concentric-tube systems are formed by assembling two or three stainless steel tube lengths one inside another with a spacer, often a spiral wire, in each inter-tube space to maintain them concentric. The multiple tube is then wound into a coil, for assembly into an outer cylindrical housing for hygienic enclosure and mechanical protection. The two-tube system is used for simple heating and cooling, where the product flows in the centre tube and the heating or cooling medium flows in the annulus, and it is also used for regenerative heat exchange where product flows in both spaces.

The triple-tube system is used for the final stage of heating to the sterilization temperature, where the product flows in the annulus between the two tubes and the heating medium flows in the centre tube and in the outer annulus. In this way, the available heat transfer area is doubled and the rate of heating is increased. Triple tubes may also be used in final cooling sections, especially where cooling rates are restricted by high products viscosity so that increased transfer area is used to compensate for reduced transfer coefficients.

A Typical flow diagram for an indirect UHT plant using concentric tube heat exchanger sections is show in Figure 10.3. The product is pumped from the balance tank (1) by a centrifugal pump, (2), through the steam-heated sterilizing heater which, during products sterilization, is inactive. The product is first heated in the regenerative heater (4) by the outgoing product. Connections are made to the homogenizer (5) at a suitable temperature point within the regenerator.

A valuable characteristic of all tubular heat exchangers is that the tubes carrying the product are strong enough to withstand the full pressure required for effective homogenization. It is therefore possible to instal the high-pressure reciprocating pump of the homogenizer before the sterile section of the plant and quite independently of the position of the homogenizer valve. The hygiene problems of installing a homogenizer after sterilization lie with the pump, not with the homogenizer valve, and the extra complication and cost of a sterile homogenizer arises from modifications that have to be made to the pump to avoid bacterial contamination. If the homogenizer pump can be included before the sterile section, it need not be of aseptic design. Furthermore, if the heat exchanger withstands the full homogenizing pressure, the homogenizing valve can be put at any point downstream from the pump, even after the sterile section. In fact, in a plant of the type shown, two homogenizing valves may be fitted, one at (5) in association with the homogenizer pump and one after the sterile section where the product is at a suitable temperature during cooling, *e.g.* at (5a). The product can therefore be homogenized before sterilization, after sterilization, or both.

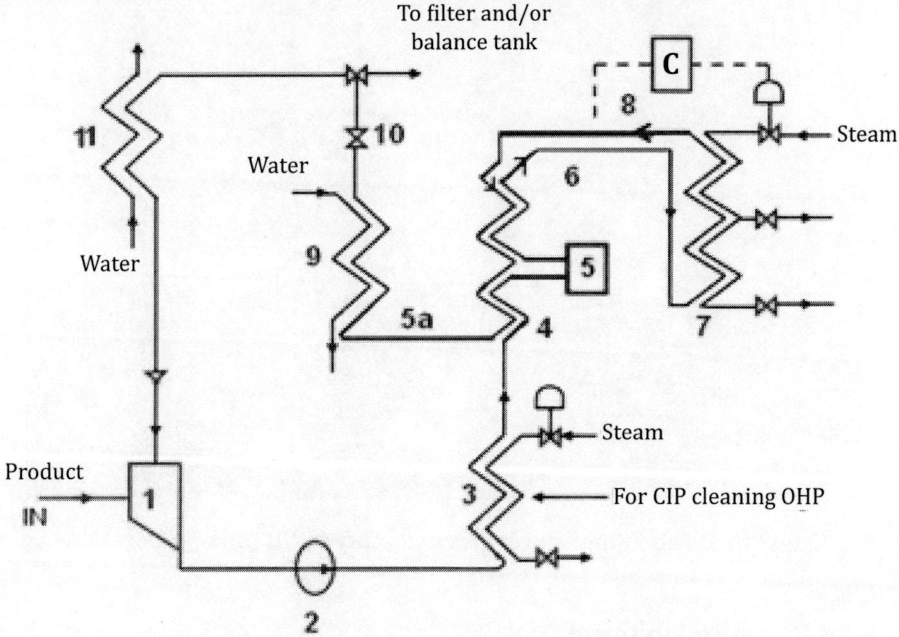

Figure 10.3. Concentric-Tube Type Indirect UHT Plant

After the first homogenizing point at (5), heating continues in the regenerator (6) and then in the stem-heated section (7) to the final sterilizing temperature. After the holding tube, (8), the product returns to the regenerator for cooling. Final cooling to the outlet temperature is in a water-cooled section (9). Back pressure is maintained by an orifice plate or back pressure valve at (10).

The required sterilization temperature in the holding tube is maintained by a controller (C) which senses the temperature in the holding tube and varies the steam supply to the final steam heating section (7) as necessary.

In industrial practice, the tubular heat exchangers are arranged in a bundle with a common chamber with insulation. The ends are provided with swing bends, so as to enable the variations such as capacity variations, holding time variations as well as position of homogenizer (Figure 10.4)

Figure 10.4. Tube Bundle provides flexibility in UHT Plant Operations

Direct Heating Systems

There are certain disadvantages in the indirect heating system. To overcome this, a system where the heating medium *viz.* Steam, is directly brought in contact with product to be heated, has been developed. This eliminates the following disadvantages of Indirect heating system.

1. Thickness of heat exchange wall slows down the rate of coming up time, cooling rate.
2. Fouling of heat transfer surface, and thereby reducing the operating time.

The product is first raised to 80-85 °C by Indirect heating system, and then the product is raised to sterilizing temperature of 140-150 °C by direct injection of steam. The temperature rises rapidly, but also causes dilution of product due to steam condensing into the milk, almost to an extent of 10 per cent or more. Hence the holding section coming later will have to be designed for greater capacity to this extent. After the designed holding time at the sterilization temperature, expansion cooling follows. The vacuum in the expansion vessel will normally be controlled at a level corresponding to a boiling temperature little above that of the product before mixing with steam. This will result in removal of exactly the amount of water added as condensed steam.

The general sequence of operations in Direct Heating Systems is shown in Figures 10.5–10.7

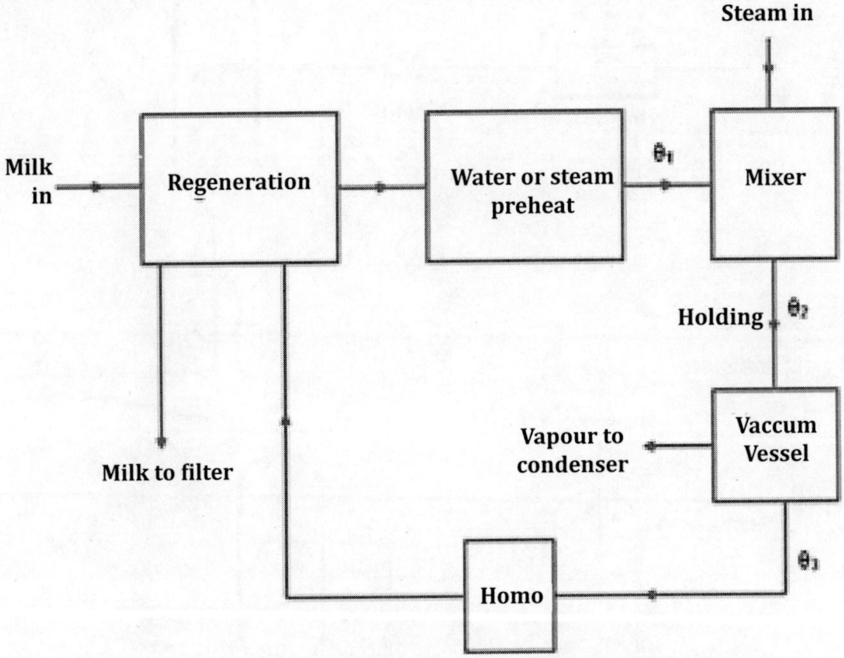

Figure 10.5. Sequence of Operations in Direct-Heating System

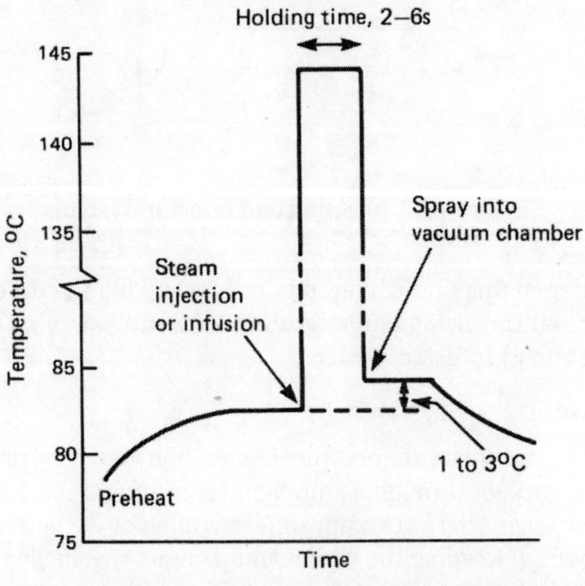

Figure 10.6. Time-temperature Relationship during Direct Heating and Cooling in Vacuum Chamber

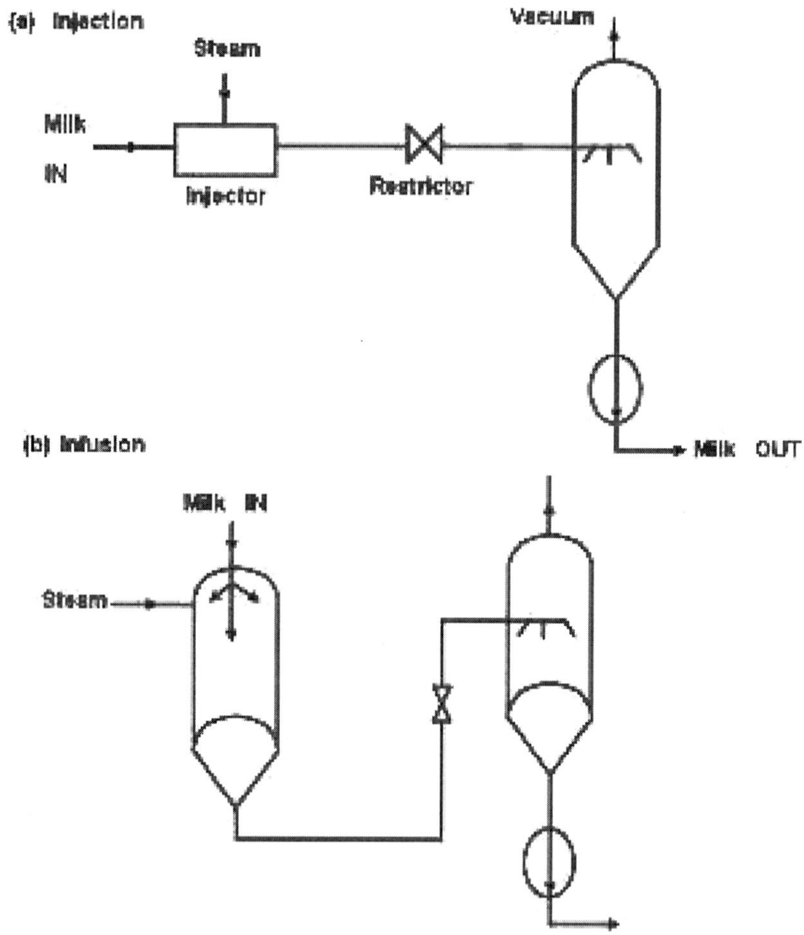

Figure 10.7. Injection and Infusion Systems

However, depending on the method used for mixing the product with steam to achieve the final sterilizing temperature, the heating systems are classified as 1) Injection system 2) Infusion system.

Injection System

In this system, the steam at a pressure higher than that of the product is injected into the product stream through a suitable nozzle, and is condensed to give the required temperature. There are many different injector designs, and meeting the basic requirement of keeping the steam and product thermally separated as far as possible until they reach the mixing zone. This will prevent surface fouling of the injector.

The important requirement in these systems is the rapid condensation of steam, to prevent the passage of bubbles of uncondensed steam into the holding tube, which would reduce the effective holding time at the sterilization temperature. The rapid condensation is achieved by introducing the steam into the liquid in the form of small bubbles or in the form of a thin sheet. It is also essential that sufficient back pressure in the liquid is maintained, above that needed to prevent boiling, at the steam injector.

Another design requirement is to keep the pressure difference between liquid and steam to be as small as possible, as this will require lower pressure of steam, and consequent savings in production of steam. Figures below show three possible designs of the injectors. In the first(a), the product tube takes the expanding part of a venturi. The steam is supplied through four tubes which surround the product tube. A series of small orifices along the lines of contact of the steam tubes with the product tube allow the steam to enter the product in small bubbles. Some of the orifices are drilled radially, and some may be drilled at an angle to encourage swirl and turbulence in the product to increase the rate of condensation.

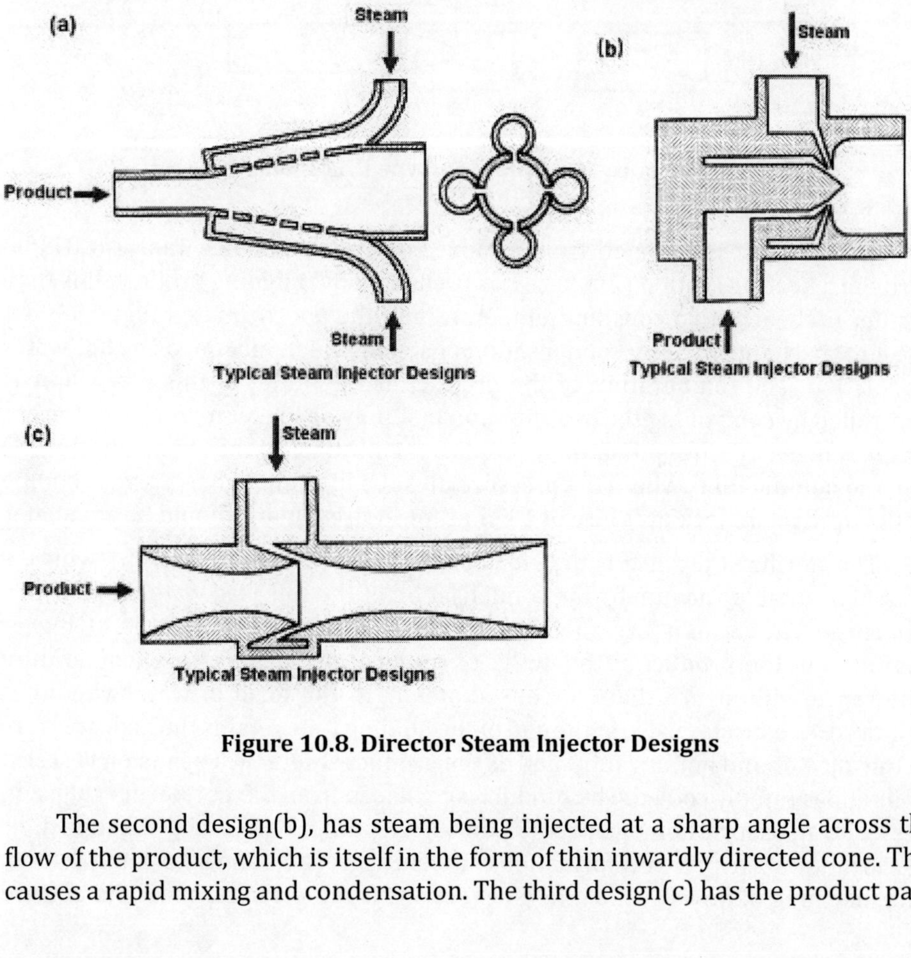

Figure 10.8. Director Steam Injector Designs

The second design(b), has steam being injected at a sharp angle across the flow of the product, which is itself in the form of thin inwardly directed cone. This causes a rapid mixing and condensation. The third design(c) has the product pass through a venturi and the steam is injected in the expansion section of the venturi where the product pressure is rising. The steam is injected in a thin annulus around the product. A second venturi section causes further fall and rise of pressure within the injector and enhances condensation and rapid heat transfer.

The product flow and arrangement of sections in a typical steam-injection type of UHT sterilizer is shown in Figure 10.9

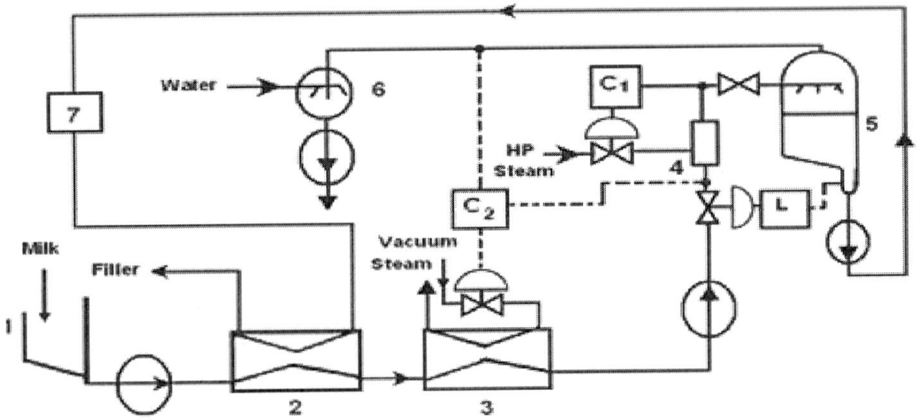

Figure 10.9. Injection Type UHT Plant

The product is pumped from a Float Controlled Balance Tank (FCBT)(1) through a heating section (2)where it is preheated by outgoing product. It is then further preheated to a constant temperature in the approximate range 75-85 °C in a heat exchanger (3) by condensation of steam under vacuum or by hot water circulation. The temperature of the product at the outlet of this pre-heater is controlled by controlling the heating vapour supply or the temperature of the hot water. A high-pressure pump then supplies the product to the steam injector (4) and the holding tube, where the pressure must be high enough to prevent boiling of the product. A back pressure of 1 bar as for indirect plant should be adequate.

The sterilized product is then lead to Expansion cooling vessel (5), which is kept at a suitable vacuum by the condensation of the released water vapour in a condenser. The vacuum corresponds to a boiling temperature of about 1-2 °C higher than that of the product at the outlet of the final pre-heater. This temperature difference will ensure that condensed steam, in the form of water vapour is removed. A restrictor valve at the end of the holding tube retains the high pressure in the injector and holding tube, and as the product enters the expansion vessel, it boils and is rapidly cooled. The product is then lead from the bottom of expansion vessel, to an aseptic homogenizer (7) through a pump. The homogenization after sterilization stage prevents formation of casein aggregates and associated defects. The matching of flows in the different sections is important to meet the defined holding time for the process. It is essential to maintain balance between water addition as steam and water extraction as vapours from the expansion vessel, by maintaining adequate difference between the temperature of the product before mixing with steam and after expansion cooling.

A closed circuit is used for plant cleaning and sterilization, in which the cleaning solutions return to the balance tank through a sterilizing water cooler and restrictor valve. By not applying vacuum to the expansion vessel during this operation, a full sterilizing temperature is achieved.

Infusion System

This system is similar to injection system in all aspects except for the method of mixing of product with steam (Figure 10.10). The infusion is done by dropping heated product into a steam pressure vessel with a conical base. The size and proportions of the vessels differ, as do the methods adopted for distribution of the product into the steam. Two designs of infusers are common.

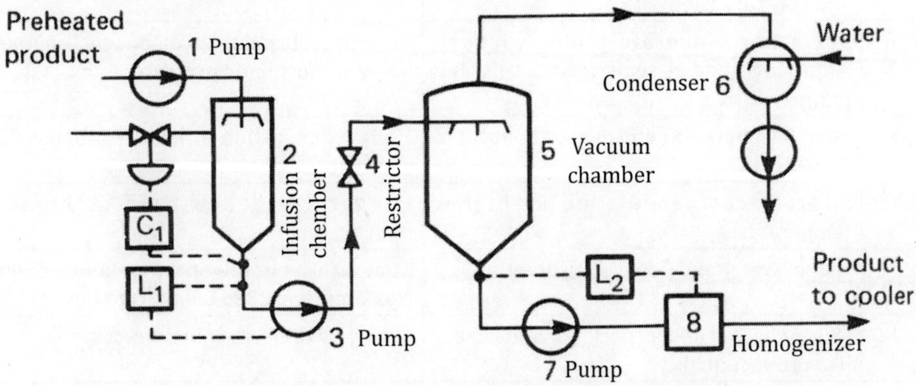

Figure 10.10. Infusion-Type UHT Sterilizer

In the first one the product is supplied to a hemispherical bowl with a loose circular disc closing the top. The product flows from the space between the bowl and the loose disc to form a thin umbrella with access to the heating steam from both sides. If the product has particles, the loose disc can lift to allow particles to pass out of the bowl with liquid. The other distributor has product flow in a series of parallel and horizontal distribution tubes. These have thin slits along the bottom, from which the product falls in thin, laminar free-falling films.

Shape of the infuser contributes to the volume of pool of product in its base, and this increases the holding time as determined by the dimensions of the holding tube and the flow rate through it. Further, for operational stability the steam pressure at the infuser should be at least 0.5 bar above that equivalent to the sterilization temperature.

Comparison of Direct Heating and indirect Heating

Table 10.1. Comparison of Direct and Indirect Systems of Sterilization

	Direct Heating System	*Indirect Heating System*
1	Higher rates of heating and cooling	Lower rates of heating and cooling
2	Quality of milk obtained is better	Slightly lower quality comparatively
3	Viscous products can be handled	Viscous products cannot be handled in PHE heating
4	Comparatively lesser fouling, except at restrictor	Fouling of heat transfer surface is a problem
5	Longer period of continuous operation	Comparatively lesser period of operation due to fouling of PHE
6	Aseptic Homogenizer required	No need of Aseptic Homogenizer
7	Initial cost is higher due to non-standard equipment, expansion vessels, more pumps and control systems	Comparatively lower cost, almost half
8	Some parts operate under vacuum, requiring special seals for vessels and pumps	No such problem, as all components operate above atmospheric pressure
9	Poor Regeneration efficiency, about 50 to 60 per cent	High regeneration efficiency, almost 90 per cent
10	Shorter residence time at higher temperatures	Longer residence time at higher temperatures
11	Higher water and power consumption	Comparatively lower water and power consumption, 25 to 50 per cent lower
12	Only preferred under special circumstances like product quality	Generally preferred over direct system

Aseptic Filling

The product from any UHT sterilizer has to be filled into individual containers for storage, distribution and sale, for commercial use. This process of filling should not lower the sterility level. Satisfactory aseptic filling is an essential part of any system to give a commercially sterile product from a UHT process. In fact, most of the spoilage is due to contamination post UHT filling.

Scientists suggest that an overall spoilage level after incubation of all filled containers of 1 in 5000 should be attainable in commercial practice. It is observed that most of the contaminations are due to failure in aseptic filling system. It is found to be more difficult to design and operate a good and reliable aseptic filling system than a good and reliable UHT sterilizer.

An aseptic filling system must satisfy the following aspects:

1. The container and method of sealing must be suitable for aseptic filling, and must not allow the passage of organisms into the sealed container during storage and distribution.
2. The container, or that part of which comes into contact with the product, must be sterilized after it is formed and before being filled.
3. No contamination should occur while container is being filled.
4. If any sealing is needed, it must be sterilized immediately before it is applied.
5. The closure must be applied and sealed within the sterile zone.

Methods of Container Sterilization

The sterilization performance required for different containers depends on the probable number of organisms existing in the container before sterilization. A sterilizing process giving about 3 to 4 decimal reductions of resistant spores is adequate to give a single survivor in about 5000 containers of 0.5 to 1 litre capacity.

Different methods of sterilizing agents are there, like saturated steam, Dry heat, Hydrogen Peroxide, Other chemical sterilants, UV irradiation, and combination of above.

Saturated Steam

It is one of the most reliable sterilant with minimum of objections. It has however to be at higher pressure to be at high temperatures that are required here. A pressure chamber must therefore be used, with the container or container material to be sterilized entering and leaving the chamber through suitable valves. Air, if present in a steam pressure chamber interferes with the transfer of heat from the steam to the container surface. Hence, any air entering the containers must be removed and not allowed to accumulate. Condensation of the steam during heating of the container surface produces condensate which may remain in the container and dilute the product.

In addition to cans, filling systems using steam have also been designed in which polystyrene thermoformed cups and polypropylene preformed cups are sterilized by steam under pressure.

Dry Heat

Dry heat can be applied either in the form of a hot gas or as hot non-aqueous liquid such as glycol. Dry heat can reach high temperatures at atmospheric pressure, which simplifies the mechanical design of sterilizer system. However, dry heat desiccates microorganisms and makes them more resistant. Much higher temperatures are therefore needed for thermal inactivation by dry heat than for inactivation in similar time by wet heat. Temperatures of the order of 200 °C may be needed if the sterilization time is not to be very long. The high temperatures needed for satisfactory sterilization by dry heat mean that containers which are heat sensitive cannot be used.

Hydrogen Peroxide

It is well known that Hydrogen peroxide is lethal to microorganisms, including heat-resistant spores. Many manufacturers are using a combination of hydrogen peroxide and heat for the sterilization of the surface of the container material. Sterilizing performance increases with both peroxide concentration and temperature. Some aseptic filling systems use a bath of hot peroxide to sterilize the container or the container material. However, the majority of the systems apply the peroxide solution (usually at 30-35 per cent concentration) to the surface of the container material by dipping or in a finely dispersed spray. The surface is then subjected to heat, either from radiant heating elements or from hot air jets. The peroxide solution on the surface is therefore heated and evaporated, to sterilize the surface at the same time remove the peroxide solution to prevent it contaminating the product after filling. Hydrogen peroxide is a poison, and in some countries there are strict limits to the concentrations which may remain in the filled container, and which are allowed in the atmosphere surrounding an aseptic filler in operation and which could be inhaled by operators.

Ultraviolet (UV) Irradiation

The optimum wavelength for UV radiation to be about 250 nm to be effective. The effectiveness falls rapidly away at shorter or longer wavelengths than the above. The effect seems to arise from direct absorption of the radiation by DNA of the bacterial cell. However, UV irradiation is not as effective as Hydrogen Peroxide.

Types of Aseptic Filling System

The principal controlling factor in design of aseptic filling system is the type of container which is to be filled, with subsidiary factors being container material and whether the container is preformed one or whether it is formed during the aseptic filing process.

Cans

Though they are the earliest aseptic fillers to be used, now they are relatively expensive, with cheaper alternatives available. It is especially so, for a low-cost product such as milk. They are also bulky to transport and store before use.

The cans may be of tinplate or drawn aluminium, with the solder being of higher melting point than normal to withstand the can sterilization temperatures. The cans are sterilized in the tunnel at atmospheric pressure by steam at 200 to 220 °C, super heated with gas flames. The sterilizing time is about 40 seconds. The can lids are sterilized, by superheated steam, in a separate unit. When the cans have passed through the sterilizing tunnel, they continue through the filling chamber where they are filled, often using a simple in-line filler of the slit or multi-port type.

The filling quantity is determined by the product flow rate and speed of travel of the can under the filler.

Cartons

These are commonly used for aseptic filling systems for milk, cream, soya-based milks and fruit juices. The filling systems are mostly in which the carton is formed within the filler from a continuous reel of material, though cartons are also supplied as preformed blanks, folded flat, and assembled into cartons in the filler. The carton material is a laminate of paperboard coated internally and externally with polyethylene, which makes the carton impermeable to liquids and allows thermal sealing of both the internal and external surfaces. An oxygen barrier is provided by a thin aluminium foil incorporated in the laminate. The structure of paperboard laminate is shown below:

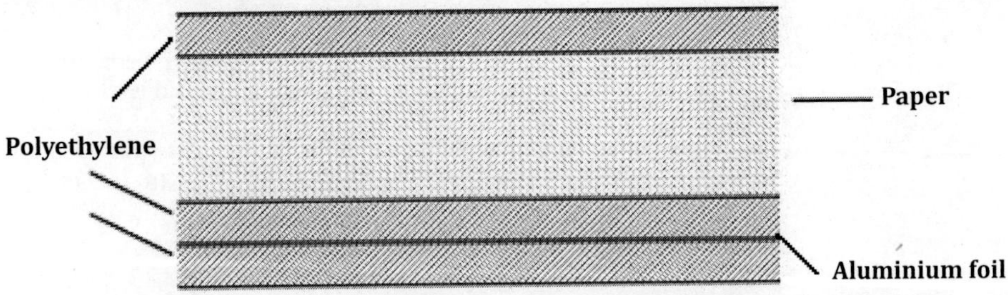

Figure 10.11. Layers of Paperboard Laminate for Aseptic Filling

In the fillers, the container material moves continuously downward in a strip, and is formed by shaping rolls into a cylinder. An overlapping longitudinal seal is formed by heat sealing. An additional thin polythene strip is heat bonded along the inside of the longitudinal seam at the same time. As the continuous cylinder moves downwards, a series of transverse heat seals are made by jaws which move down at the same speed as the cylinder. These seals have the effect of closing the bottom of the cylinder, so that it can be filled with product. This is done through a filling tube from above, and a float-operated filling valve at the outlet of the tube maintains the liquid level above the sealing level (as shown in diagrams in Figure 10.12a-b).

Different forms of seals were evolved, the earlier ones giving shape of tetrahedral, while the later ones are moulding into rectangular blocks.

In some of the designs a headspace is provided in the carton by arranging a float-controlled filling valve to give a product level below the top transverse seal. The filled volume is then determined by the carton dimensions and the position of the valve. The cartons are sterilized with hydrogen peroxide and heat, before they are filled as above. Depending upon the application of Hydrogen Peroxide, and stage at which it is exposed to sterile air, it is done in two alternative methods, as shown in Figure 10.13a-b.

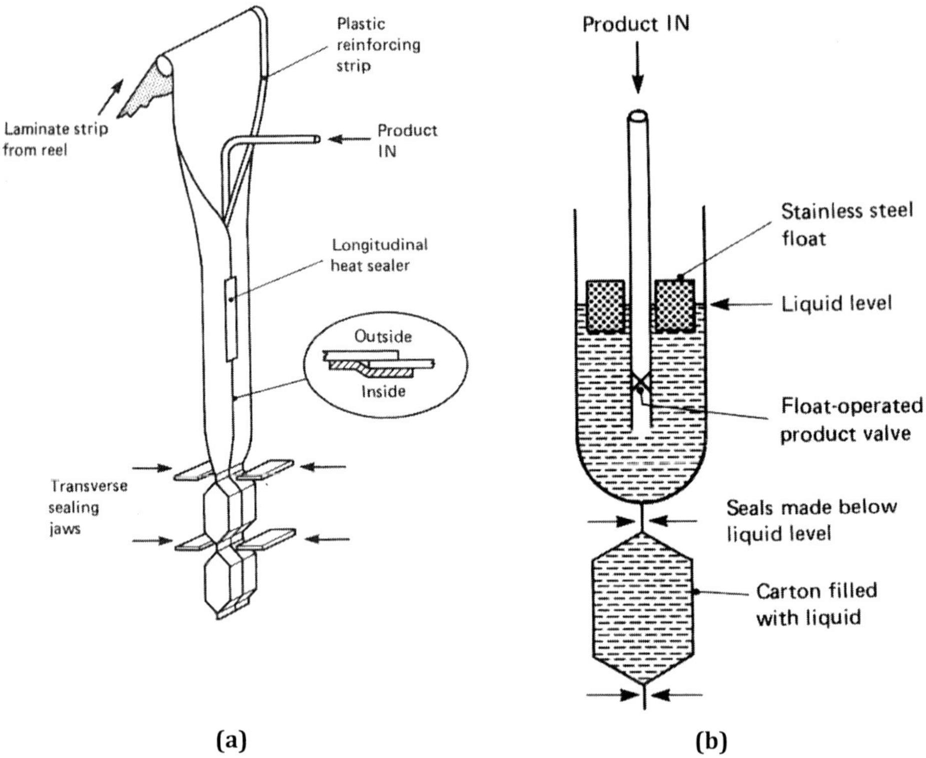

Figure 10.12. Carton Forming and Filling in Aseptic Filling Equipment

UHT Plant operational care and maintenance

1. Raw milk quality should be properly tested, and only milk of highest quality must be accepted for UHT processing. The factors of importance are microbiological, flavor related, and adulteration.
2. As much as possible, electrical failure should be avoided, and if in doubt, the plant should be started and operated on standby electrical generator itself.
3. All the precautions for the plant hygiene and personnel hygiene must be observed, in the lines of latest standards of hygiene.
4. Test must be carried out periodically for the plant hygiene, and corrective measures should be immediately initiated.
5. Care must be taken to maintain operative temperatures and pressures inside the heat exchangers, so that there is no formation of vapours and subsequent formation of scale. About 3-4 bar pressure is normal. If 6-8 bars are approached, then stop the plant.

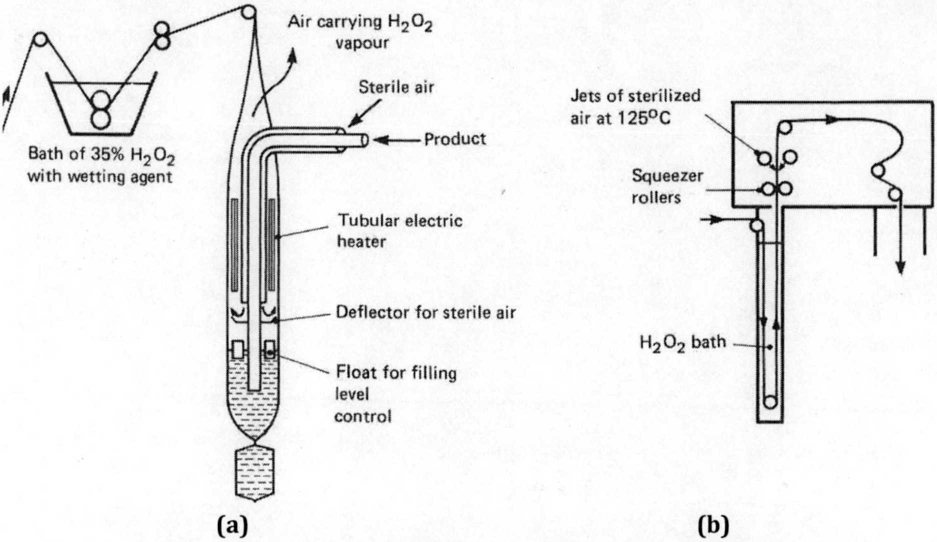

Figure 10.13. Methods of Sterilizing of Carton Material

6. Also special gaskets will be needed. Nitrile rubber-138 °C; Ethylene Propylene Diene Methylene 155 to 160 °C is for UHT plant. Right type of fixing material and choice has to be better. Also mechanical methods are used for holding.
7. New methods of cleaning and solutions are being used for complete and effective cleaning of deposits especially in heating section.

REFERENCES

"Ultra-high temperature processing of milk and milk products" by H. Burton (1988), Pub: Elsavier Applied Science.

Chapter 11

Filling Equipment

Filling Operation: Principles and Working of different Types of Bottle Fillers and Capping Machine

Introduction

After the process systems, the filler is the heart of any milk handling line in a dairy plant. Whether it's a simple 6-head rotary filler, or a high-speed multilane linear machine with a hepafiltration chamber, the filler is the centre of line activity. Filling department is so far a labour intensive one, though shifting more towards mechanization and automation.

The nature of the filler business has changed dramatically in recent years, as more and more beverages (including single serve milk) are being sold in plastic. Packaging materials and package sizes have changed, and many manufacturers are now offering different lines of equipment for use with more than one kind of packaging. Machines are faster and achieve higher levels of hygiene for longer shelf life and ambient storage.

Companies have included blow moulded plastic in their lines of packaging and equipment. Some companies have got linear aseptic filler for HDPE bottles in commercial use for low acid food and beverages. Customers can now commercially fill low acid products into HDPE plastic containers and distribute them without refrigeration.

The latest filling machines have features like 64-valves with 32 pocket star wheels and designed to fill plastic and glass bottles with a maximum diameter of 3.0' (76 mm) at speeds to 800 bpm. Other models are available for larger containers. Special features include demountable valve flanges, electronic level control system, filler driven in-feed conveyor, automated lubrication system, automated cleaning and sanitizing system, and digital control system with touch screen operator interface.

A new line of multi-lane inline equipment developed from a company which can fill in both Ultra-Clean and Aseptic processes. The machine can receive product/s from one of many different UHT processors. Because it is multi-lane, it may be used to fill more than one type or flavour of product simultaneously. Production rates will vary, depending on the number of lanes, the size of the container and the viscosity of the product. Fill time is the major limit on cycle time, but the bottles can be filled in multiple stages to shorten the cycle time.

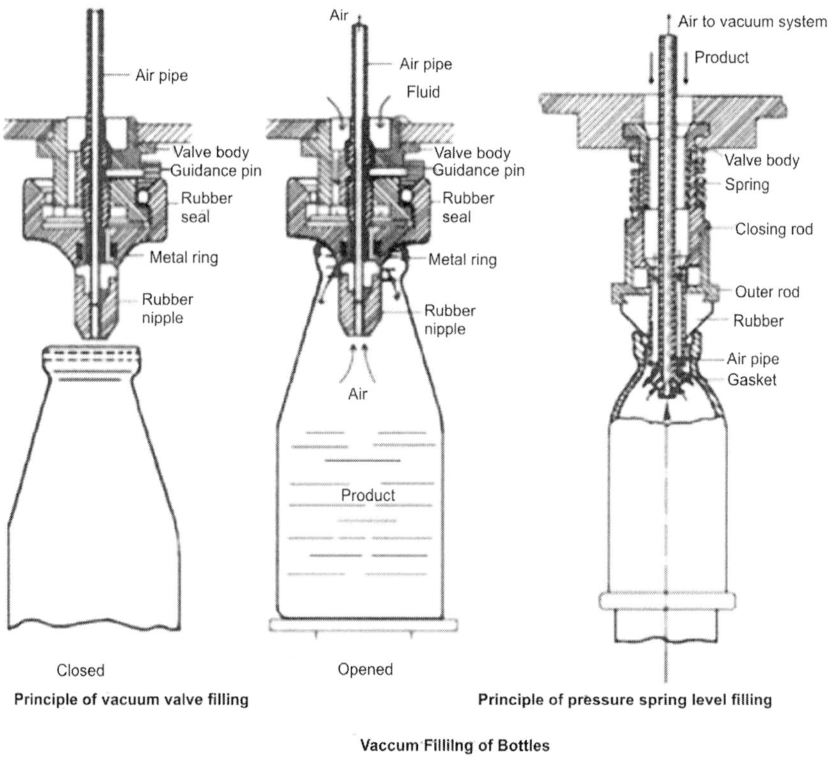

Figure 11.1. Vacuum Filling of Bottles

Vacuum Filling of Bottles

The bottles after bottle washing machine are fed to the bottle filler by conveyor. The bottles are made to feed on to individually by a star wheel on to a Platform (Figure 11.1). The platform will rise and fall making it to meet the filler valve positioned above. The mouth of the bottle is pushed against the rubber seal of the filling device thus making it air tight. The air in the bottle is removed due to the vacuum which exists in the upper part of the filling vessel, which will also be rotating along with the platform of bottles. Simultaneously, the opening between the rubber seal and filling tube is released by the raise of the bottle against the rubber seal, and milk enters the bottle, due to low pressure existing in bottle. Once the milk has risen to the level of the hole of the filling tube, no more air can be removed from the bottle, and the filling operation is terminated.

This type of filling has the advantages of rapid filling, reduced milk losses. If the mouth of the bottle is damaged, it cannot properly raise the rubber seal, and milk will not be filled in that bottle. This automatically reduces the product loss. The vacuum system also ensures foam free filling.

Pouch Filling Machine, Form-fill-seal

The Form Fill Seal machines are a genre of filling equipment that can fill in a flexible packing material. The product should be free flowing type, either liquid or even granular. Various operations in the equipment may be controlled electronically or mechanically.

The process involves certain steps, which will take place cyclically in auto operation mode.

There is option for variation in size and quantity of the product packed. To a large extent, the market milk is now being sold by packing in these machines.

The following are the operations that go cyclically.

1. Forming of tube of packing material from film in rolled state
2. Simultaneous operation of filling and sealing
3. Movement of film to form next package
4. Simultaneous separation of filled and sealed packet while filling of next packet.

Sequence of Operation

With reference to the Figures 11.2 and 11.3, the film roll is loaded at the backside of the filling machine on a sliding platform. The film edge is passed over end of role contact lever, dancer roller, UV tube and brought to the front side over to forming plates. The forming plates roll the flat film in to tube with a certain band of overlap. Within the tube is the fluid filling pipe enveloped. The tube then passes over to vertical seal jaws that are engaged and disengaged with the help of an air operated piston, or in some machines by mechanical means. In between the jaws, the overlapped part of the film tube passes. The set of jaws have one stationary and one moving jaw. The moving jaw has a Ni-chrome rod, supplied with variable voltage such that the heat is generated when the current passes intermittently. During the period when the current does not pass, when the jaws are disengaged, the cooling water being circulated in the moving jaw, cools it and prevents continuous over heating of the sealing rod. The film is supported by Teflon cloth and rubber cushion, as well as protected by Teflon cloth from sealing rod. This arrangement prevents electricity passing on to the film and other parts, while allowing only the heat to pass on to the film and partly melting and fusing the vertical joint.

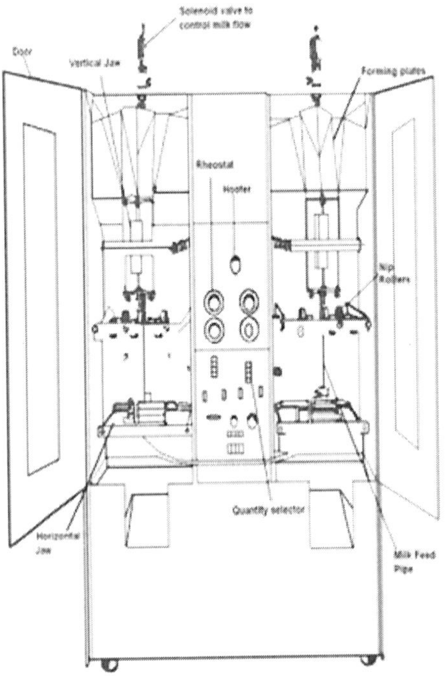

Figure 11.2. Form Fill Seal Machine

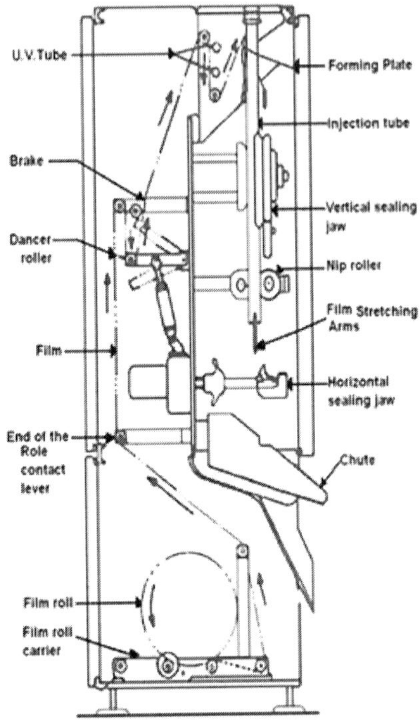

Figure 11.3. Film Feed Circuit

Lower down the film tube, there are a pair of nip rollers giving a holding and pulling down action, when the jaws are disengaged, making the film to move to seal the next portion of vertical overlap, by a predetermined length.

Further lower down the film tube is engaged by horizontal jaws, at a sufficiently below the lower edge of fluid filling pipe. This arrangement allows the formation of lower seal of the packet, while the fluid is being filled to a known quantity. The quantity of flow is controlled by a valve operated by a rod which is lifted by a solenoid coil positioned at the top of machine, just at the feeding line from the over head tank carrying the fluid to be filled. While filling is taking place, a pair of flat blades operated by spring keeps the film perfectly flat at horizontal edge so that there is no folds and horizontal seal is perfect.

When the filling of fluid and the horizontal sealing is complete, the horizontal jaws (as well as vertical jaws) get disengaged, and the nip rollers start rolling to bring the next length of film tube to be filled for next packet. While the second packet is being filled, the first packet already filled will be getting the horizontal seal of top portion of the filled packet. When the next time the jaws open, the first packet drops down by its own weight and weakened connection to the rest of the tube.

The above cycle of operation is repeated when the controls are in automatic operation, while single action takes place when in manual operation during initial adjustment of time and temperature combination for obtaining proper seal.

Controls

The Form Fill Seal Machine has various controls for the following operations.

1. Adjusting the temperature of sealing rod by controlling the electric supply, to match to the thickness of the film to be sealed.
2. Adjusting the timing for the jaws to be engaged and simultaneously filling operations to take place, with a known quantity of fluid.
3. Adjusting the quantity of fluid to be filled when jaws are engaged
4. Adjusting the timing for the jaws to be disengaged and allow time for movement of film to the required length of package.

The other useful instrumentation are the end of film indicator (gives audio signal), so that the new film roll can be changed easily, fine adjustments for the quantity of fluid filled in few grams range, fine adjustment of timings, *etc.*

The equipment will require water for cooling the jaws at a fixed flow rate and of low temperature. Also, compressed air is required which is at required pressure and free of condensed moisture.

The above filling equipment is available in number of variations to cater to different sizes, speeds, and flow properties of product being filled.

Aseptic Filling Bulk Handling System, Care and Maintenance

Introduction

With increasing consumption of long life foods like milk, fruit juices and other liquid foods, as well as certain pharmaceutical applications, the bulk handling of foods aseptically is taking prominence.

The above operation is possible with improvement in packing material and related technologies. The large size packages that can withstand mechanical damage and ability to have perfect seal are being manufactured. The packing containers are heat sealable like the milk pouch film sealing. There is also both one way as well as returnable containers being manufactured.

Bulk Filling Equipment

The Bulk fillers like individual package fillers, are located very near to the UHT plant, and the product is not stored in between. The product is filled as it comes out of UHT plant.

Standard equipment (Figure 11.4) basically has an arm protruding out like a cantilever, from a support base. The arm carries (Figure 11.7) a long pair of horizontal jaws for sealing, supplied with heating element. The arm also carries a valve sealing equipment that exactly matches with the nozzle of the packing bag. Once the valve is connected to the nozzle of the packing bag (Figure11.5), it creates an air tight grip on to it. The arm is connected with vacuum line, which when activated, draws away the air from the bag to be filled. The option of circular valve or horizontal jaws depends on the type of bag that the bulk filling is done. This action is similar to the vacuum cleaner (Figure 11.6).

Figure 11.4. Bulk Filler

Figure 11.5. Aseptic Closure Assembly

Once the bag is cleared of the air, the product flows through another pipe connected to the arm, and a known quantity is filled into the bag. The choice of the quantity, the time of sealing, the temperature *etc.* is programmed. The sample packing for smaller bags can be done manually. If the bag is large, the arm has adjustment facility to lift the bag to the required height and hold it till the product is filled, so that there are no empty pockets in the bag.

The controls can be simple touch screen type and has all the machine modes and also any alarm that are required for indicating any process deviations to the operating personnel. There is a control for the sealing time, temperature, change in volume *etc.* The controls have both the Manual and Auto operation facility, so that filling operation goes on continuously, once the process parameters are achieved. The manual operation is usually done for sample package sealing that is required for the quality control laboratories from time to time. The touch screen controls can also be custom programmed to give the information in addition to the operation, so that history of filling operations, shift basis or day basis.

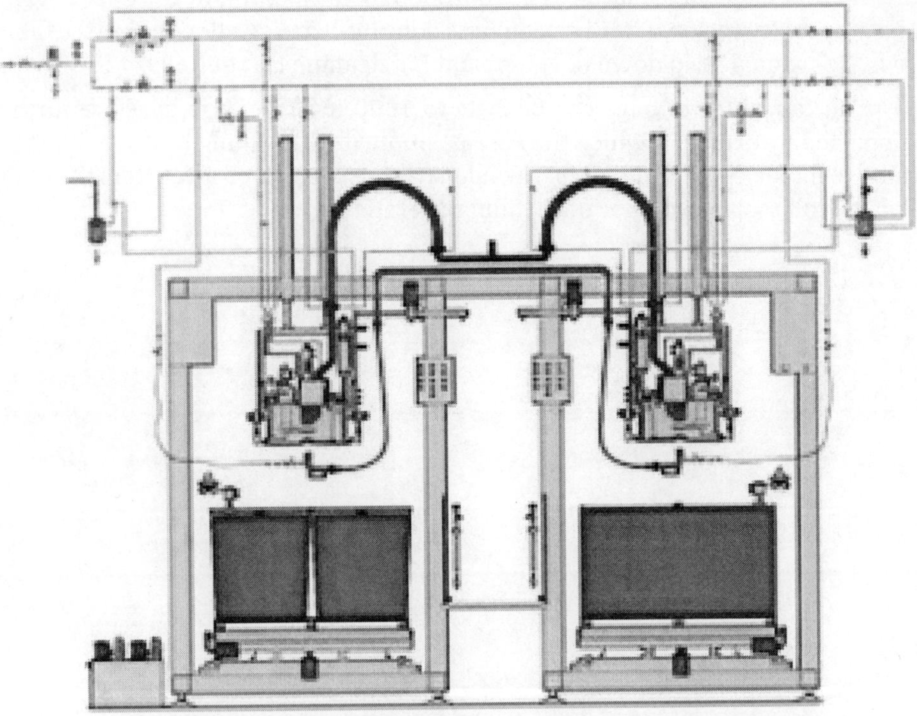

Figure 11.6. Schematic Diagram and Bulk Filler

Figure 11.7. Package Material Holder cum Filling Tube

All product contact surfaces are of AISI 316, for maintaining proper hygiene and prevent corrosion. The CIP cleaning is done by fitting an adopter into the filling jaws. No manual strip down of equipment for cleaning is required.

The capacities of bags can be 5 lts to 1000 lts. The large bags are further supported by drums in which they are positioned while filling it. The equipment can also fill variety of containers. The filled packets are moved away from the filling arm, by roller conveyors for quick filling operations.

Chapter 12

Mixing and Agitation Equipment

Introduction

In most of the processing operations in dairy plant, the process fluid may have to be heated or cooled. It would need some kind of agitation. Similarly mixing various ingredients like that for Ice cream mix, is also common.

Agitation

It is the induced motion of a material in a specified way usually in a circulatory pattern inside some sort of container. *e.g.* Agitation of milk in a storage tank, agitation of ghee in ghee boiler.

Mixing

It is random distribution, into and through one another, of two or more initially separate phases. *e.g.* Mixing of Ice cream ingredients before freezing.

Agitation of Liquids

The purpose of agitation can be many, depending on the process objective.

1. To distribute solid particles more uniformly. *e.g.*: Ingredients in ice cream mix
2. Blending miscible liquids *e.g.*: Gerber test shaking before Centrifuge
3. Dispersing gas through the liquid. *e.g.*: Agitation in silos by compressed air
4. Dispersing immiscible liquids to form emulsion or suspensions of fine drops *e.g.*: Butter oil in Recombined Milk

5. Promoting heat transfer between the liquids, *e.g.*: agitation of cream in ageing tanks.

Agitation Equipment

1. Vessels are usually cylindrical, slightly round bottom to avoid pockets of stagnant liquid. The top of the vessel may be closed or open. The height of liquid approximately equal to diameter of vessel, or any specific suitable ratio.
2. Motor with shaft of sufficient length, directly connected or through reduction gear. Shaft has an impeller and may have support.

Types of Agitators

There are different types of agitators, to suit to the varied process operations. Below are some of them. The types a) and b) Are of Turbine types, while the c) and d) type are propeller type

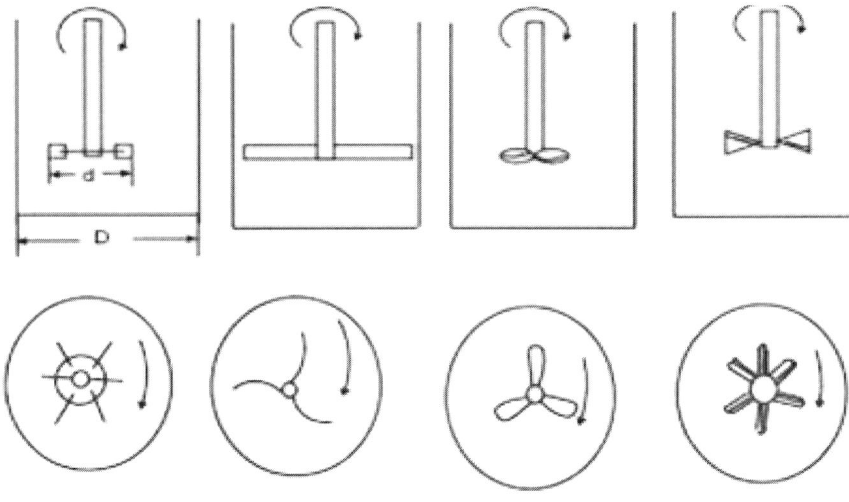

Figure 12.1. Types of Agitators.

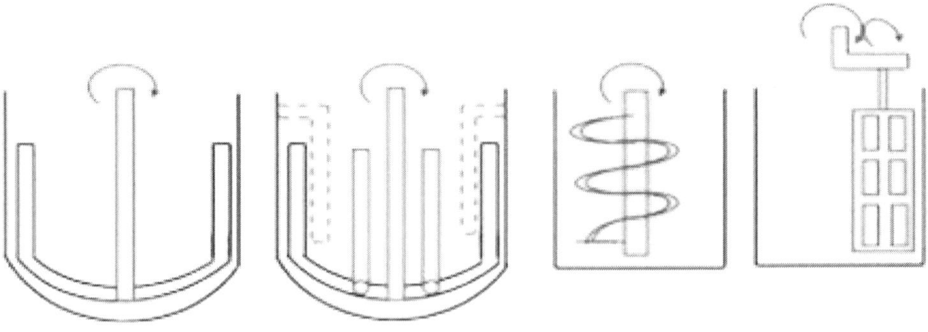

Figure 12.2 Methods of preventing vortex formation

Methods to Avoid Formation of Vortex

Action of agitation sometimes leads to formation of Vortex, which can suck air into the fluid being agitated (Figure 12.2). Methods by which this objective is achieved are: i)Tilted impeller shaft ii) Mounted on side (eccentrically) of tank iii) Baffles, which are vertical strip perpendicular to the wall of tank. Generally, four baffles are sufficient to prevent vortex formation. It's width can vary from 1/12 to 1/18 of the diameter of the vessel. Another method is the propeller type agitators, either inclined or off-centre to the tank.

Types of Agitators

There are number of types of agitators, to meet the specific requirements of agitation of fluids.

i) Paddle	ii) Propeller	iii) Turbine
Open	Bladed	Vertical
Straight	Disk	Curved blade
Blade		

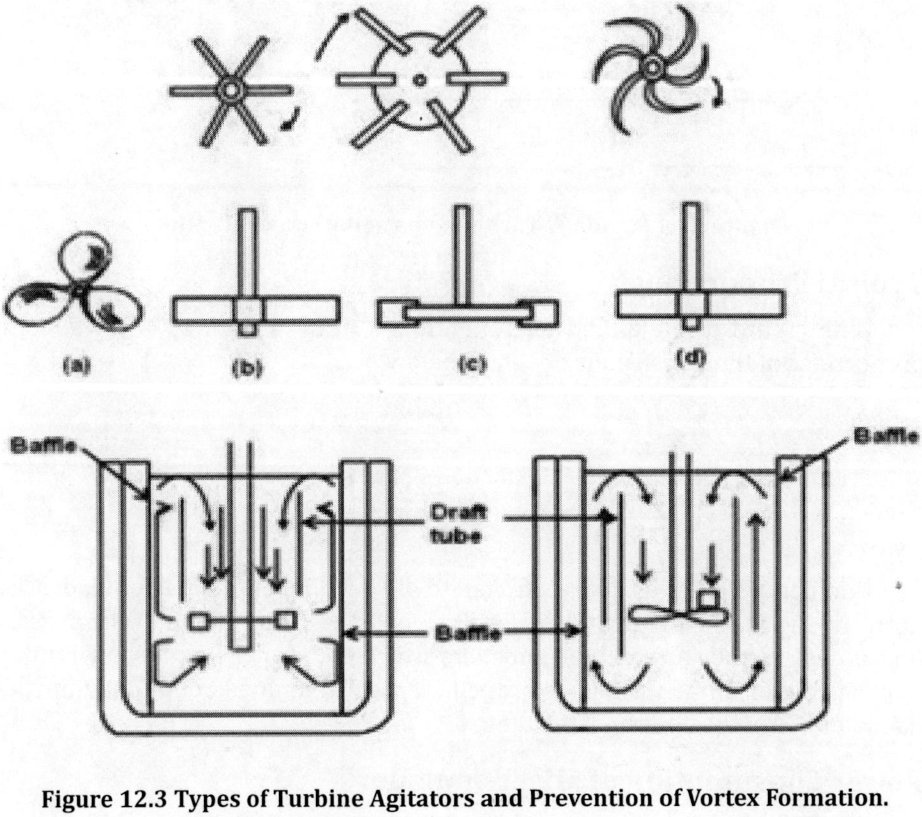

Figure 12.3 Types of Turbine Agitators and Prevention of Vortex Formation.

Draft Tubes

Draft tubes are used when direction and velocity of flow to the suction of the impeller to be controlled. They are useful when high shear at impeller itself desired. *e.g.*, Emulsions. Also, they are used when solid particles tend to float on surface. The disadvantages are i) Draft tubes add to the fluid friction ii) For a given power input, reduce flow rate.

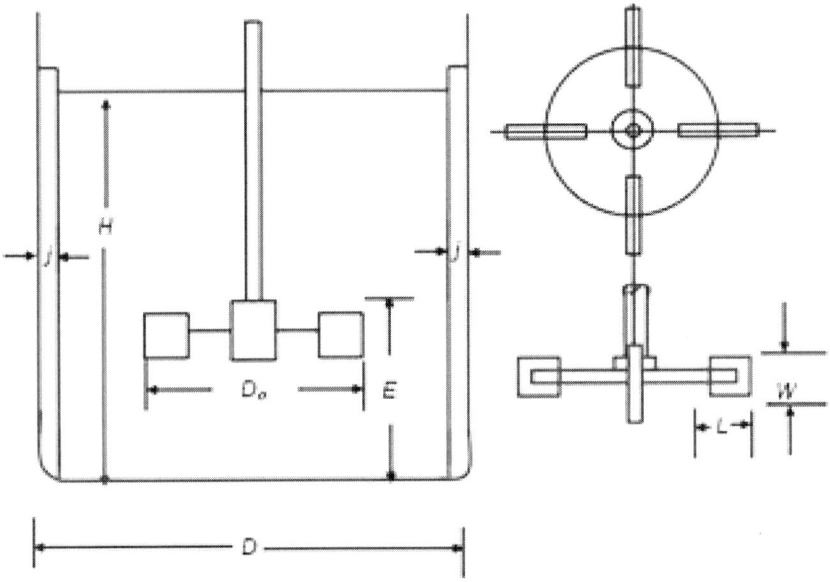

Figure 12.4 Standard Turbine type agitator and notations

Typical Proportions

With regard to the dimensions mentioned in the Figure 12.4, the general proportions of the dimensions are given below:

$$\frac{Da}{Dt}=\frac{1}{3} \quad \frac{H}{Dt}=1 \quad \frac{J}{Dt}=\frac{1}{12}$$

$$\frac{E}{Dt}=\frac{1}{3} \quad \frac{W}{Da}=1 \quad \frac{J}{Da}=\frac{1}{4}$$

- Number of Baffles usually restricted to 4. The number of impeller blades are normally in the range 4 to 16 but usually 6 or 8. If greater depth of the process fluid is desired, then two or more impellers are mounted on the same shaft. Bottom impeller, either of the turbine or propeller type, is mounted about one impeller diameter above the bottom of the tank.

Power Consumption of Mixer-Impeller

Introduction: The operation of agitation has to be effective in achieving the objective, at a minimum cost as well as in a short time. With increasing cost of electricity, the energy consumption of agitation and mixing has to be carefully audited for energy savings. In the analysis of electrical energy consumption, techniques of optimization, dimensional analysis is done to minimize the cost of installation and operation.

Circulation, Velocities and Power Consumption in Agitated Vessels

Volume Circulated by impeller must be sufficient to sweep out the entire vessel in reasonable time. Velocity of the stream leaving the impeller must be sufficient to carry the currents to the remotest points. Turbulence in moving stream often governs the effectiveness of the operation. It is caused by large velocity gradients.

Volumetric flow rate, "q" is proportional to the speed and the cube of diameter of impeller. Another important parameter is the Flow Number, a dimensionless number.

Flow number, $N_Q = \frac{q}{nD_a^3}$

where,

n = rpm or rps; D_a = Diameter of the agitator; q = Volumetric flow rate, m^3/min or m^3/s

Flow number is constant for each type of impeller

For standard flat – blade turbine, in a baffled vessel, $N_Q \approx 1.3$

For Flat – blade turbines, then total flow, estimated from the average circulation time for particles is

$$q_T = 0.92 n D_a^3 \left(\frac{Dt}{Da} \right) \tag{Eq.12.1}$$

For a $\frac{Dt}{Da}$ =3 q_T=2.76 nD_a^3 or 2.1 times the value at the impeller (N_Q=1.3). The above equation should be used only for Dt/Da ratio between 2 and 4

Power Consumption

When the flow in the tank is turbulent, the power requirement can be estimated from the product of the flow 'q' produced by the impeller and the kinetic energy Ek, per unit volume of the fluid.

$$Q = nD_a^3 n_Q \ \& \tag{Eq.12.2}$$

$$Ek = \frac{\rho (V^1{}_2)^2}{2gc} \tag{Eq.12.3}$$

Where V^1_2 = Actual velocity of the particle at the impeller blade tip, which is slightly smaller than the tip speed, μ_2

Power requirement, then is = $nd_a^3 N_Q . \frac{\rho(\alpha \pi n Da)^2}{2gc}$ (Eq.12.4)

$$= \frac{\rho n^3 Da^5}{2gc} \left(\alpha^2 \frac{\pi^2}{2} N_Q \right) \quad \text{(Eq.12.5)}$$

where ,

$$\alpha = \frac{V^1{}_2}{\mu^2}$$

Solving Problems

Example 1

A flat-blade turbine with six blades is installed centrally in a vertical tank. The tank is 6 ft (1.83 m) in diameter; the turbine is 2 ft (0.61m) in diameter and is positioned 2 ft (0.61m) from the bottom of the tank. The turbine blades are 5 in (127 mm) wide. The tank is filled to a depth of 6 ft (1.83 m) with a solution of 50 percent caustic soda, at 150 °F (65.6°C), which has a viscosity of 12 cP and a density of 93.5 lb/ft³ (1498 kg/m³). The turbine is operated at 90 rpm. The tank is baffled. What power will be required to operate the mixer?

Solution

Curve *A* in Figure 12.5, applies under the conditions of this problem.

$S_1 = D_a/D_t$; $S_2 = E/D_t$; $S_3 = L/D_a$; $S_4 = W/D_a$; $S_5 = J/D_t$ and $S_6 = H/D_t$

The above notations are with reference to the diagram below against the dimensions noted. The notations are valid for other examples following also.

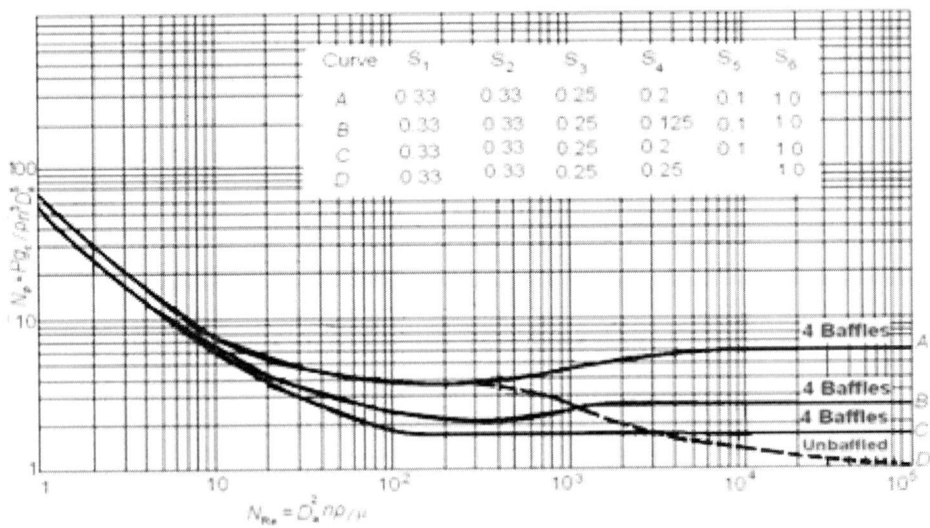

Figure 12.5. Turbine Blade Performance Chart (from MacCabe and Smith).

The Reynolds number is calculated. The quantities for substitution are, in consistent units,

Da = 2ft n = $\frac{90}{60}$ = 1.5 rps

μ = 12 x 6.72 x 10^{-4} = 8.06 x 10^{-3} lb/ft-s

ρ = 93.5 lb/ft^3 g= 32.17 ft/s^2

Then $P = \frac{N_P n^3 D_a^5 \rho}{gc}$

$$N_{RC} = \frac{D_a^2 n\rho}{\mu} = \frac{22\times1.5^3\times93.5}{8.06\times10^{-3}} = 69{,}600$$

From curve *A* (Figure 12.5), for N_{RC} = 69,600, N_P = 5.8, and from Eq. of P.

$$P = \frac{5.8\times93.5\times1.5^3\times2^5}{32.17} = 1821\,\text{ft-lbs}$$

The power requirement is 1821/550 = 3.31 hp (2.47 kW).

Example 2

What would the power requirement be in the vessel described in example 1. If the tank were unbaffled?

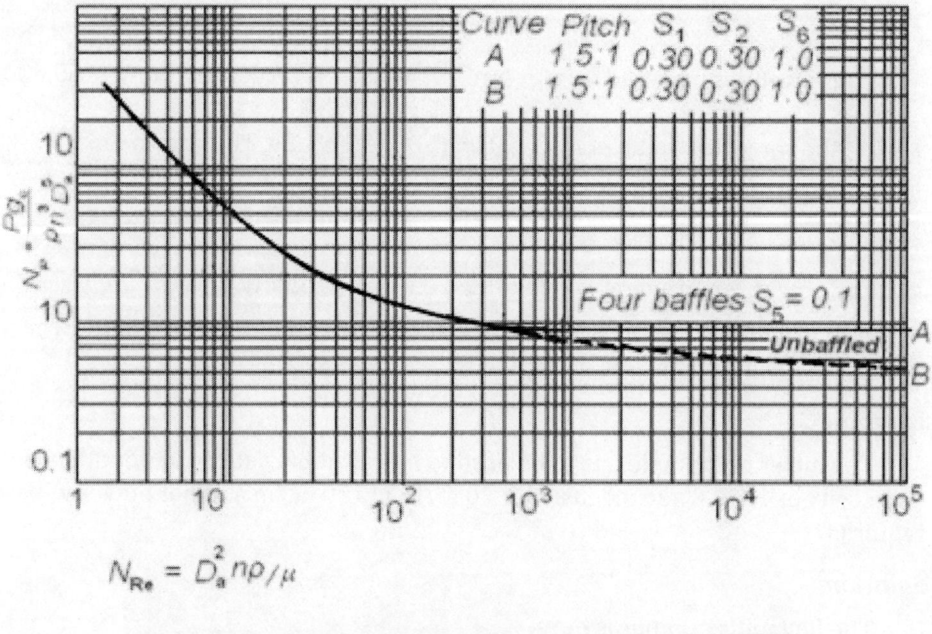

Figure 12.6. Flat Blade Turbine (from MacCabe and Smith).

$$N_P = \frac{Pg_c}{n_3 D_a^5 \rho}$$

Curve *D* of Figure 12.5 now applies. Since the dashed portion of the curve must be used, the Froude number is a factor; its effect is calculated as follows:

$$N_{FR} = \frac{n^2 D_a}{g} = \frac{1.5^2 \times 2}{32.17} = 0.14$$

Constants a and b in $m = \frac{a - \log_{10} N_{\text{Re}}}{b}$ are given in the table below

Figure	*Line*	*a*	*B*
12.5	D	1.0	40.0
12.6	B	1.7	18.0

The constants a and b for substitution into Eq. $m = \frac{a - \log_{10} N_{\text{Re}}}{b}$ are

a = 1.0 and b = 40.0 from eq. $m = \frac{a - \log_{10} N_{\text{Re}}}{b}$

$$m = \frac{1.0 - \log_{10} 69{,}600}{40.0} = -0.096$$

The power number read from Figure 1 $P = \frac{N_P n^3 D_a^5 \rho}{gc}$, curve *D*, for N_{RC} = 69,600, is 1.07; the corrected value of N_p is 1.07 x $0.14^{-0.096}$ = 1.29. Thus, from Eq.

$$P = \frac{1.29 \times 93.5 \times 1.5^3 \times 2^5}{32.17} = 406\,\text{ft - lbs}$$

The power requirement is 406/550 = 0.74 hp (0.55 kW).

It is usually not good practice to operate an unbaffled tank under these conditions of agitation.

Example 3

The mixer of example1. is to be used to mix a rubber-latex compound having a viscosity of 1200 P and a density of 70 lb/ft^3 (1120 kg/m^3). What power will be required?

Solution

The Reynolds number is now

$$N_{RC} = \frac{2^2 \times 1.5 \times 70}{1200 \times 0.0672} = 5.2$$

This is well within the range of laminar flow.

Values of constants K_L and K_T in Eqs. $N_P = \frac{K_L}{N_{\text{Re}}}$ and $N_P = K_T$ $N_p = \frac{K_L}{N_{\text{Re}}}$ are

Given in the table below for baffled tanks having four baffles at tank diameter

Type of Impeller	K_L	K_T
Propeller, three blades		
Pitch 1.0[1]	41	0.32
Pitch 1.5[2]	55	0.87
Turbine		
Six-blade disk [2] (S_3 = 0.25, S_4 = 0.2)	65	5.75
Six curved blades[1] (S_4 = 0.2)	70	4.80
Six pitched blades [3] (45°,S_4 = 0.2)	-	1.63
Four pitched baldes[2] (45°,S_4 = 0.2)	44.5	1.27
Flat paddle, two baldes[1](S_4 = 0.2)	36.5	1.70
Anchor [2]	300	0.35

K_L = 65; from Eq., N_P = 65/5.2 = 12.5, and

$$P = \frac{12.5 \times 70 \times 1.5^3 \times 2^5}{32.17} = 2938 \text{ft - lbs}$$

The power required is 2938/550 = 5.34 hp (3.99 kW). This power requirement is independent of whether or not the tank is baffled. There is no reason for baffles in a mixer operated at low Reynolds numbers, as vortex formation does not occur under such conditions.

Note that a 10,000-fold increase in viscosity increases the power by only about 60 percent over that required by the baffled tank operating on the low-viscosity liquid.

REFERENCE

Unit operations of Chemical Engineering by McCabe W.L., Smith J.C and Harriott P., 1993 McGraw-Hill Book Co.

Chapter 13

Evaporation

Dairy Industry acts as a balancing wheel between the times of surplus milk supply to the times when it is less available, by converting the milk constituents into a dried form and butter. This also makes economical transportation of these products from areas of surplus milk production to areas where it is scarce or higher in demand. In this conversion process care is taken to minimize the heat damage to the milk constituents and preserve its flavor. It is done in two stages:

1. Evaporation of the moisture in the milk to obtain a concentrated product with 50 per cent total solids
2. Drying of the concentrated milk to a final moisture content of 2 to 4 per cent.

This chapter deals with the equipment employed for concentration and drying of milk and methods of their operation.

However, we must be aware that concentration of milk causes elevation of boiling point, and promotes chemical reactions that can result in browning of and flavor and physical defects to the concentrated product.

Basic Principles

Evaporation is a process by which a liquid is brought to its boiling point by application of external heat and transforming it to vapour, which leaves the surface of the liquid. The rate of evaporation depends primarily on:

a) Heat transfer rate at the interface of heating surface and the liquid
b) Area of the heating surface in contact with the liquid
c) Rate of removal of vapors formed

The evaporation of water from milk requires attention because of its heat sensitivity and the process being energy intensive. For this reason, the evaporation has to be carried out under vacuum to:

a) Reduce the boiling point to a temperature where heat damage to the milk components is minimum, (especially the proteins) and
b) Enable multi-stage evaporation at different vacuum levels.

The water contents of the liquid dairy products, their concentrates for spray drying and dried powders are given in Table 13.1. From these figures it is obvious that a substantial part of the water is removed by evaporation and remaining by spray drying and possibly fluidized bed after-drying, if employed.

Table 13.1: Moisture Content of Milk and its Products

Product	*Moisture Content Per cent*		
	Unprocessed	*Concentrate*	*Powder*
Whole Milk	87-88	48-52	2-3
Skim milk	91-92	48-52	3-4
Sweet whey	94-95	40-60	2

Beside those listed above, other dairy products processed by evaporation and spray drying are buttermilk, acid whey, concentrates and permeates obtained from ultrafiltration, special formulations *e.g.* milk formulas and ice cream mixes.

There are two main reasons for using evaporation prior to spray drying:

a) It has a positive influence on many qualitative properties of the finally produced powder,
b) It is a far more economical from energy stand point to remove water by evaporation than by spray drying.

Consequently, removal of highest possible amount of water by evaporation makes the process economic from the energy usage stand point.

Guided by above reasons the design of an evaporator are based on:

a) Using vacuum level, where the boiling temperature of milk is below the temperature that would cause heat damage.
b) Providing sufficient heat transfer surface for the liquid to achieve high heat transfer and consequently evaporation rates so that exposure time to heat is reduced.
c) Keeping a low temperature difference between the heating surface and boiling point of milk, to avoiding local product concentration and scorching.

The main components of an evaporation plant are:

1. Heat exchanger for preheating the milk either indirect or direct
2. Pasteurizing system including holding tubes
3. A system for uniform distribution of milk in calandria
4. Calandria(s)
5. A liquid vapour separator for separation of the vapour from the concentrated product
6. Vapour recompression systems
7. Vacuum equipment
8. Flash coolers
9. Sealing water equipment for milk pumps
10. Cooling towers
11. Milk pumps

The Table 13.2 below explains differences between boiling and evaporation.

Table 13.2. Differences between Boiling and Evaporation

Boiling	*Evaporation*
Boiling is change of liquid phase to vapour phase. It occurs when the liquid is at saturation temperature at prevailing pressure. It requires application of heat or mechanical energy	Evaporation is a type of vaporization which occurs at liquid surface due to difference between vapour pressure at the interface of liquid and surroundings. Vaporization takes place even at temperatures below the boiling point
Rapid process	Evaporation is generally a slower process
Vapour bubbles are formed at the heating surface and rise from boiling	During evaporation, the liquid vaporizes only from the surface of the liquid

General terminologies and aspects of the evaporator/evaporation:

(i) **Concentration:** The initial and final concentration of solute in the solution should be considered. As the concentration increases, the boiling point of the solution increases.

(ii) **Foaming:** Few products have tendency to foam, which reduce heat transfer. It makes controlling level of liquid difficult which ultimately increases the product (entrainment) losses.

(iii) **Heat sensitivity:** Milk, and other food products, are sensitive to high temperatures. If time of exposure to heat is more, damage to milk proteins occurs.

(iv) **Scale formation/Fouling:** Deposition of solids on the heat exchanger surface is a common-phenomenon. However, the scale forming tendency can be reduced with low temperature difference and relatively clean and smooth heat transfer surface. The flow velocity of the product has a significant effect on scale formation. Scale reduces rate of heat transfer and makes cleaning difficult.

(v) **Specific heat:** It increases with concentration of solution. More heat is required to be supplied at high specific heat values.

(vi) **Viscosity:** There is an increase in viscosity of solution during evaporation which increases time of contact and hence chances of burning or damage the product.

(vii) **Capacity:** It is the amount of water evaporated per hour. It depends on the surface area of heat transfer, temperature difference and the overall heat transfer co-efficient.

(viii) **Economy:** It is the amount of water evaporated per kg of steam used. It increases with number of evaporation effects.

Types of Evaporators

The types of evaporators commonly used are:

- Batch type (Vacuum pan)
- Forced circulation
- Natural circulation
- Wiped film
- Rising film
- Plate evaporators
- Falling film
- Rising/falling film

Vacuum Pan

Concentration of milk with batch pan (Figure 13.1) is one of the oldest methods. It is mostly outdated, but used in a few limited applications, such as the concentration of jams and jellies, where whole fruit is present.

With the Vacuum pan evaporator, the product residence time is several hours. Therefore, it is essential to boil the product which is heat sensitive or thermo-degradable at low temperatures under vacuum. The batch pan is either jacketed or has internal steam heating coils. Heat transfer areas normally are small due to vessel shapes. The heat transfer coefficients (HTCs) tend to be low under natural convection conditions. Low surface areas together with low HTCs generally limit the evaporation capacity of the vacuum pans. Heat transfer is improved by agitation of the product within the vessel. In many cases, large temperature differences cannot be used to increase the heat transfer rate for the risk of rapid fouling of the heat transfer surface. The low evaporation capacity, caused by the above factors, therefore, limits the use of vacuum pan.

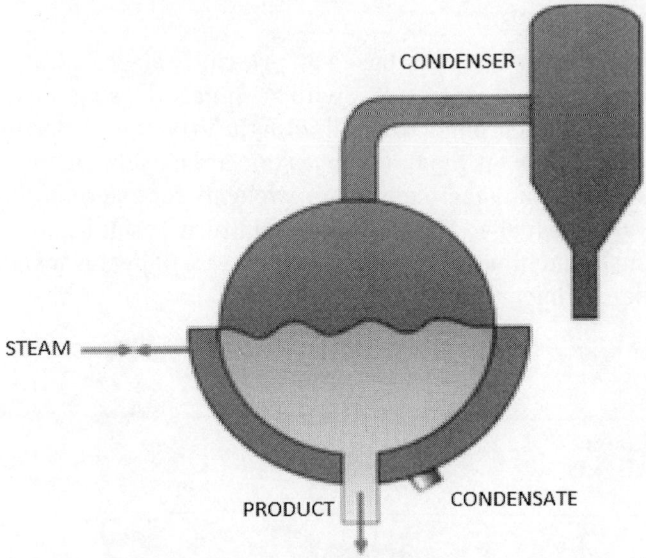

Figure 13.1. Vacuum Pan

Natural Circulation

Natural circulation evaporator is illustrated in (Figure 13.2) is achieved through the use of having an external shell and tube heater outside of the main vessel. The most common application for this type of unit is as a reboiler at the base of a distillation column.

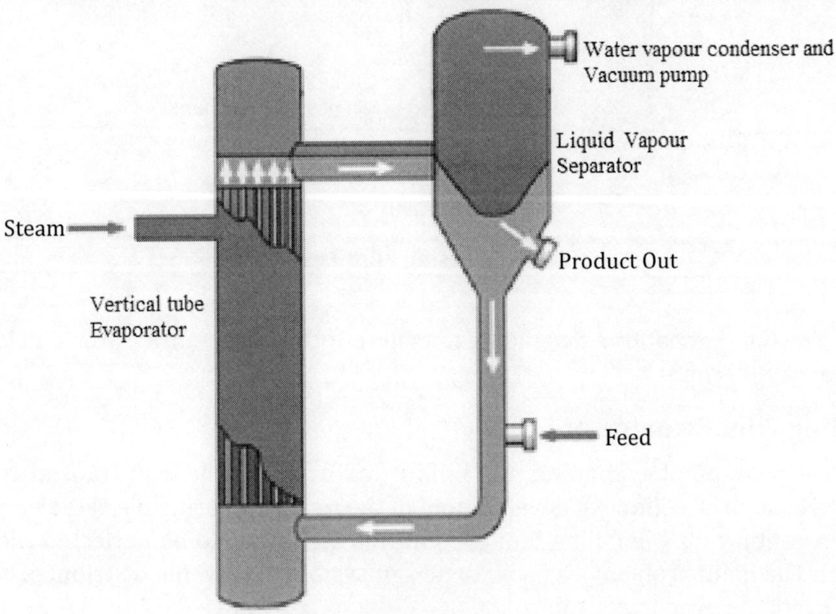

Figure 13.2. Natural Circulation Evaporator

Rising Film Evaporator

The rising film evaporation is based on principle (Figure 13.3) of the product flowing upwards inside vertical tubes with steam condensing outside the tubes. Liquid inside the tube is brought to a boil, with the vapor generated forming a core in the center of the tube. As the fluid moves upwards inside the tube, more vapor is formed resulting in a higher central core velocity forcing the remaining liquid to the tube wall. The higher vapor velocities, in turn, result in thinner and more rapidly moving liquid film. This provides higher rate of heat transfer and shorter product residence time.

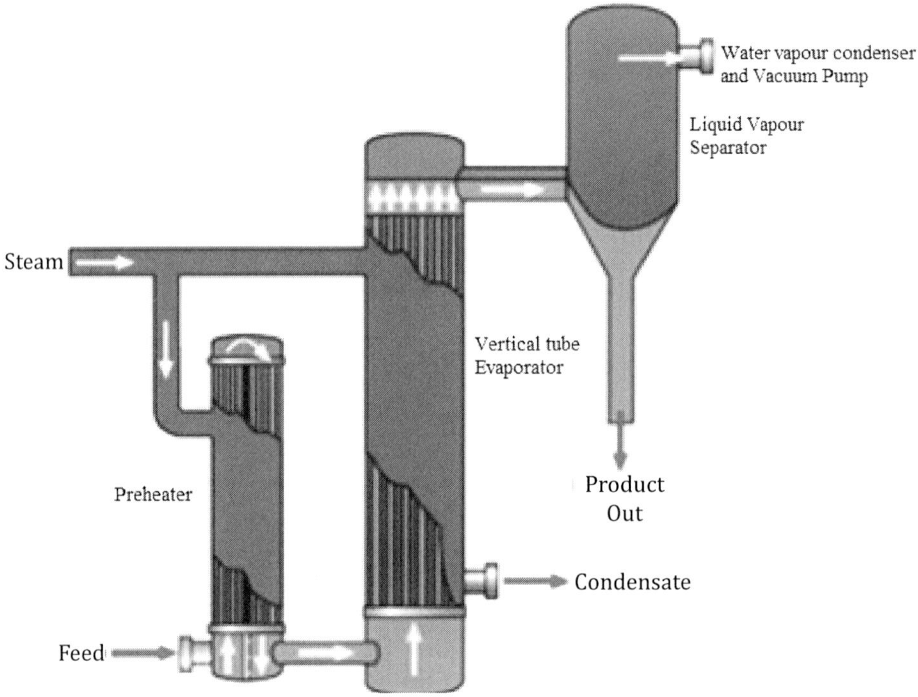

Figure 13.3. Rising Film Evaporator

The higher heat transfer rate results in reduced heat transfer area and lower initial capital costs.

Falling Film Evaporator

In this type of evaporator, the milk is fed from top the heat transfer tubes, called calandria. Following development of the rising film principle, it took almost half a century for the falling film evaporation technique to be perfected (Figure 13.4). The main problem was how to design system that evenly distributes liquid to each of the evaporator tubes.

While the technique varies with each manufacturer, the distribution generally is based around use of a perforated plate positioned above the top of tube plate of the calandria. The problem of spreading the liquid to each tube is sometimes further enhanced by flash vapor generated at this point. The falling film evaporator does have the advantage of the film 'going with gravity' instead of against it. This results in a thinner, faster moving film and gives rise to an even shorter product contact time and an improvement to heat transfer.

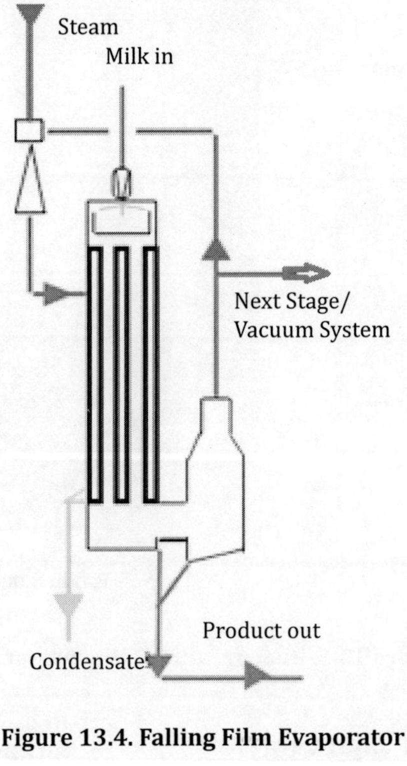

Figure 13.4. Falling Film Evaporator

To establish a well-developed film, the rising film unit requires a driving film force, typically a temperature difference of at least 14°C between condensing steam and evaporating temperature of the product. In contrast, the falling film evaporator does not have a driving force limitation–permitting use of a greater number of evaporator effects within the same overall operating limits.

Rising/Falling Film Evaporator

The rising/falling film evaporator (Figure 13.5) has the advantages of the ease of liquid distribution of the rising film unit coupled with the requirement of lower head room. The tube bundle is approximately half the height of either a rising or falling film evaporator for similar evaporation capacity, The vapor/liquid separator is positioned at the bottom of the calandria.

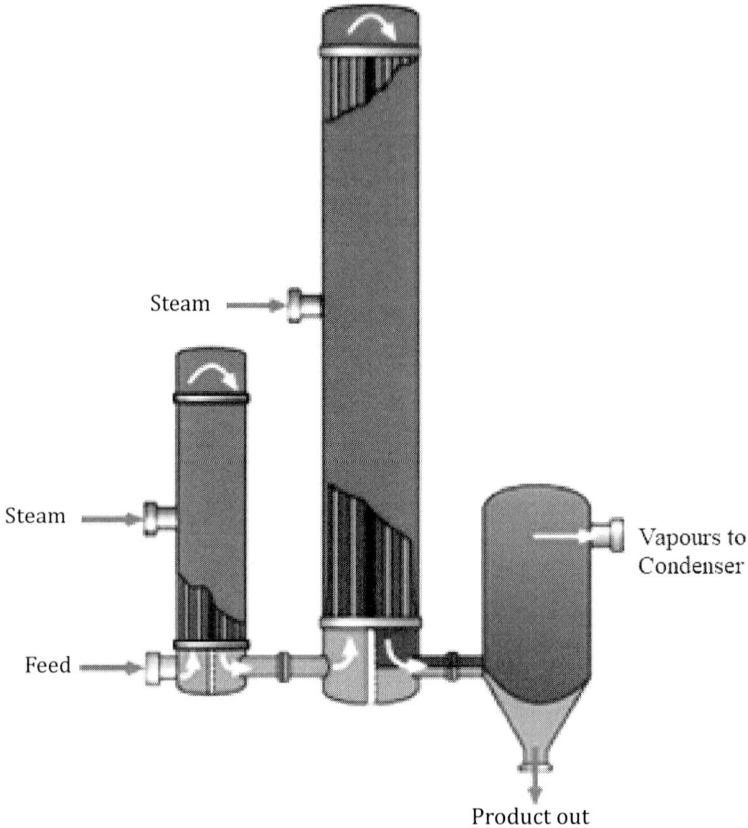

Figure 13.5. Rising/Falling Film Evaporator

Forced Circulation Evaporator

The forced circulation evaporator (Figure 13.6) was developed for processing liquids which are susceptible to scaling or crystallizing. Liquid is circulated at a high rate through the evaporator, with boiling being prevented within the unit by the hydrostatic head maintained above the top tube plate. As the liquid enters the evaporators where the absolute pressure is slightly less than in the tube bundle, the liquid flashes to form vapors.

The main applications of forced circulation evaporator are the concentration of inversely soluble materials, crystallizing operation, and in the concentration of thermally degradable products which result in the deposition of solids. In all these applications, the temperature rise across the calandria is kept as low as possible, (2-3°C). This results in a high recirculation ratio. The high recirculation rates result in high fluid velocities through the tube which help in minimizing the buildup of deposits or crystals on the heating surface. Forced circulation evaporators are more expensive than film evaporators because of the need for large size circulating pipework and large recirculation pumps. Operating costs also are higher.

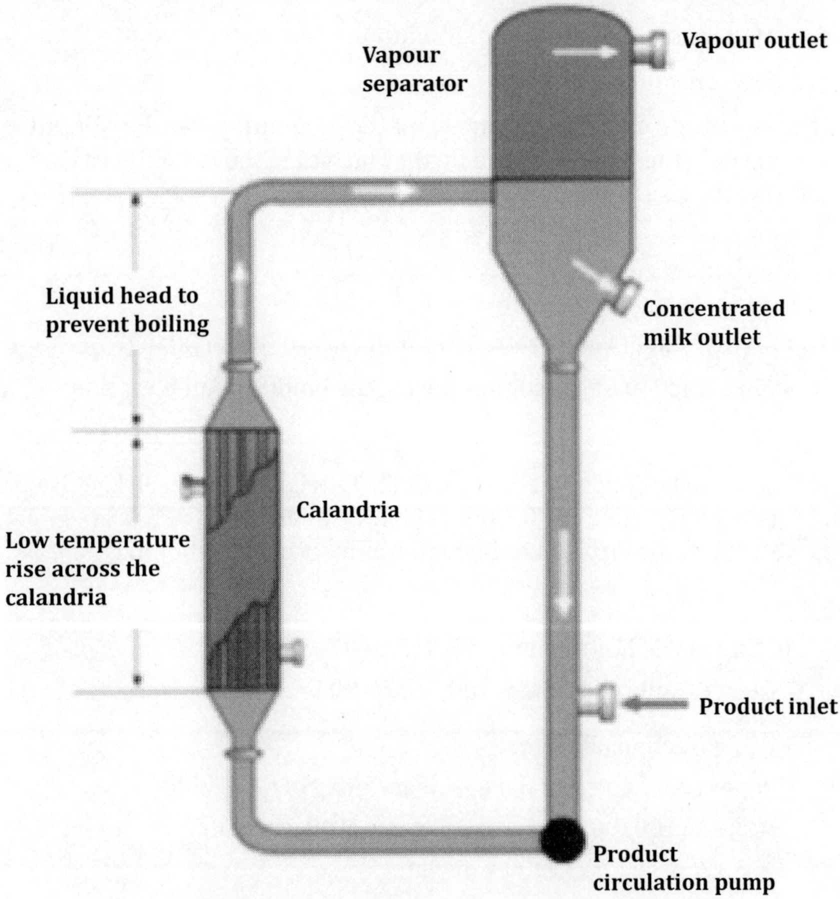

Figure 13.6. Forced Circulation Evaporator

Single Effect Evaporation design calculations

Boiling Point Elevation

Any solute present in the product, will affect the Boiling point and Freezing point, depending on its molecular weight and its concentration. The boiling point elevation is given by the equation:

$$(\lambda_v/R_g)\{(1/T_{AO}) - (1/T_A)\} = -\ln X_A \quad \text{(Eq.13.1)}$$

where,

λ_v = Latent heat of vaporization of water (kJ/mole)

T_{AO}= Boiling point of pure water (K) at a given pressure

T_A= Boiling point of the solution (K) at the same pressure

X_A = Mole fraction of water in the solution

R_g = Gas constant (J/mole K)

If the solution is a dilute one, the above equation can be simplified, by considering only the first term, and further including the molality. The simplified equation for the rise in the boiling point, ΔT_B is:

$$\Delta T_B = R_g (T_{AO})^2 X_B m/(L_v 1000) \qquad \text{(Eq.13.2)}$$

where,

L_v = Latent heat of vaporization per unit mass of water(kJ/kg)

X_B = Mole fraction of the solute causing the boiling point elevation

Example 1

Evaluate the boiling point of Skim milk, with composition as Lactose 5.0 per cent, Protein 3.5 per cent, Fat 0.5 per cent and Minerals and Ash 0.5 per cent, and rest is water. Assume lactose has a greater influence on the boiling point.

Solution

1) The total solids amount to 5.0+3.5+0.5+0.5 =9.5

   Hence Water per cent = 100 – 9.5 = 90.5

2) Mole fraction of Lactose:

   5 per cent Lactose = 5.0 g lactose/100 g of Skim Milk

   = 5.0 x (100/90.5) =5.525 g lactose/100 g of water

   = 55.25 g lactose/1000 g of water

   Mole fraction of lactose, m = 55.25/342 (Molecular weight of lactose = 342)

3) Rise in Boiling point due to dissolved lactose

   $\Delta T_B = R_g (T_{AO})^2 X_B m/(L_v 1000)$

   ΔT_B = (8.314 J/mole K) (373 K)^2 (18) (0.162)/(4.0626 x 10^4 J/mole) (1000)

   = 0.083K or °C

   Hence, the boiling point of skim milk is (100+0.083) = 100.083°C

4) Calculate the boiling point of concentrated milk to 40 per cent Total Solids.

   Mole fraction of Lactose:

   Percentage of Lactose in concentrated milk: 5.0 x (40/9.5) = 21 per cent

   Molality of lactose, m:

21 per cent lactose = 21.0 g of lactose/100 g product

= 21 x (100/40) = 52.63 g lactose/100 g of water

= 526.3 g of lactose/1000 g of water

M = 526.3/342 = 1.538

5) As the solute is not a dilute one, the equation used is

$- \ln X_A = (\lambda_v/R_g) \{(1/T_{AO}) - (1/T_A)\}$

Mole fraction of Water, $X_A = (60/18)/\{(60/18) + (21/342)\}$

= 3.33/(3.33 + 0.06)

= 0.9823

$-\ln(0.9823) = [(4.0626 \times 10^4/8.314) \{(1/373) - (1/T_A)\}]$

On solving, T_A = 373.64 K or 100.64°C

6) If the concentrated product is assumed to be dilute,

$\Delta T_B = \{(8.314 \text{ J/mole K}) (373^2 \text{ K}) (18) (1.538)\}/\{(4.0626 \times 10^4 \text{ J/mole}) (1000)\}$

= 0.788 °C

Or, the Boiling point = 100.788°C

Retention Time

Concentration with Vacuum Pan, may be assumed as a continuous operation, provided the feed is introduced and mixed with the product undergoing concentration in a continuous manner. An approximation may be made assuming ideal mixing of concentrated product and feed, is given below:

$$\text{Per cent Product Removed} = [1 - \exp(-t/R)] \times 100 \qquad \text{(Eq.13.3)}$$

where,

R = Ratio of evaporator volume to rate of product removed

t = Retention time

The above equation gives the product removal rate from the evaporator for a given retention time. The retention time is an important factor as it indicates the time of exposure of heat sensitive components in the feed to the evaporation temperature.

Determination of Heating Surface Area

Saturated steam is only used exclusively as the heating medium for the milk evaporators. The essential aspect to consider when designing a milk evaporator is to estimate the heating surface area. Generally, it must be large enough to secure the required heat transfer rate but not excessively large to keep good coverage of the over-all tube surface by the evaporating liquid. It is calculated by following equation:

$$Q = U A (T_s - T_m) \quad \text{(Eq.13.4)}$$

where,

A is the heating surface in m^2,

Q is heat transfer rate required kcal/s or W,

T_s is the condensing steam temperature °C,

T_m is temperature of the evaporating milk temperature °C,

U is the overall heat transfer coefficient in kcal/h.m^2. °C or W/m^2. K

The heat transfer rate 'Q' is calculated from the amount of water to be evaporated 'w' (kg/h) and the latent heat of evaporation under the conditions of vacuum pan operation (vacuum pan operating pressure and temperature).

Heat Balance

The Mass and Heat balance can be shown for a Single Effect evaporator as follows: It is assumed that there is no loss of either mass or heat from the system:

$$\text{Mass balance: } F = V + P \quad \text{(Eq.13.5)}$$

$$\text{Material balance: } X_F F = X_p P \quad \text{(Eq.13.6)}$$

where,

F = Mass flow rate of Feed entering the evaporator; P = mass flow rate of concentrated product leaving the evaporator; V= mass flow rate of Vapours leaving the evaporator

X= Total solids fraction, with subscripts F and P for feed and concentrated Product respectively.

Energy Balance

$$F C_{pF} (T_F - 0) + w_s H_s = VH_v + P C_{pP} (T_P - 0) + w_s H_c \quad \text{(Eq.13.7)}$$

where,

H = the specific enthalpy values, with subscripts s,v, and c refer to steam, vapour and condensate respectively.

w_s = Mass flow rate of steam supplied to the evaporator at saturated temperature and pressure of T_s and P_s respectively

C_p= Specific heat with second subscripts F and P are that of Feed and Product respectively

T = Temperature °C

The base temperature of 0°C is utilized so that enthalpy values from Standard Steam tables can be used. Sub-cooling of condensate is neglected here.

Heat to be added to the product in order to evaporate specified quantity is, neglecting subcooling of condensate,

$Q = w_s H_s - w_s H_c$ (Eq. 13.8)

A term, Steam Economy, is often used in evaluating the performance of an evaporator. It is the measure of steam utilized to for required amount of water in the product. It is given by,

Steam Economy = V/w_s (Eq. 13.9)

Example 2

A fruit juice with 10 per cent total solids is being concentrated in single effect evaporator with a feed rate of 12000kg/h at 25°C. The evaporator is being operated at a vacuum to evaporate moisture from product at 70 °C while supplying saturated steam at 169.1 kPa absolute. The final concentration of the product is 45 per cent total solids. Calculate the steam requirement and steam economy for the process when the condensate leaves at 80°C. If overall heat transfer coefficient is 1600 W/m² K, calculate the required heat transfer area for the above operation. Take the specific heat of feed as 3.81 and that of product as 2.83 kJ/kg K.

Solution

Given Data: $X_f = 0.10$; $X_p = 0.45$; $m_f = 12{,}000$ kh/h or 3.33 kg/s; $T_f = 25$°C

$P_s = 169.1$ kPa; Ts = 115°C (from steam tables); $T_c = 80$°C

U = 1600 W/m² K; Cpf = 3.81 kJ/kg K; Cpp = 2.83 kJ/kg K

Data from Steam tables:

Enthalpy of saturated steam at 169.1 kPa (115°C) = 2699.0 kJ/kg

Enthalpy of condensate at 80°C = 334.91 kJ/kg

Enthalpy of Vapour at 70 °C. = 2626.8 kJ/kg

a) Mass balance equation

   $X_f \times m_f = X_p \times m_p$

   0.1 x 12000 kg/h x h/3600 s or 0.1 x 3.33 = 0.45 x m_p or $m_p = 0.74$ kg/s

   $m_f = m_v + m_p$ or $3.33 = m_v + 0.74$ or $m_v = 2.59$ kg/s

b) Enthalpy balance equation

   $(m_f \times H_f) + (m_s \times H_s) = (m_p \times H_p) + (m_v \times H_v) + (m_s \times H_c)$

   [3.33 kg/s x 3.81kJ/kg. K x {(25+273) – (0+273)}] + [m_s s 2699]

   = [0.74 kg/s x 2.83 kJ/kg K x {(70+273)- (0+273)}] + [2.59 kg/s x 2626.8 kJ/kg K] + [m_s x 334.91 kJ/kg]

   $317 + 2699\, m_s = 156.5 + 6803.4 + 334.91\, m_s$

   $m_s = 2.81$ kg/s

c) Steam Economy = $m_v/m_s = 2.59/2.81 = 0.92$.

d) Heat Transfer area:

Rate of heat given up by the Steam = Rate of heat transferred to the product (Neglecting heat losses)

$m_s (H_s - H_c) = U A (T_s - T_p)$

$2.81 (2699 - 334.91) = (1600/1000) \times A \times (115\text{-}70)$

$6643.1 = 72 \times A$

Area of heat transfer, $A = 92.3\ m^2$

Example 3

A single effect evaporator is required to concentrate a solution from 10 per cent solids to 30 per cent solids for a feed rate of 250 kg per hour. If the pressure in the evaporator is 77 kPa absolute, and if saturated steam is available at 200 kPa gauge, calculate the quantity of steam required per hour and the area of heat transfer surface if the overall heat transfer coefficient is 1700 J $m^{-2}\ s^{-1}\ {}^{o}C^{-1}$.

Assume that the temperature of the feed is 18°C and that the boiling point of the solution under the pressure of 77 kPa absolute is 91°C. Assume, also, that the specific heat of the solution is the same as for water, that is 4.186×10^3 J $kg^{-1}\,{}^{o}C^{-1}$, and the latent heat of vaporization of the solution is the same as that for water under the evaporation conditions.

From steam tables, the saturation temperature of steam at 200 kPa (gauge) [300 kPa absolute] is 134°C and latent heat 2164 kJ kg^{-1}; the saturation temperature at 77 kPa (abs.) is 91°C and latent heat is 2281 kJ kg^{-1}.

Mass Balance (kg h^{-1})

	Solids	*Liquids*	*Total*
Feed	25	225	250
Product	25	58	83
Evaporation			167

Heat Balance

Heat available per kg of steam

= latent heat + sensible heat in cooling to 91°C

$= 2.164 \times 10^6 + 4.186 \times 10^3(134 - 91)$

$= 2.164 \times 10^6 + 1.8 \times 10^5$

$= 2.34 \times 10^6$ J

Heat required by the solution

= latent heat + sensible heat in heating from 18°C to 91°C

= 2281 x 10^3 x 167 + 250 x 4.186 x 10^3 x (91 - 18)

= 3.81 x 10^8 + 7.6 x 10^7

= 4.57 x 10^8 Kg J h^{-1}

Now, heat transferred from steam = heat required for necessary evaporation.

Therefore, quantity of steam required per hour = (4.57 x 10^8)/(2.34 x 10^6)

= 195 kg h^{-1}

Quantity of steam/kg of water evaporated = 195/167

= 1.17 kg steam/kg water.

Heat-transfer Area

Temperature of condensing steam = 134°C.

Temperature difference across the evaporator = (134 - 91) = 43°C.

Writing the heat transfer equation for q in joules/sec,

$q = UA\,\Delta T$

(4.57 x 10^8)/3600 = 1700 x A x 43

A = 1.74 m^2

Area of heat transfer surface = 1.74 m^2

(It has been assumed that the sensible heat in the condensed (cooling from 134°C to 91°C) steam is recovered, and this might in practice be done in a feed heater. If it is not recovered usefully, then the sensible heat component, about 8 per cent, should be omitted from the heat available, and the remainder of the computations adjusted accordingly).

Example 4

Tomato juice is to be concentrated from 12 per cent solids to 28 per cent solids in a climbing film evaporator having tubes of 3 m high and 4 cm diameter. The maximum allowable temperature for concentrating tomato juice is 57°C. The juice is fed to the evaporator at 57°C and at this temperature the latent heat of vaporization is 2366 kJ kg^{-1}.

Steam is used in the jacket of the evaporator at a pressure of 170 kPa (abs). If the overall heat-transfer coefficient is 6000 J m^{-2} s^{-1} °C^{-1}, estimate the quantity of tomato juice feed per hour. Assume heating surface as 3 m long x 0.04 m diameter.

Mass balance: basis 100-kg feed

	Solids	*Liquids*	*Total*
Feed	12	88	100
Product	12	31	43
Evaporation			57

Heat balance

Area of evaporator tube $=\pi d\, h$

$= 3.14 \times 0.04 \times 3$

$= 0.38\ m^2$

Saturation temperature at 170 kPa (abs) = 115°C from Steam tables.

Carrying out heat balance across the evaporator

$q = UA\,\Delta T$

$= 6000 \times 0.38 \times (115 - 57)$

$= 1.32 \times 10^5\ J\ s^{-1}$

Heat required per kg of water evaporation in feed

$= 0.57 \times 2366 \times 10^3$

$= 1.34 \times 10^6\ J$

Rate of evaporation = $(1.32 \times 10^5)/ 1.34 \times 10^6)$

$= 0.1\ kg\ s^{-1}$

Rate of evaporation = 360 kg h^{-1}

Quantity of tomato juice feed per hour = 360 kg

REFERENCE

APV Handbook, 4th Edn., Invensys Co, USA

Chapter 14

Multiple Effect Evaporators

The single effect evaporators discussed in the previous chapter, evaporates the water from milk using steam, and also expends energy to condense the vapours generated in the process. It is necessary to maintain vacuum in the evaporator. Hence, multiple effect evaporators have been introduced to save energy used for evaporation. In this system, the vapours generated in the first effect, which is at a higher temperature, are used to concentrate the product in the second effect, which is pumped from first effect. However, compared to first effect the degree of vacuum is higher in second effect and hence the boiling temperature is lower than that of first effect. This process repeats from second effect to third effect and so on to the last effect.

Main Components of the Multiple Effect Evaporator

The main components are:

1. Heat exchanger for preheating the liquid either indirect or direct
2. Pasteurizing system including holding tubes
3. System for uniform product distribution in the tubes
4. Calandria(s) with tubes
5. Liquid vapour Separator for separating of the vapour from the concentrated product
6. Vapour recompression system(Thermocompressor)
7. Vacuum equipment
8. Flash coolers
9. Water Sealing equipment

10. Condensers
11. Cooling towers
12. Milk pump

Heat Exchangers for Preheating

As the milk that is pasteurized or raw is at storage temperature, it needs to be brought up to evaporation temperature, in addition to any other high heat treatment, that the process may need. Hence, the milk is preheated in heat exchangers. The preheaters are mostly of two types: a) Spiral – tube pre heaters and b) Straight tube preheaters.

Spiral-Tube Preheaters

The spiral tubes are placed on shell-side of the calandria surrounding the falling-film tubes, and are heated by the steam or water vapours produced in the evaporator. Though the system is simple, it does not offer the possibility of inspection for deposits or leakages. Except for few smaller capacity evaporators, they are not used anymore (Figure 14.1).

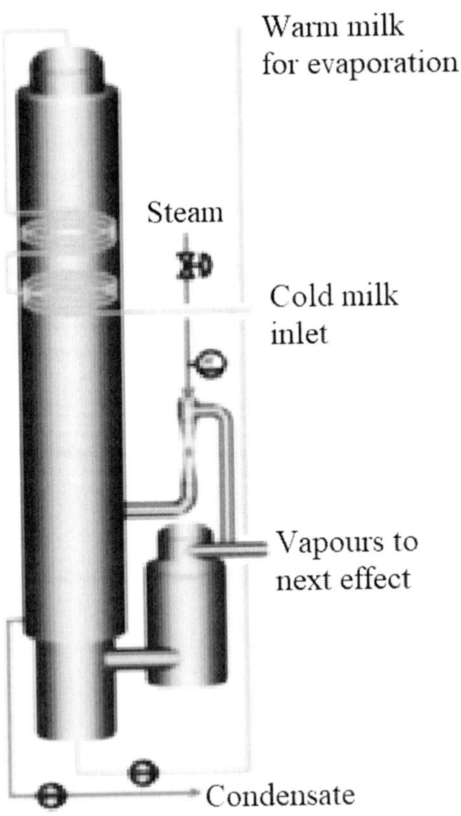

Figure 14.1. Spiral Tube Pre-heater

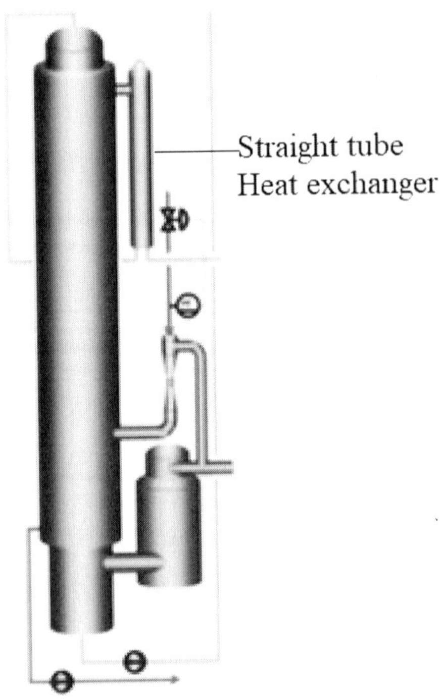

Figure 14.2. Straight Tube Pre-heater

Straight-Tube Preheaters

The straight-tube preheaters are basically shell-and-tube heat exchangers placed vertically outside the evaporator and heated by the water vapours from the corresponding calandria. The vapour connection to the preheater is at its top so that non-condensable gasses can easily be extracted. This ensures an optimum utilization of the heating surface of the evaporation tubes. In this system, inspection and manual cleaning are possible. The Figure 14.2 shows straight tube pre-heater.

Heat transfer surface in the preheater is arranged in groups of parallel tubes of small diameter to make the surface area large. Each group of tubes is connected by normal dairy fittings at the terminal end. Due to flow being parallel, the holding time is very short. The viscosity of the final concentrate is therefore lower in evaporators equipped with straight-tube preheaters.

The large surface of the preheaters and the temperature level prevailing during operation (rise from 5 to 65°C) offers optimal conditions for growth of mesophilic and thermophilic bacteria. After 14-16 h of operation a bio-film is formed on the inner surface of the preheaters, where spores can be formed. Unless attention is paid to this problem, one cannot expect a 20-hour production without increase of mesophilic and thermophilic bacteria and their spores particularly during the last 4-5 hours of a 20-hour production period. To ensure production of milk powder without microbial problems the following types of preheaters should be used:

Direct Contact Regenerative Preheaters

By using a direct contact regenerative preheater of similar design as the direct contact regenerative flash chambers, the heating from 5°C to 40°C and from 40°C to 70°C can be done rapidly without heat transfer surfaces where biofilms can be formed.

The milk is pumped into the direct contact preheater(s), where the water vapour from one of the calandrias is introduced by means of live steam through a small thermo-compressor (Figure 14.3). By applying this method, it is possible to operate the evaporator plant continuously for 20 hours or more without growth of mesophilic and/or thermophilic bacteria and their spores at reduced steam consumption. The direct contact regenerative preheaters are also used where Direct Steam Injection (DSI) for pasteurization is used.

Duplex Preheating System

By installing duplex preheaters, it is possible to have a continuous run of evaporator for 20 hours, as the preheaters can be cleaned before the critical level of biofilm growth is reached. Cleaning and effluent treatment add to the costs. Although, the investment is higher, the actual direct production costs and time are not affected.

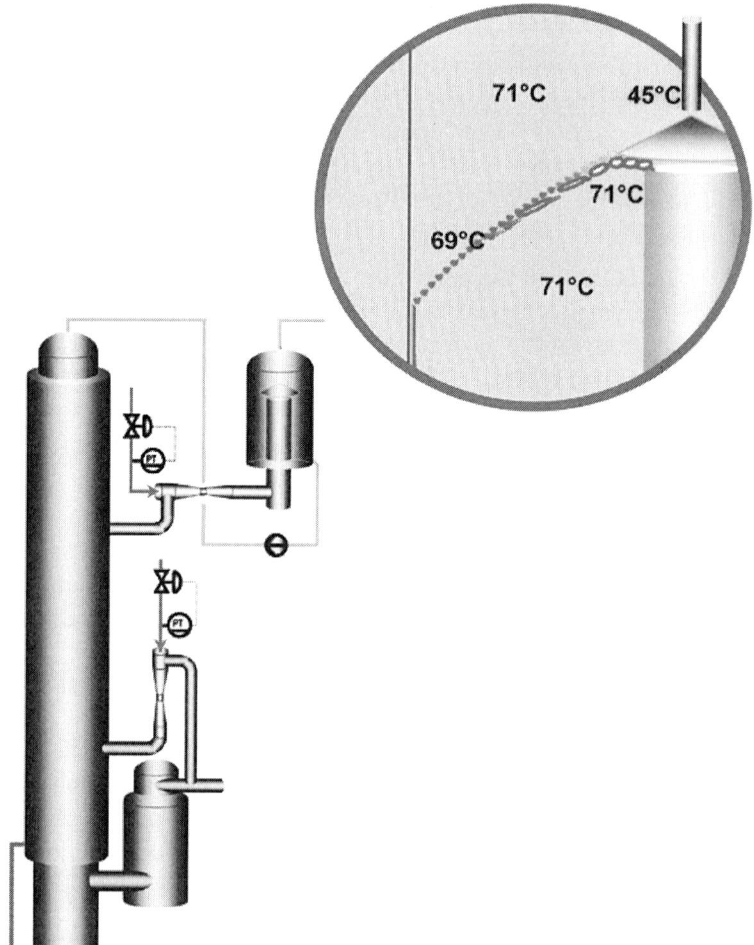

Figure 14.3. Direct Contact Regenerative Heater

Pasteurizing System

As mentioned earlier, the milk requires certain preheat treatment for destroying heat resistant bacteria and to prevent gelation of the concentrated product. Though, it is the most important function, there are other reasons like need to meet the requirements of final product, such as the type of milk powder to be manufactured using drying operation. More detailed information can be obtained from literature on Milk Powder Technology. (Ref. Vegn Westergaard 2010).

The heat treatment is done in two ways, i) Indirect pasteurization ii) Direct pasteurization.

Indirect Pasteurization

For a better finished product, the milk has to be preheated to high temperatures, before concentration by evaporation. This can be done either by indirect heaters or by direct injection of steam (DSI) systems. The indirect heaters are heat exchangers such as the plate, straight-tube and spiral-tube (Figure 14.4). If temperatures up to 110°C are required, it is recommended to have two heaters. While one is in operation the other can be cleaned.

The advantage of the indirect heating is that the product will not be mixed with the condensing steam and diluted. The disadvantage is that it takes a long time to heat the product from 80°C to 110°C resulting in a concentrate with high viscosity. This is because the whey proteins, when unfolded due to heat treatment will react with each other and also with k-casein. For improved energy efficiency one or more regeneration systems can be incorporated.

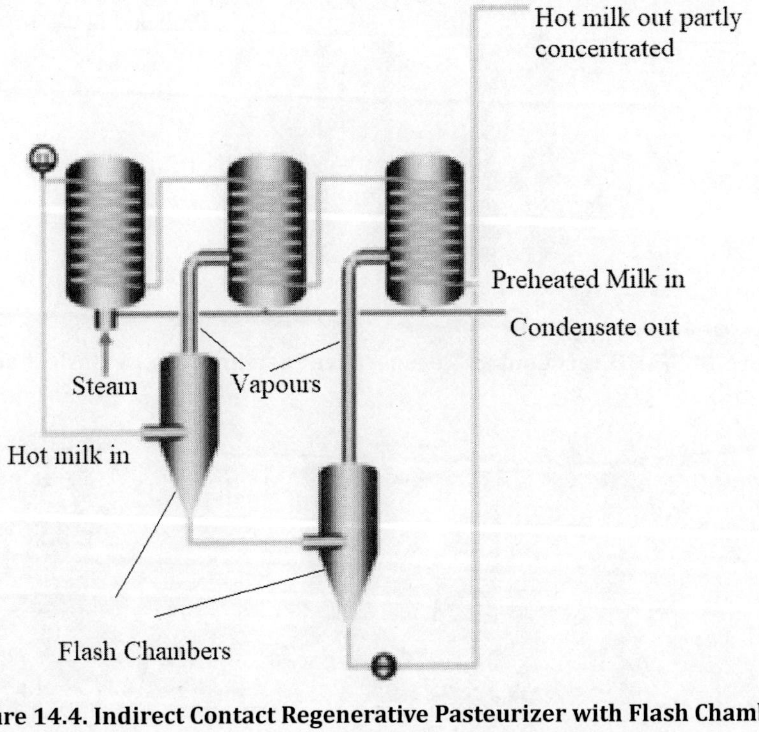

Figure 14.4. Indirect Contact Regenerative Pasteurizer with Flash Chamber

Direct Pasteurization

Direct pasteurization is done in two different ways. One method is by direct steam injection, where the steam is mixed with the milk using a Tangential Swirl Heater (TSH), (Figure 14.5 b). It offers controlled and short residence time with no mechanical impact, even at temperatures above 120°C. It can operate for 20 hours or longer without cleaning. In the other method milk is sprayed into an atmosphere of steam (infusion) at sufficient pressure. The steam must be of culinary type, Products made with culinary steam are fit for human consumption. Culinary steam boiler (vapours condensed from evaporation of milk condensate) uses an indirect coil-type heater and steam as heating medium. The milk condensate (vapours condensed from evaporation of milk) is converted to steam in the culinary boiler. The advantage of direct pasteurization is the short time taken to reach the desired temperature.

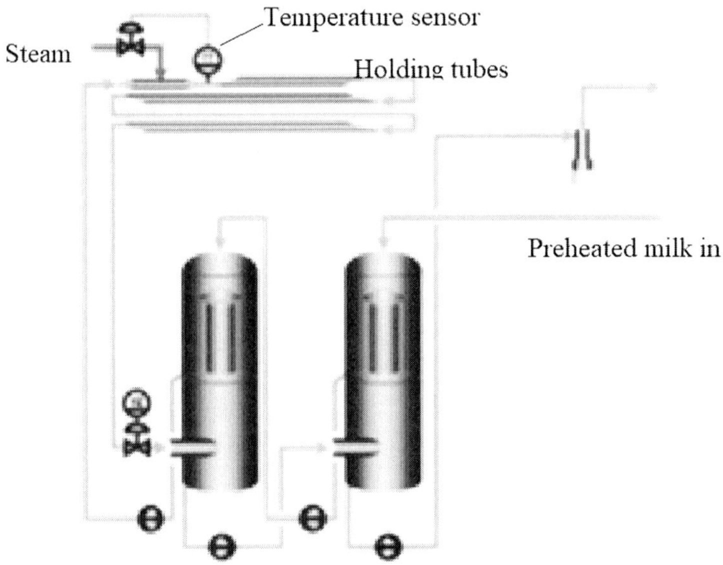

Figure 14.5(a). Direct Contact Regenerative Pasteurizer with Flash Chamber

Figure 14.5(b). Tangential Swirl Heating System for Injection of Steam into Milk

In comparison to indirect heating method, the direct heating has a less pronounced effect on the denaturation of the whey proteins for the same pasteurization temperature/time.

In indirect preheating, regenerative flash chambers are used when high pasteurizing temperatures are needed. The temperature of milk will drop due to the evaporation of water, and the vapours produced are used for preheating prior to pasteurization. The regenerative flash chamber can be either indirect as shown in Figure 14.4 or direct contact type as shown in Figure 14.5a, b. The direct contact regenerative system is preferable for its short residence time and due to absence of heat contact surface where deposits can develop.

The pasteurization temperature has influence on the total steam consumption, which will increase by increasing temperature. For the same pasteurization temperature, the direct pasteurization will result in higher steam consumption compared to indirect pasteurizing due additional energy required for evaporation of the extra water formed by the steam condensation. However, the additional steam used is partly offset by the flashing off vapours used as heating medium in the subsequent Calandria.

Holding Tubes

The holding of pasteurized product is always carried out in horizontally placed tubes, with specific length and diameter to give the desired holding time. There are usually several tubes of the same length in the holding section, combining appropriately of which enables the variation of holding time.

Product Distribution System

It is essential that the product to be evaporated is distributed uniformly into all the tubes of the calandria for effective use of heat transfer surface. The distribution system is therefore given a special attention when designing an evaporator. In principle there are two different systems: i) Dynamic distribution system ii) Static distribution system.

Dynamic Distribution System

In the dynamic distribution system, the necessary kinetic energy for distribution of the product is obtained by from pressure drop of the product over a full-cone nozzle. As the product is superheated in relation to the pressure inside the tubes, flash vapour will instantaneously is formed. The mixture of product and vapour is sprayed into the inlet of the tubes thus being covered uniformly by the product. This system is very inflexible with regard to capacity variation. It is not used in dairy evaporators designed for production of milk products with different solids content.

Static Distribution System

In the static distribution system, the incoming superheated product is first separated into flash vapour and product. The product flows to a distributor plate placed inside an open cone as the product enters the calandria. The cone is placed above a distributor bowl with a number of holes. Here a certain level of product is maintained. The product flows through the holes in the plate by gravity. Each hole is placed above the area between the tubes. Thus, the product flows onto the tube plate and then over the edge down along the inside surface of each tube. The flash vapour also enters the tubes and pushes the product against the inner surface of the tubes and transfers its kinetic energy. This gives the product nitial downward velocity.

This distribution system is much more flexible with respect to capacity, as an increase in the level in the distributor bowl - as a result of increased capacity - will make the product flow through the holes at a higher velocity, thus maintaining the level.

During CIP operation of the evaporator and especially the pasteurizing equipment, some jelly lumps of milk protein deposits may cause blocking of the holes in the distributor plate. To avoid this problem, a self-cleaning hydro cyclone is installed in the product line between the discharge from the flash vessel of the regenerative pasteurizer and the inlet to the first calandria.

Calandria(s) with Boiling Tubes

The liquid to be evaporated is evenly distributed on the inner surface of the evaporator tubes. The liquid will flow downwards forming a thin film, from which the evaporation will take place. Heat energy required for evaporation is obtained from condensation of steam on outside tube surface. The steam condensate flows downwards the outer surface of the tube. A large number of tubes are placed side by side to form an assembly. The tube ends are fixed to the tube plates. Finally the entire assembly is enclosed by a jacket. The steam is introduced through the jacket. The space between the tubes is the heating section. The inner side of the tubes is the evaporating section. Together they are called calandria. The concentrated liquid and the vapours leave the calandria from the bottom part, from where the major proportion of the concentrated product is discharged.

The remaining portion of the concentrate together with the vapours enter liquid- vapour separator tangentially. The separated concentrate is finally discharged (usually by means of the same pump used for the concentrate from the calandria). The vapours leave the separator from the top. The heating steam condenses on the outside surface of the tubes and is collected as condensate from the bottom part of the heating section. The condensate is discharged by means of a pump.

In order to understand the heat and mass transfer principles forming the basis for the evaporation process it is necessary to define specific parameters. Assume the mass flow rate of liquid product (Feed) entering the evaporator as A. The rate water evaporated and mass flow rate of vapours leaving the evaporator is B. The mass flow rate of concentrate is C. The total solids per cent defined as (Mass of solids in feed)/(Mas of feed) will increase from Total solids per cent in feed (TS_{feed}) to Total solids per cent in concentrate ($TS_{concentrate}$). And thus:

$$A=B+C \qquad \text{(Eq. 14.1)}$$

Mass flow rate of solids entering the evaporator = mass flow rate of solids leaving the evaporator

$$\text{Thus, } A\,(TS_{feed}) = C\,(TS_{concentrate}) \qquad \text{(Eq. 14.1a)}$$

$$e = A/C = TS_{concentrate}/TS_{feed} \qquad \text{(Eq. 14.2)}$$

The evaporation ratio (e) is a measure for the evaporation intensity and can be defined either as the ratio between the flow rates of feed and concentrate or the ratio between the total solids (TS) percentages in the concentrate and in the feed. If the concentrations or the evaporation ratio are known the quantities A, B or C can be calculated, provided one of them is known.

Milk, is a heat-sensitive product due to its protein content, and evaporation (*i.e.* boiling) at 100°C will result in denaturation of proteins to such an extent that the final product is considered unsatisfactory. The evaporator is therefore operated under vacuum, which means that the boiling/evaporation takes place at a lower temperature than saturation temperature corresponding to the normal atmospheric pressure. The vacuum is created by a vacuum pump prior to starting-up of the evaporator and is maintained by condensing the vapours by means of cooling water. A vacuum pump or steam ejector is used to evacuate non-condensable gases from the milk.

At 100°C the enthalpy of water is 539 Kcal/kg and at 60°C it is 564 Kcal/kg. As the milk has to be heated from 6°C to the boiling point, and as approx. 20 Kcal/kg of energy is required for maintaining a vacuum corresponding to a boiling point of 60°C, the following energy consumption can be estimated, assuming heat losses at 2 per cent:

Table 14.1. Comparison of Evaporation Process at Atmospheric Pressure and under Vacuum

	Energy Required (in kcal/kg)*	
	At 100 °C	*At 60 °C*
Heating	94	54
Evaporation	539	564
Vacuum	-	20
Net Energy consumption	633	638
Heat loss (approx.)	15	15
Total energy consumption	648	653

* Corresponding to about 1.1 kg of steam/kg of water evaporated.

Vapour Separator

The purpose of vapour separator is to separate the vapour from the concentrated product leaving the evaporator. Milk evaporators operate exclusively with centrifugal type vapour separators.

Separators with Tangential Vapour Inlet

As the vapours generated from the evaporation are used as heating media in the "next" calandria, any product present in the vapours must be separated, since it would otherwise contaminate the condensate and further represent a loss.

The major portion of the concentrate is discharged from the bottom of the calandria below the tube bundle. Due to the high vapour velocity some of the concentrate is carried along with the vapour as small droplets. The separation of product droplets from vapours is done in a separator with a tangential vapour inlet connected to the calandria below the tubes (Figure 14.6).

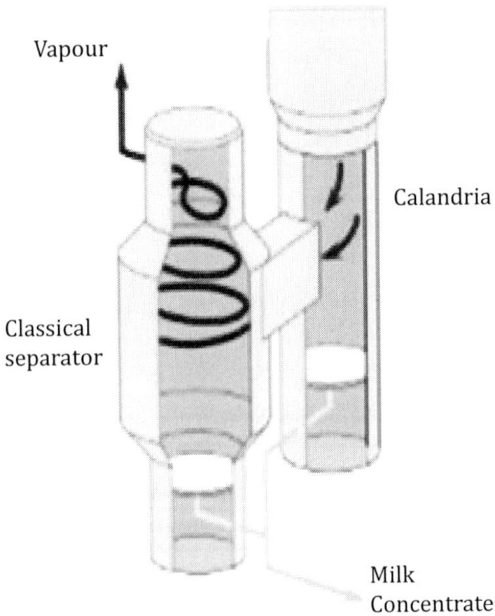

Figure 14.6. Vapour Separator with Tangential Vapour Inlet

Special care is taken in the design of the separator to avoid product carry-over at lowest possible pressure drop, as a drop in the pressure causes drop in enthalpy of saturated vapours generated in the following calandria (next effect) with an overall drop in the efficiency as a result.

Wrap-around Separator

To reduce space requirements, the Wrap-around separator was developed, see

Figure 14.7. It is integrated into the bottom portion of the calandria. It has efficiency as high as the classical separator, but with a low pressure drop. It is typically used on large calandrias operating with MVR compressors connected with a very short vapour duct to minimize the pressure drop. The saving in floor space is typically around 30 per cent.

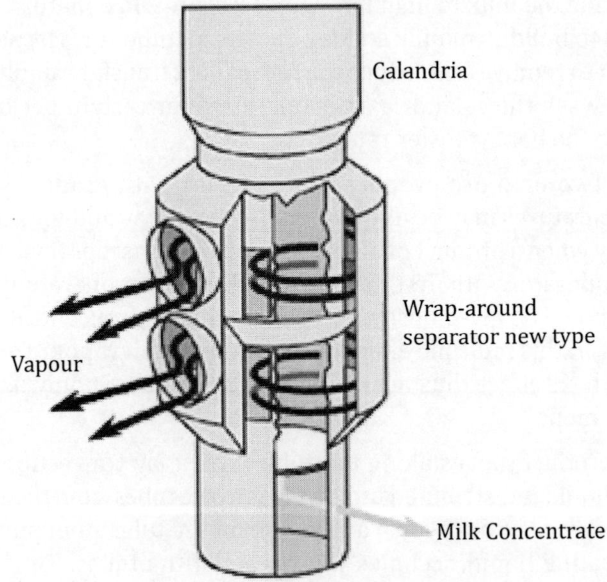

Figure 14.7. Wrap-around Separator

Evaporator Operation and Various Feeding Systems

Introduction

The most commonly used evaporation plant in the dairy industry is the falling film evaporating plant. It consists of the following components which are assembled together in the required manner.

a) Heat transfer tubes or calandria
b) Liquid/vapour separation system
c) Vapour removal and condensing system and vacuum control system
d) Ancillary equipment such as pumps for transferring milk, cooling water pumps, valves, gauges, thermometers *etc.*

a) Heat Transfer Surface

Efficient heat transfer from steam to the liquid product is vital; both to the process and to the quality of the final product. Chemical, physical and biological changes in the product depend partly on process time and temperatures. In general, high temperature and short heat treatment time produce less chemical and higher biological effect than lower process temperatures and long times. For economy, it is important to keep the temperature as high and heating time as short as possible. Heat transfer rates and flow of product through the evaporator system creates conditions specific to each type of equipment.

During heating the milk to high temperatures, protein denaturation will take place. This leads to buildup of milk solids deposits on tube surface, which become hard and difficult to remove and reduce the rate of heat transfer rapidly. Thus, there a need arises to wash the calandria after operation for certain period to remove scale and restore the heat transfer rate.

The use of two or more evaporators in series with gradually decreasing evaporator temperatures may be used for energy economy and improved product quality, especially when a product of higher viscosity approaches final density. Thus, a triple effect evaporator with first, second and third effect operating respectively at 70 °C, 57 °C and 44 °C gives desirable product quality, concentration level and energy economy. In this multiple effect evaporator system vapour from one effect is reused in the next effect thus achieving greater heat economy as well as less severe heat treatment.

The liquid product moves along the tube surface by convection and vapour propulsion. The liquid is distributed uniformly into the tubes from the top by means of weir or other device so as to form a thin film on the tube inner surface. Vapour from the evaporating liquid occupies the center of the tubes. Thus, only a thin quickly moving film is in contact with the heating surface. This evaporator system is called falling film evaporator. It provides a gentle heat treatment.

b) Liquid/Vapour Separation System

The separation of vapours from the concentrated liquid is possible by giving the liquid-vapour mixture, centrifugal or rotary motion at the entry of the separator. The adaptation of cyclone principle eliminated the need for deflecting plates, baskets and entrainment separators in the vapour space. The trend is to reduce the vapour system in size, by achieving higher efficiency of separation.

c) Vapour Removal and Vacuum Control System

The key component of the evaporator is the vacuum system, on which the success of operation, ease of operation and final product quality depend. It is important to remove besides the water vapour evaporated from the concentrated product, the non-condensable gases, which enter through leaks in the system or are entrained in the product.

d) Ancillary Equipment

The heat economy in evaporation is obtained by the continuous re-use of vapours generated during product concentration. This vapour may be recompressed by live steam as in Thermo Vapour Recompression System (TVR). The use of preheaters, interstage heaters and raw product heaters using condensate, all provide for economical operation. The use of heaters in this way requires pumps to transfer the liquid product. These pumps are often operating under vacuum, and therefore require water seals to maintain the vacuum.

Overall Heat Transfer Coefficient

The Overall heat transfer coefficient (U) is the most critical factor in operating the evaporator at its full capacity. The numerical values of U are influenced by a number of external factors as well as the properties of the evaporating liquid at any stage of the process (*i.e.* density, viscosity, surface tension, temperature, boiling point elevation, thermal conductivity, specific heat) properties of the heating medium, tube wall thickness and its thermal conductivity tube surface treatment and cleanliness, velocity of the product film flow and thickness of the product film.

The numerical value of the overall heat transfer coefficient varies between 3000 and 100 $W/m^2/^oC$ for low viscous liquids and clean surfaces and high viscous liquids and fouled surfaces respectively. Therefore, under continuous operating conditions in a multi-effect evaporator the overall heat transfer coefficient decreases from stage to stage due to rising viscosity and formation of deposits (mainly calcium phosphates and precipitated proteins) on the heating surfaces. The overall heat transfer coefficient for skim milk is about 2500 $W/m^2/^oC$ in the first effect and drops down to below 1000 in the last effect. For whole milk values are about 15 per cent lower.

Wetting of the Heat Transfer Area

It is a common problem occurring in evaporators operation, that the tube surfaces are not completely covered with the product due to low average flow of product per tube or due to poor product distribution in the tubes. The increased demand for large multi-effect evaporators require larger heating surface which can only be met by using increasing number of tubes. This would, however, mean that less liquid is entering each of the tubes, and the product film generated in the tubes becomes too thin. This will result in some product particles obtained from drying discolored and scorched. These particles will not dissolve when the powdered product is reconstituted. In extreme cases the tubes of the evaporator are blocked completely. In such situation manual cleaning of the tubes is necessary.

The design of the evaporator system therefore operates with the so-called coverage coefficient which is defined as:

Product flow rate at the lower end of tubes/Periphery of the tubes

The problem of complete wetting is overcome by recirculating part of the feed from the outlet of the calandria to its inlet of same, thus increasing the amount of product flow sufficient to cover surface of all the tubes.

However, this solution is not ideal, as the residence time of the product is increased, which will result in increased viscosity and protein denaturation.

This problem is solved by dividing the effects with low coverage coefficient in two or more separate calandrias or splitting the calandria into two or more sections with the same evaporating temperature and often with one liquid-vapour separator.

Hence, it is now common in the industry to find calandria with longer tubes in order to obtain more heating surface per tube and to use them in calandrias with two or more "splits" and maintain the coverage coefficient at a safe level. Earlier the evaporators were equipped with tubes of 3-4 m long and operated with a temperature difference of about15°C. Currently evaporators with tube lengths up to 14 m and operating with a temperature difference as small as 2°C are common. This also results in less product passes needed to obtain enough wetting of tubes, fewer pumps, and reduced residence time. The disadvantage is that there will be an increased pressure drop of the vapour over the longer tubes, and (that is on the account of) overall evaporation capacity. Any further increase in tube length results in increased vapour velocity and carryover of concentrate droplets.

The design of evaporator or spray dryer is optimized mostly for a particular product though it may also be adopted for other similar products with slightly lesser efficiency. Further, vapours generated from first effect is used as heat source in second effect resulting in saving of 50 per cent of energy. As the number of effects increase within the operational limits, the energy use will decrease for same amount of water evaporation.

The Velocity at the Bottom of Calandria

As the product flows under gravity down the tubes in the form of a film the amount of vapour increases as evaporation progresses and influences the flow of the film. The vapour reaches its maximum velocity at the bottom of the tubes shortly before the discharge end. The velocity can be calculated as illustrated below:

Velocity, V = Volumetric flow rate (m^3/s)/Cross sectional area of a tube (m^3)

= Mass flow rate x Specific volume (m^3/kg)/Cross sectional area of a tube

$$V = (\pi \times d \times h \times U \times \Delta K \times \upsilon)/\{L \times (\pi/4)\, d^2\} \qquad \text{(Eq. 14.3)}$$

where,

d = diameter of the tube in calandria, m

h = height of the tube, m

U =Overall heat transfer co-efficient, $kJ/m^2\,K$

ΔK = Temperature difference between evaporating temperature and the condensing steam

υ = Specific volume, m^3/kg

L = Latent heat of vaporization, kJ/kg

Then, the Eq.14.3 reduces to

$V = (4 \times U \times \Delta\theta \times \upsilon \times h)/L \times d$ (Eq. 14.4)

Example: The following data, $U = 2200\ W/m^2\ K$; $\Delta\theta = 10\ K$; $\upsilon = 7.68\ m^3/kg$

L (at 60 °C) = 2358 kJ/kg; h =6 m; d = 0.032 m

Velocity, $V = (4 \times 2200\ J/s\text{-}m^2\text{-}K \times 10K \times 7.68 m^3/kg \times 6m)/2358 kJ/kg \times 10^3\ J/kJ \times 0.032m$

= 53.7 m/s

This relatively high velocity, which can be influenced by the choice of the diameter of the tubes, the length of the tubes, the temperature difference and the level of the vacuum, has an accelerating effect on the already concentrated film of liquid particularly in the lower part of the evaporator.

The following example shows to what extent the drop of pressure in this type of the flow influences the boiling point.

Reynolds Number, Re. = Velocity x diameter/Viscosity x Specific volume

$= (53.7 \times 0.032)/11.15 \times 10^{-6} \times 7.68 = 20{,}067$

The value of Re indicates that flow is turbulent.

The rate of flow of vapours is not as high along the whole length of the tube. On the other hand, the film of the liquid reduces the diameter of the tubes.

The pressure drop for the whole length of the tube can be expressed by :

$\Delta P = (0.3164/\sqrt[4]{Re})\ (h \times V^2/d \times 2 \times \upsilon)$

In the present example, considering h = 6 m and d = 0.032 m

$\Delta P = \{0.3164 \times 6 \times (53.7)^2\}/\sqrt[4]{20067} \times 0.032 \times 2 \times 7.68$

$(0.3164 \times h \times V2)/\sqrt[4]{Re} \times d \times 2 \times \upsilon$

= 936 Pa = 0.00936 bar.

The above drop in pressure is for saturation temperature of water vapour at 60°C. This pressure difference corresponds to a temperature difference of only 1°C. However, the rise in temperature is significant in a multiple effect evaporator, where each effect operates at very small temperature difference.

For estimating the mean thickness of the film of product, δ, the following relationship which takes into account the effects of gravity and viscosity in laminar flow can be used.

$\delta = \sqrt[3]{\{(m \times 3 \times \upsilon)/(\rho \times \pi \times d \times g)\}}$

where,

m= mass flow rate of product, kg/s

υ = kinematic Viscosity, m^2/s

ρ = Density, kg/m^3

d = Diameter of pipe, m

g = Acceleration due to gravity, m/s^2

Different Feed Arrangements in Multiple-effect evaporators

The Multiple effect evaporator, provides an opportunity to advantageously use the different effects being at a range of temperatures, to obtain final product at temperature, or concentration that is most suitable for the further processing, especially drying. The most commonly employed feeding arrangements are: 1) Forward feed 2) Backward feed 3) Mixed feed and 4) Parallel feed.

In the forward feed system, the flows of process fluids and of steam are in parallel. Forward feed has the advantage that no pumps are needed to move the product from one effect to the next effect (not applicable to modern calandria type evaporator). It has the disadvantage that all the heating of feed is done in the first effect, causing less vapour to be generated for each kg of steam supplied, resulting in lower economy. It has the further disadvantage that the most concentrated product is subjected to the lowest temperatures. Low temperatures may be helpful in preventing decomposition of organics, but high viscosity of the product will sharply reduce the heat transfer coefficient of last effect. Forward feed system is generally used for concentrating heat sensitive products.

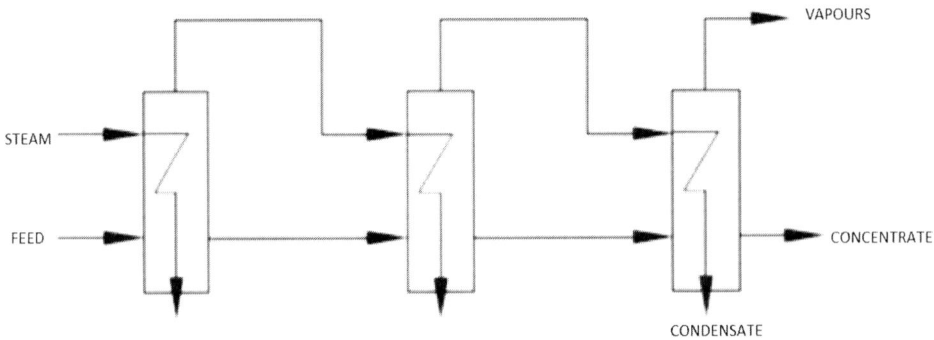

Figure 14.8. Forward Feeding System

In a backward feed system, the feed and steam are in counter flow. This system requires pumps between the effects. The feed solution is heated as it enters each effect, which results in better economy than that obtained with forward feed. The viscosity spread is reduced since the most concentrated product evaporates at the highest temperature. But this may affect the heat sensitive materials. This system is used for concentrating highly viscous product.

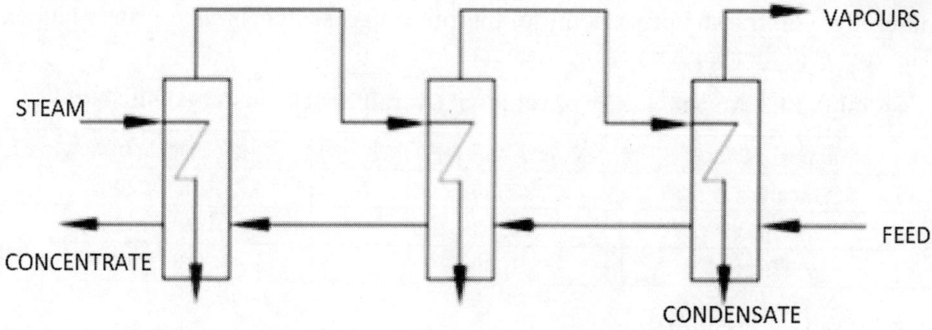

Figure 14.9. Backward Feeding System

For best overall performance, evaporators may be operated with flow sequences that combine both these methods (*i.e.* mixed feed), or each effect may be fed in parallel with fresh feed evaporating to final concentration.

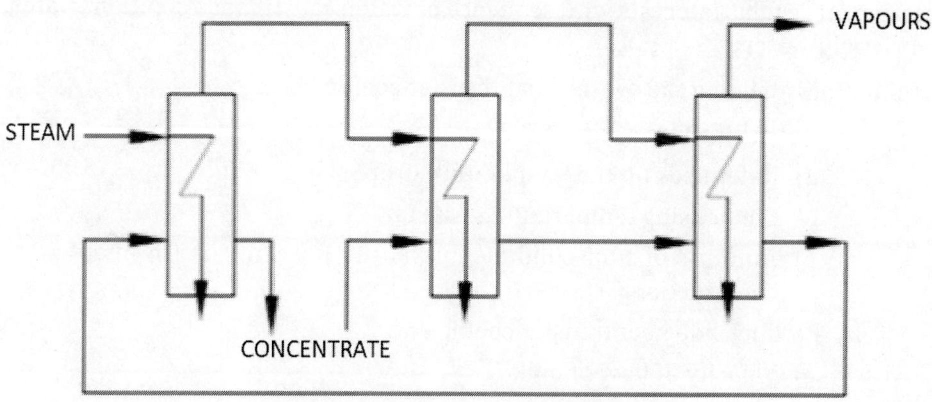

Figure 14.10. Mixed Feeding System

Capacity of Multiple – Effect Evaporators

Although the use of the multiple-effect principle increases the steam economy, it must not be thought that there are no accompanying disadvantages. Along with importance of economy of an evaporator system, is the question of its capacity. Capacity is the total evaporation of water per hour obtained since latent heat nearly remains the same over the range of pressure across the evaporator effects. Capacity can also be expressed as the total heat transferred in all the effects. If it is assumed that for both the batch type and multiple effect evaporators the overall heat transfer coefficients are the same, and that they are operating between the same temperature difference between condensing steam and evaporation temperatures and have equal heat transfer area, then their capacities do not show a significant difference. But the steam economy for multiple effect system is significantly higher. The cost is higher for having more effects.

Table 14.2. Normal Range of Values of Overall Heat Transfer Coefficients

Number of Effect	*Skim Milk (W/m^2K)*	*Whole Milk (W/m^2K)*
1st effect	2300-2600	2000-2200
2nd effect	1900-2200	1700-1900
3rd effect	1000-1200	900-1100

Note: 1 Watt = 0.86 kcal/h

During continuous operation of an evaporator, the overall heat transfer coefficient, 'U' drops due to the deposition of precipitated proteins or calcium phosphate on inside surface of the calandria tubes. Deposits may also form on the steam side if incomplete separation of the product from vapour occurs in the vapour separator and this vapour is used in the next effect. So plant must be cleaned at regular intervals with sequence of water-caustic soda solution- water-nitric acid-water.

Factors affecting the overall heat transfer coefficient:

i) Steam side film coefficient:
   a) It depends on the temperature drop
   b) Condensing temperature of steam
   c) Amount of non-condensable gases present. d) On evaporator construction.

ii) Product side heat transfer coefficient:
   a) Velocity of flow of milk
   b) Viscosity of fluid
   c) Cleanliness of the heating surface

The volume of the vapours generated also changes with the temperature and pressure in each effect.

Table 14.3. Temperature, Vapour Pressure and Specific Volume of Water Vapours

Temperature °C	*Absolute Pressure (mm of Hg)*	*Specific Volume (m^3/kg)*
75	434	2.8
70	233	4.8
60	149	7.7
50	92	12.0
40	55	19.6

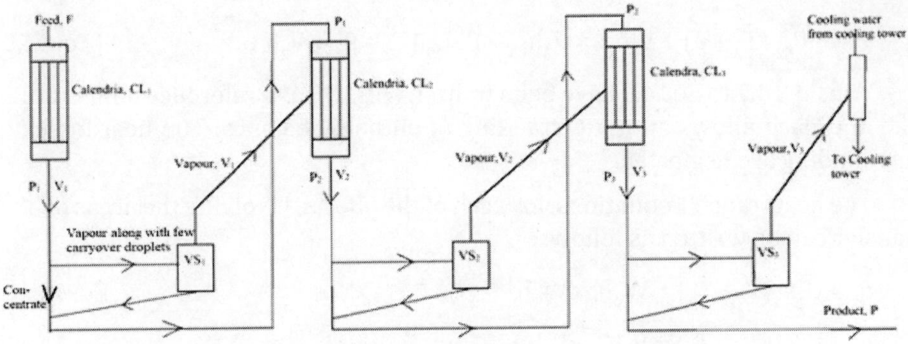

Figure 14.11. Schematic of a Triple Effect Evaporator

Analysis of Multiple Effect Evaporator

Legend:

Calandrias : CL_1, CL_2,CL_3

Vapour Separators: VS_1, VS_2,VS_3

Product from each effect (kg/hr): P_1, P_2, P_3

Vapour from each effect (kg/hr): V_1, V_2, and V3

Specific Enthalpy: h (kJ/kg), with subscripts S,V,L for Steam, Vapour and Condensate at the effects of 1,2,3 respectively

Steam supply (kg/hr): W_S

Feed: F (kg/hr); Product: P (kg/hr)

X_F and X_c are mass fraction of solids in feed and in concentrate respectively

The normal procedure is to write the mass and Energy (Enthalpy) balance equations for the entire system and for each effect.

Forward-feed triple-effect system is considered:

The mass balance equations:

$$F = V_1 + V_2 + V_3 + P \qquad \text{(Eq.14.5)}$$

$$X_F F = X_p P \qquad \text{(Eq. 14.6)}$$

The equation Eq. 14.5 and 14.6 describe the total mass balance and solids mass balance respectively.

An enthalpy balance can be written for each effect of the triple-effect system as follows:

$$F C_{PF} (T_F -0) + W_S h_S = V_1 h_{v1} + P_1 C_{P1} (T_1 - 0) + W_s h_L \qquad \text{(Eq.14.7)}$$

$$P_1C_{P1}(T_1 - 0) + V_1h_{V1} = V_2h_{v2} + P_2C_{P2}(T_2 - 0) + V_1h_{L1} \quad \text{(Eq.14.8)}$$

$$P_2C_{P2}(T_2 - 0) + V_2h_{V2} = V_3h_{V3} + P_3C_{P3}(T_3 - 0) + V_2h_{L2} \quad \text{(Eq. 14.9)}$$

The Eq. 14.7 to Eq14.9 have been written with 0°C as a reference temperature. The approach allows an understanding of enthalpy balance. The heat losses to surrounding are neglected.

The heat transfer equations, for each of the effects, involving the areas of heat transfer can be written as follows:

$$q_1 = U_1A_1(T_S - T_1) = W_Sh_S - W_Sh_L \quad \text{(Eq. 14.10)}$$

$$q_2 = U_2A_2(T_1 - T_2) = V_1h_{V1} - V_1h_{L1} \quad \text{(Eq. 14.11)}$$

$$q_3 = U_3A_3(T_2 - T_3) = V_3h_{V3} - V_3h_{L3} \quad \text{(Eq. 14.12)}$$

After computing the quantity of heating medium required, equations Eq. 14.10 to Eq. 14.12 can be utilized to compute the heat-transfer area with knowledge of overall heat-transfer co-efficient for each effect.

The usual procedure for computations utilize several assumptions such as equal heat flux, equal heat-transfer area or equal temperature gradients in each effect of the system.

The influence of product concentration on the overall heat-transfer co-efficient should not be overlooked, and should be computed or estimated for each effect of a multiple-effect evaporator.

The objective of the design of a multiple-effect evaporation system is to evaluate the efficiency as expressed by a steam economy and to compute the required heat-transfer surface area. The computation of steam economy is done by determining W_s, V_1, V_2 and V_3.

In practice, as the product at final stage is highly concentrated, for obtaining better heat transfer, a mixed feed is generally employed in the dairy industry.

Example: Concentration of Milk in a Triple-effect Evaporator

A triple effect forward – feed evaporator is being used to evaporate milk containing 10 per cent solids to a condensed milk of 50 per cent T.S. The boiling point rise of the milk (independent of pressure) can be estimated from BPR°C = $1.78x + 6.22x^2$, where x is weight fraction of T.S. in milk (K1). Saturated steam at 205.5 kPa (121.1°C saturation temperature) is being used. The pressure in the vapor space of the third effect is at 13.4 kPa. The feed rate is 22680 kg/h at 26.7 °C. The heat capacity of the milk is estimated by using the relationship (K1) $C_p = 4.19 - 2.35x$ (kJ/kg.K). The heat of milk is considered to be negligible. The coefficients of heat transfer have been estimated as $U_1 = 3123$, $U_2 = 1987$, and $U_3 = 1136$ W/m². K. If each effect has the same surface area, calculate the area, the steam rate used, and the steam economy.

Solution

Following the above steps outlined, the calculations are as follows:

Step 1. For 13.4 kPa, the saturation temperature is 51.67°C (from the steam tables). Using the equation for BPR for 3 - effect with $x = 0.5$,

$BPR_3 = 1.78x+6.22x^2 = 1.78(0.5)+6.22(0.5)^2 = 2.45$ °C, so $T_3 = 51.67+2.45 = 54.12$ °C

Step 2. Making an overall and a solids balance to calculate the total amount vaporized $(V_1+V_2+V_3)$ and L_3,

$F = 22680 = P_3 + (V_1+V_2+V_3)$

$Fx_F = 22680(0.1) = P_3(0.5) + (V_1+V_2+V_3)(0)$

$P_3 = 4536$ Kg/h

Total water evaporation rate = $(V_1+V_2+V_3) = 18\ 144$ kg/h

Assuming equal amount of water is vaporized in each effect, $V_1 = V_2 = V_3 = 6048$ kg/h. Making a total material balance on effects 1, 2, and 3 and solving.

(1) $F = 22\ 680 = V_1 + L_1 = 6048 + L_1$, $L_1 = 16\ 632$ kg/h

(2) $L_1 = 16\ 632 = V_2 + L_2 = 6048 + L_2$, $L_2 = 10\ 584$ kg/h

(3) $L_2 = 10\ 584 = V_3 + L_3 = 6048 + L_3$, $L_3 = 4536$ kg/h

Making a solids balance on effects 1, 2 and 3 and solving for x,

(1) $22\ 680(0.1) = L_1x_1 = 16\ 632\ (x_1)$, $x_1 = 0.136$

(2) $16\ 632(0.136) = L_2x_2 = 10\ 584\ (x_2)$, $x_2 = 0.214$

(3) $10\ 584(0.214) = L_3x_3 = 4536\ (x_3)$, $x_3 = 0.50$

Step 3. The BPR in each effect is calculated as follows:

(1) $BPR_1 = 1.78\,x_1 + 6.22(x_1)^2 = 6.22(0.136)^2 = 0.36$°C

(2) $BPR_2 = 1.78(0.214) + 6.22(0.214)^2 = 0.65$°C

(3) $BPR_3 = 1.78(0.5) + 6.22(0.5)^2 = 2.45$°C

$\Sigma\,\Delta T$ available = $Ts_1 - T_3$ (saturation) – $(BPR_1 + BPR_2 + BPR_3)$

$= 121.1 - 51.67 - (0.36 + 0.65 + 2.45)$

$= 65.97$°C

Making an initial assumption of distribution of available temperature difference between boiling points of first effect to last effect:

$\Delta T_1 = 12.40$ °C similarly $\Delta T_2 = 19.50$°C and $\Delta T_3 = 34.07$°C

However, since a feed at a lower temperature enters the first effect, this effect requires higher heat input.

Increasing ΔT_1 and lowering ΔT_2 and ΔT_3 proportionately as a first estimate,

$\Delta T_1 = 15.56°C, \Delta T_2 = 18.34°C, \Delta T_3 = 32.07°C$

To calculate the actual evaporation temperature of the milk in each effect,

(1) $T_1 = Ts_1 - \Delta T_1$

$= 121.1 - 15.56 = 105.54°C$

$Ts_1 = 121.1°C$ (Condensing temperature of saturated steam to effect 1)

(2) $T_2 = T_1 - BPR_1 - \Delta T_2$

$= 105.54 - 0.36 - 18.34 = 86.84°C$

$Ts_2 = T_1 - BPR_1 = 105.54 - 0.36$

$= 105.18°C$ (Condensing temperature of water vapours to effect2)

(3) $T_3 = T_2 - BPR_2 - \Delta T_3$

$= 86.84 - 0.65 - 32.07 = 54.12°C$

$Ts_3 = T_2 - BPR_2 = 86.84 - 0.65$

$= 86.19°C$ (Condensing temperature of water vapours to effect3)

However, further calculations depend on iterative steps to reach the appropriate solution.

In the dairy industry, neither the approach of i) Equal heat transfer areas ii) Equal quantity of vapour generation in each effect, are practical. This is because of i) decreasing quantity of water vapours as the concentration progresses ii) As the products gets concentrated, the temperature differences, need to be reduced since the product becomes extremely heat sensitive. It may be an approach suitable where parallel feed, and not the forward feed or mixed feed is employed.

An example of the calculations for an evaporator concentrating milk is illustrated below:

1. Capacity Calculations

A Falling Film Milk Evaporator is continuously supplying 48 per cent TS Milk Concentrate to a Spray Dryer operating at a capacity of 20 Tons/Day on 20 Hours working basis. The feed to the evaporator is skim milk with an average TS of 8.70 per cent The final output of SMP produced from spray dryer has an average 3.5 per cent Moisture content.

Determine the hourly water evaporation capacity required from the Evaporator.

Solution

a. Hourly SMP Output from Dryer: 20 x 1000/20 (Daily Working Hours) = 1000 kg/h
b. With 3.5 per cent Moisture in SMP, hourly production of Milk Solids: 1000 x 3.5/100 = 965 kg/h
c. Feed Rate to Evaporator: 965 x 100/8.70 = 11,019.95 = 11,020 kg/h

d. Hourly output of Concentrate from Evaporator at 48 per cent TS: 965 x 100/48 = 2,010 kg/h

e. Hence Water Evaporation required in the Evaporator: 11,020 – 2,010 = 9,010 kg/h.

2. Estimation of Steam for TVR

The above Milk Evaporator is designed as a Four Effect Falling Film configuration with Thermo Compression across the First Effect. The Thermo Compressor installed has an entrainment ratio of 2.3:1, which means that for each kg of Motive Steam used, the TVR is recompressing 2.3 kg of water vapours. Calculate the hourly quantity of Motive Steam required for the Evaporator and work out the Specific Steam Consumption Ratio for the Evaporator. Neglect all the heat losses.

Solution

Since the designed system configuration has Four Effects with the TVR across the first effect the solution is worked out as shown below;

Total Discharge Ratio from TVR: 1 Part Motive Steam + 2.3 Parts Suction Vapours = 3.3

Evaporation Ratio for the System: 3.3 (first effect) + 1 (second effect) + 1 (third effect) + 1 (Fourth Effect) = 6.3

Hence Motive Steam Required: 9,010/6.3 = 1430.16 kg/h

Specific Steam Consumption Ratio: 1430.16/9,010 = 0.15873

Water Evaporation in First Effect: 1430.16 x 3.3 = 4719.53 kg/h

Water Evaporation in Second Effect: 1430.16 kg/h

Water Evaporation in Third Effect: 1430.16 kg/h

Water Evaporation in Fourth Effect: 1430.16 kg/h

Total Evaporation: 4719.53 + (3 x 1430.16) = 9,010 kg/h

3. Determination of Heat Transfer Area (HTA) for each Calandria

The above Evaporation System is designed with the following Temperature Profile:

First Effect Jacket: 75°C

First Effect Evaporating Temp: 71 °C

Second Effect Evaporating Temp: 66 °C

Third Effect Evaporating Temp: 58 °C

Fourth Effect Evaporating Temp: 48 °C

Considering following average Overall heat transfer coefficients (U values) for different Effects, calculate the HTA required and the number of Tubes with 50.8 mm OD x 1.5 mm Thickness x 14,000 mm Length required for each Effect.

First Effect: 1700 kCal/m^2 h °C

Second Effect: 1500 kCal/m^2 h °C

Third Effect: 1300 kCal/m^2 h °C

Fourth Effect: 1000 kCal/m^2 h °C

Solution

From Steam Tables we can find the Latent Heat of vaporization at above evaporating temperatures as shown follows:

First Effect at 71°C: 556.7 kCal/kg

Second Effect at 66 °C: 559.6 kCal/kg

Third Effect at 58 °C: 564.4 kCal/kg

Fourth Effect at 48 °C: 570.1 kCal/kg

Equation Used: Q=U x HTA x ΔT or

HTA = Q/U ΔT

Q for an effect = Water evaporated in the effect x latent heat of vapors

HTA for First Effect: 4719.53 x 556.7/1700 x (75-71) = 386.4 m^2

HTA for Second Effect: 1430.16 x 559.6/1500 x (71-66) = 106.7 m^2

HTA for Third Effect: 1430.16 x 564.4/1300 x (66-58) =77.6 m^2

HTA for Fourth Effect: 1430.16 x 570.1/1000 x (58-48) = 81.5 m^2

Internal Diameter of Tube: (50.8 – 3.0)/1000 = 0.0478 m

HTA per Tube: 0.0478 x 3.146 x 14 = 2.105 m^2

Number of Tubes First Effect: 386.4/2.105 = 184

Number of Tubes Second Effect: 106.7/2.105 = 51

Number of Tubes Third Effect: 77.6/2.105 = 37

Number of Tubes Fourth Effect: 81.5/2.105 = 39

4. Working out Product Wetting Rates

Work out Product Wetting Rates from each Calandria and find out if it should be designed in Single or Multiple Passes.

Solution

Note: Product Wetting Rates are always worked out based on product flow exiting from the bottom of each Calandria.

Output product from First Effect: 11,020 – 4720 = 6300 kg/h

Output product from Second Effect: 6300 – 1430 = 4870 kg/h

Output product from Third Effect: 4870 – 1430 = 3440 kg/h

Output product from Fourth Effect: 3440 – 1430 = 2010 kg/h

Hence Wetting Rates can be worked out as below:

First Effect flow rate/tube: 6300/184 tubes = 34.24 kg/h/tube

Second Effect flow rate/tube: 4870/51 tubes = 95.49 kg/h/tube

Third Effect flow rate/tube: 3440/37 tubes = 92.97 kg/h/tube

Fourth Effect flow rate/tube: 2010/39 tubes = 51.54 kg/h/tube

Output Concentration Levels for product coming out from the bottom of each Calandria are:

Total input Milk Solids: 11,020 x 8.70/100 = 958.74, Say 960 kg

Concentration of solids in product exiting each effect is as follows:

First Effect: 960/6300 = 15.24 per cent

Second Effect: 960/4870 = 19.71 per cent

Third Effect: 960/3440 = 27.90 per cent

Fourth Effect: 960/2010 = 47.80 per cent

Wetting rate per tube has direct bearing on the sticking of milk solids on the internal surfaces of evaporator tubes causing tube choking/reduced evaporation rates *etc.* Based on their past experience different designers of Evaporator choose certain minimum wetting rate for different Calandria for different configurations. This minimum figure is chosen based on product flow ability, product concentration level, viscosity, operating temperature and also the Protein content levels, *e.g.* SMP has higher Protein content than WMP or Dairy Whiteners. In the above calculated figures it is found that First and Fourth Calandria wetting rates are much lower than the values preferred to keep fouling minimum. It is preferred to design these two Calandria in multiple passes. For example in this case it would be good to split the First in two Passes (102 + 82) tubes and the Fourth also in two passes (21+18) tubes.

5. Condenser Design

For the above Evaporator, the final and Fourth Effect water vapours are condensed in a Surface Type Tube Heat Exchanger. Calculate the HTA required for this Condenser and also work out the Cooling Water Flow Rate and Cooling Tower Capacity.

Solution

Quantity of water Vapours coming out from Fourth Effect: 1430 kg/h

Hence Total Heat to be rejected: 1430 kg/h x 570.1 kcal/kg = 815,243 kCal/h

Note: Under Indian conditions and considering a maximum wet bulb temperature of 28 °C in humid rainy season, Cooling Towers can cool the water at best to 32 °C. The return water temperature for a Condenser is generally kept about

5-6 degree C below the condensing vapour temperature. Hence we can choose inlet water temperature of 32 °C to the Condenser and 42 °C as the outlet/return temperature to the Cooling Tower.

Based on the above assumptions Cooling Water Circulation Rate can be worked out as follows:

815,243/(42-32) = 81,524.3 kg/h. However, we always choose at least around 15 per cent extra when specifying the cooling water circulation/cooling tower capacity as over a time the cooling water becomes dirty and also there is scaling occurs in the Condenser Tubes causing reduced overall heat transfer coefficient/ heat rejection capacity.

Therefore in this case we can specify 90,000 LPH as Cooling Water Flow Rate.

Cooling Tower Capacity: 90,000 x 8 (LMTD)/3024 = 238 TR Say 250 TR

HTA for Condenser: 815243/1000 (U) x 10.20 (LMTD) = 79.93 m^2, Say 80 m^2

Note: To obtain higher flow velocity and better HTC, Condenser tubes are generally chosen with smaller diameter and lower wall thickness. In this particular case e 25.4 mm OD x 1.2 mm thick x 14 m long tubes can be selected.

HTA for each tube works out as: (25.4 - 2.4)/1000 x 3.146 x 14 = 1.013 m^2

Total Number of Tubes Required = 80/1.013 = 78.97, Say 80 divided in two passes of 40 tubes each.

Volumetric Flow per tube: 90/40/3600 = 0.00625 m^3/s

Cross Section Area/Tube = 0.00416 m^2

Flow Velocity through Tube = 0.00625/0.00416 = 1.502 m/s, which is satisfactory.

6. Working Out amount of Steam required for Direct Injection for Preheating the Feed Milk

In the above Evaporator, the feed milk @ 11,020 kg/h flow rate is coming out of the last Pre-Heater at 70 °C. It needs to be heated to 90 °C by direct Steam Injection. Culinary quality steam is available at 2.0 bar (gauge) pressure. Work out the total quantity of steam required for the purpose.

Solution

Heat Load: 11,020 x 1 kcal/kg-°C x (90-70) C = 220,400 kCal/hr

L.H. of Steam @ 2.0 bar (g) = 517.2 KCal/kg from Steam Table

Steam Required = 220,400/517.2 = 426 kg/h, Say 425 kg/h

(Sensible Heat of the Steam is ignored).

7. Waste Heat Recovery

In the above Evaporator, how much waste heat going out of the system in the form of vapours to the Condenser that can be recovered for pre-heating of the Feed Milk in a Tubular Heater.

Solution

Since the vapours are available at 48 °C and the Feed Milk coming out of the storage tank is generally at 4 °C, a portion of the total vapours can be used to pre-heat the feed milk to around 42 °C (that is 6 °C less than the Vapour Temperature). This will result in a portion of the vapours condensing in this pre-heater. Thus the load on the Surface Condenser is reduced by the amount of heat rejected from vapours to milk which ultimately reduces the size for the Condenser and also lower Cooling Water flow Rate/lower size of Cooling Tower.

Assuming a U value of 1000 KCal/m^2 - °C and Specific heat of milk as 1 kCal/kg - °C Required HTA of Pre-Heater: 11,020 kg/hr x 1 kcal/kg - °C x (42-4) °C/1000 kCal/m^2 - °C x 19 °C = 22.04 m^2

Number and size of tubes= 10 Tubes of 50.8 mm OD x 1.5 mm Thickness x 14 m Long. It should be designed in Single Pass.

The following chart shows the operating data of a multiple effect evaporator concentrating milk.

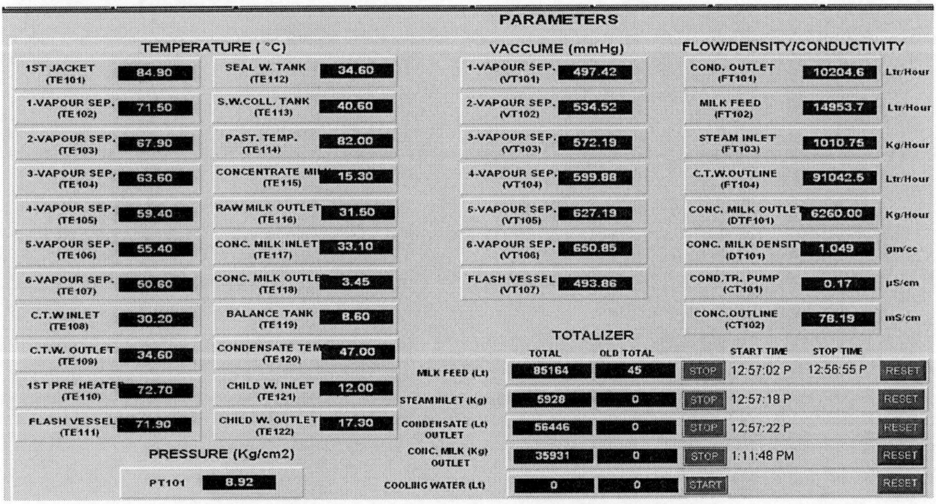

Figure 14.12. Chart Showing the Operating Data of a 6-effect Evaporator.

REFERENCES

"Food Engineering and Dairy Technology" by H.G. Kessler (1981), Pub: Verlag A. Kessler, Germany

"Milk Powder Technology: Evaporation and Spray Drying" by Vagn Westergaard, GEA Niro, Copenhagen, Denmark

Chapter 15

Auxillary Equipment for Evaporators

Vapour Recompression Cystem

The vapours leaving the vapour separators have considerable amount of energy, and by increasing its temperature, they can be reused. For increasing the temperature, the vapours need to be compressed. Two types of vapour compressors are used with the evaporators. They are i) Thermo Vapour Re-compressor and2) Mechanical Vapour Re-compressor

Thermo Vapour Recompression (TVR)

The thermo-compressor increases the temperature/pressure level of the vapour, by compressing the vapour at a lower pressure to a higher pressure by employing high pressure steam. The thermo-compressor has no moving parts and steam flows through it at high velocity. The construction of thermos compressor is simple, its size is small, and the cost is low.

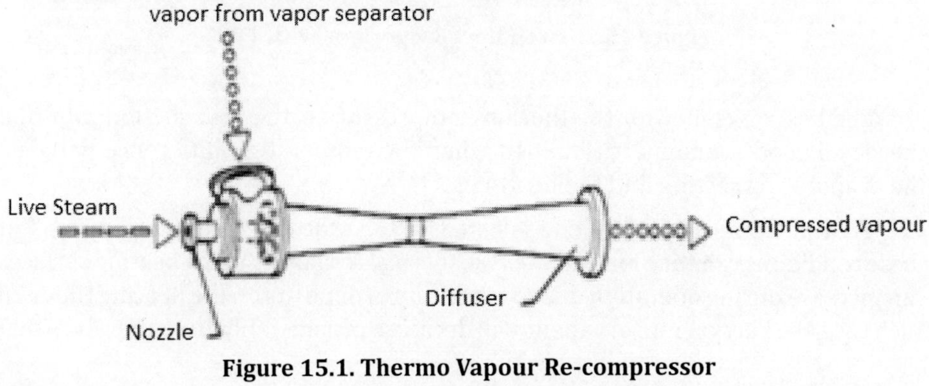

Figure 15.1. Thermo Vapour Re-compressor

In the nozzle the pressure of the inflowing steam is converted into velocity. A high velocity jet is therefore created at nozzle exit. A portion of the vapour is drawn from the vapor separator. In the diffuser the velocity of mixture of steam and vapours, is converted into pressure (pressure and temperature increase). This recompressed vapour can now be used as heating medium for the evaporator. The Figure 15.2 shows a double-effect evaporator with thermo-compressor.

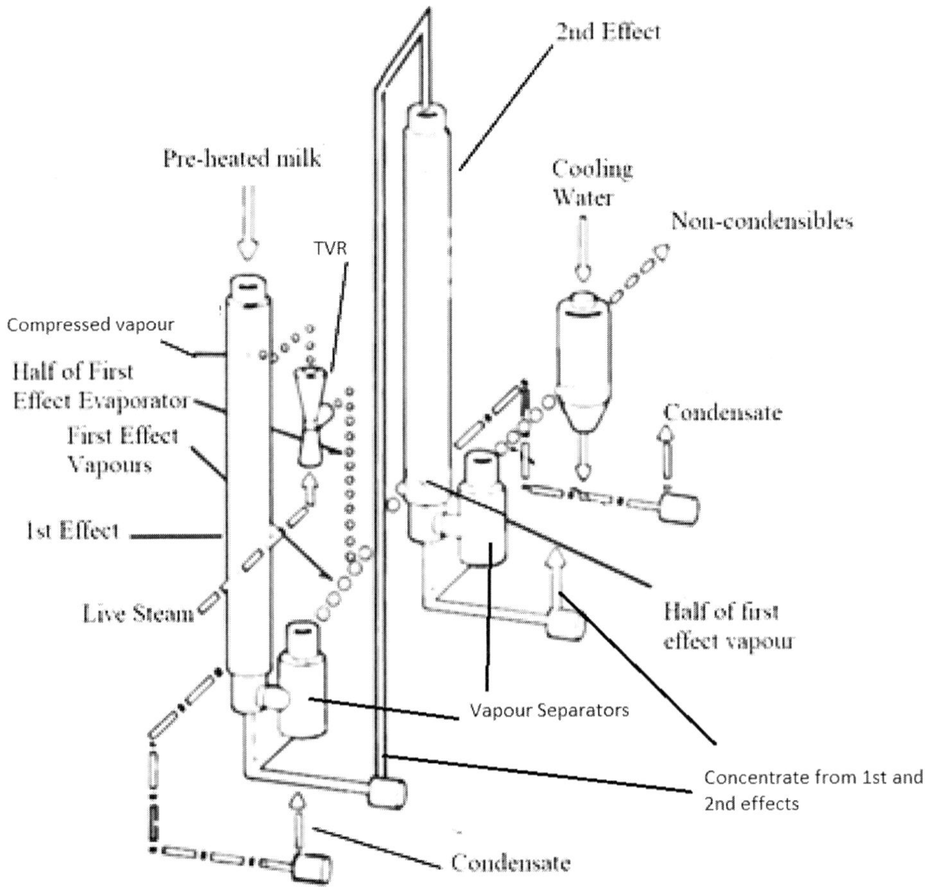

Figure 15.2. Two Effect Evaporator with TVR

The best efficiency for the thermo-compressor, *i.e.* the best suction rate, and thereby a good economy, is obtained when the temperature difference between the evaporating section and the heating section is low.

Thermo-compressor must be selected for the exact operating conditions. But these conditions can vary, for example, the thermal resistance of the heating surfaces can increase during operation due to formation of deposits on the heating tubes. In such case, the suction rate of vapour will decrease considerably. In evaporators that have to meet varying capacities a number of thermo compressors with different characteristics are used. Further, a thermo-compressor, which is designed for use with a higher live steam pressure, will draw a larger amount of vapour from the separator than one built for a lower steam pressure. For simplification we will assume an efficiency of 1:2, but latest thermo-compressors under certain conditions can operate with an efficiency of 1:3, *i.e.* for one kg of live steam, 3 kg of low pressure vapour is compressed by life steam.

By adding a thermo-compressor to a three-effect evaporator 1kg live steam evaporates 5 kg of water, *i.e.* the steam economy is as great as that obtained by addition of two effects. Dividing a given total Δt between the first and last effect in multi-effect evaporators requires an enormous heating surface and consequently makes the installation expensive.

Mechanical Vapour Recompression - MVR

As an alternative to the thermo-compressor, during the last few decades the mechanical vapour compressor has found extensive use with evaporators in the dairy industry. The energy used for the compressor is usually electricity, but diesel engines and gas turbines are also used. In situations where processes require steam at low pressure, and the compressor could be driven by a steam turbine which essentially acts as a pressure reducing valve. The exhaust low pressure steam from the turbine can be used for other processes. The economics of the system is dictated by local pricing policy for energy. As a rule of thumb, an MVR solution is economical, if the cost/kW is less than three times the price/kg steam. However, the selection of compressor is influenced by the end product quality - the milk powder. In the MVR evaporator the residence time is resulting in lower viscosity of the concentrate.

The mechanical vapour compressor is a high speed (~3000 rpm) fan capable of operating under vacuum. At low boiling temperatures the volume of the vapours generated is enormous. Consequently, a limit is imposed on the lowest temperature that can be used in practice. As the energy applied to the compressor is utilized most efficiently by low compression ratios, the temperature/pressure increase obtained is limited. Therefore, a large heat transfer surface for the evaporator is required tending to increase the capital costs of the equipment. As it is essential to operate an MVR evaporator at a low overall temperature difference between the vapour evolved from the product and the heating medium as a result of the compression, it is necessary for the boiling point elevation of the product is kept at a minimum, as this would otherwise even further minimize the temperature difference available for the evaporation. Figure 15.3 illustrates a single-effect MVR evaporator. The incoming cold milk is first preheated by the concentrate then by condensate from the calandria followed by a final pasteurization by means of live steam. The vapour is compressed by the high speed blower and then enters the evaporator.

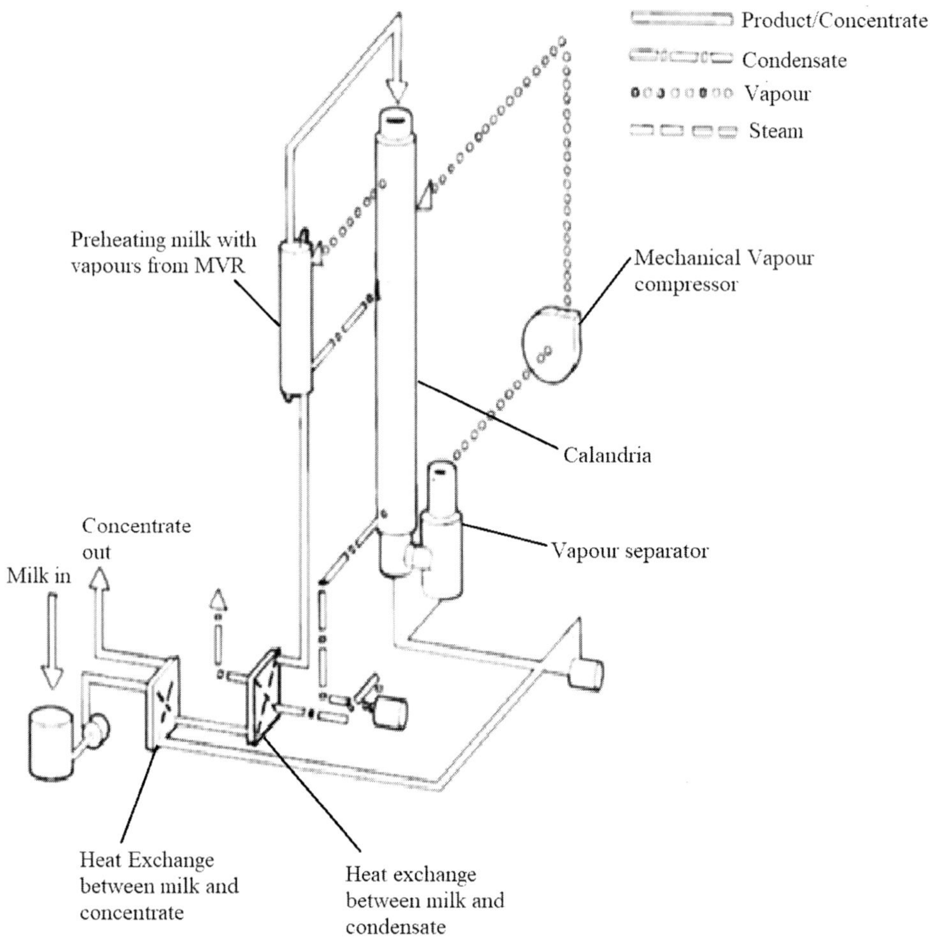

Figure 15.3. Single Effect MVR Evaporator

A vacuum pump, together with a water-cooled condenser, maintains the desired vacuum in the system. As it can be seen, no energy is rejected from the plant in the form of warm condensate, except a minor part via the cooling water (depending upon the pasteurization temperature desired). The MVR Evaporator is often used as pre-condenser of milk products for transport purposes, where the required solids content is in the range of 30-35 per cent. Thus, the boiling point elevation is limited. With the concentrate leaving the plant at low temperature, this kind of installation is competitive to membrane process.

The vapour from the separator is drawn by the blower and the compressed vapour is de-superheated by spraying water into the outlet of the blower. The compressed vapour is condensed on the heat exchanger surface in the subsequent calandria, where it is discharged as condensate. Simultaneously, water is evaporated from the milk and separated as vapour in the separator.

The MVR evaporator offers much better capacity variation/turn-down capability, as speed of the fan only needs adjustments for such variations.

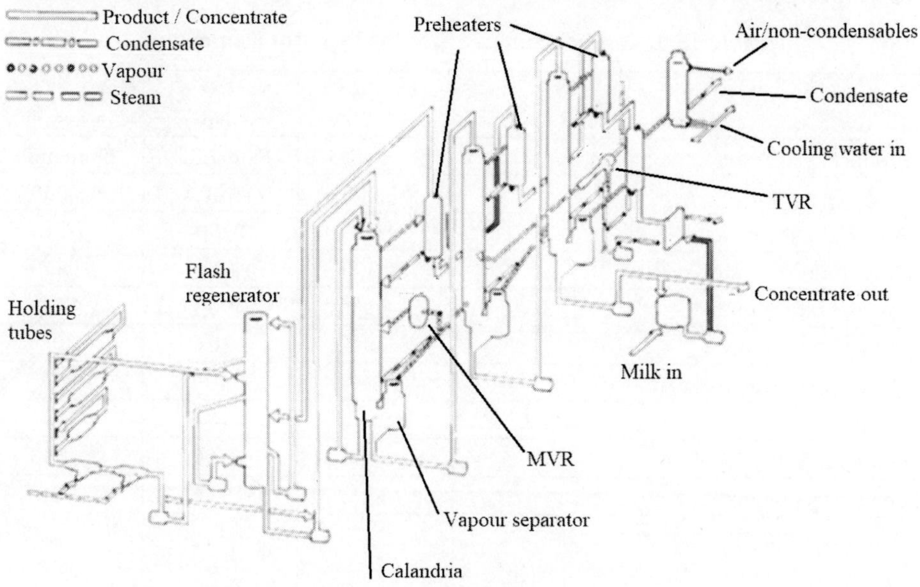

Figure 15.4 Combined MVR/TVR Evaporator

Usually, the MVR evaporator is combined with a TVR unit, if a concentrate with solids content suited for a spray drying is planned for production (Figure 15.4). The steam consumption per kg of water evaporated is of course less than in a multi-effect evaporator, but if the MVR unit is driven by an electric motor, the electrical energy consumption will be higher. Since cooling water requirement is minimal, this system is very attractive. However, a higher investment should be anticipated. When electrical energy cost is lower, it is advantageous to replace the TVR unit with an additional MVR unit to compress the vapour over the last effect. It is therefore essential that each case is studied carefully taking into consideration local pricing of steam, electricity and cooling water.

With reference to the Figure 15.4, the uncondensed milk enters the balance tank, and is pumped to the first preheater where condensate from last effect heats the incoming milk. The milk is then pumped through the preheaters connected with each of the effects, from the lowest temperature ones to highest. It is then pumped in to high heaters, and held for a predetermined time, at this temperature, to obtain specific properties in the final product and product not to show signs of gelling. The milk at high temperature is made to come down in temperatures instantaneously through Flash regenerator. The milk is then sequentially passed through first effect to the last one. To improve upon the thermal efficiencies of the whole system, TVRs and MVR are employed, either exclusively or together.

Comparison of Energy Consumption in Multiple Effect and MVR/ TVR Evaporators

Table 15.1. Comparison of different Types of Evaporator

	5-Effect TVR	*7-effect TVR*	*1-effect MVR/ 2-effect TVR*
Product	Skim milk	Skim milk	Skim milk
Capacity, kg/h	15,000	15,000	15,000
Solids in/out, per cent	9/50	9/50	9/50
Evaporation of water, kg/h	12,300	12,300	12,300
Pasteurization temp, °C	90	90	90
Holding time, sec	30	30	30
Steam Consumption, kg/h	1,610	1,190	375
Steam pressure, bar	10	10	10
Condensate, kg/h	13,400	1,400	12,800
Condensate temp, °C	54	51	22
Power consumption - MVR - Pumps, kw	 75	 75	 150 50
Cooling water., m^3/h	32	3.5	2*
Cooling water temp. °C in/out	28/35	28/35	12/50
Power consumption. cool. tower, kw	10	2.5	—
Residence time, min	12	18	6

* To be used only if the temperature of the raw milk is above 5 °C.

Vacuum Producing Devices

Production and maintenance of vacuum in the evaporator is essential for the smooth and efficient operation of evaporators. The vacuum producing devices used in dairy industry also includes water vapour condensers which are 1) indirect type condenser 2) Direct type condenser

Indirect Type Condenser

The surface condenser is an indirect type heat exchanger in which water at ambient temperature is on one side, causes the vapour coming from the product to condense on the other side. The most common among these types is the vertical shell and tube type condenser, which is similar to the calandria, but less in height.

Direct Type Condenser

In this type of condenser, the cooling water is sprayed as a jet (Figure 15.5b), into the chamber in which the water vapour from the evaporator is condensed.

These are further classified as parallel flow and counter flow type. The parallel flow condenser is normally operated as a wet condenser and the counter current flow condenser as a dry condenser. In parallel flow condenser, non-condensable gases and water are removed at the same temperature, where as in counter current condenser, the non-condensable gases are removed at the temperature of incoming water.

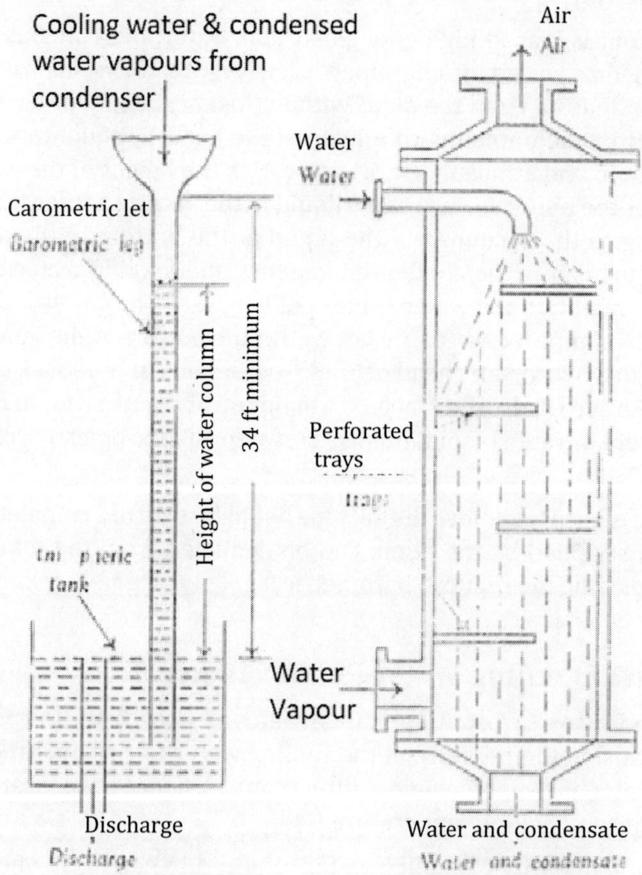

(a) Barometric Leg (b) Tray Direct Type

Figure 15.5. Types of Condensers

In case of dry condenser, the cooling water is removed by one pump and the non-condensable including the air are removed by a vacuum pump. In a wet condenser, condensed vapour and non-condensable are all removed together.

A jet condenser will use cooling water amounting to 20 to 50 times the weight of vapour. Thus the vapour being condensed and the cooling water are mixed. A jet condenser is normally used in milk evaporating operation in preference to a surface condenser. The surface condenser is more expensive.

The quantity of water required is less for a counter flow type of condenser. Another advantage is that the vapour with air need not enter at the top of the unit as is in parallel flow. The quantity of air removed from the evaporator system is about 15 – 25 per cent of the volume of cooling water. Leaks in the system can cause the quantity of air to be considerably higher, resulting in expensive operation.

Barometric Leg

This system is placed high enough so that water and condensate from the condenser are discharged by a barometric leg (Figure 15.5a). In order to remove water and condensate from the plant without losing vacuum it is necessary that a leg of liquid is maintained with a hydrostatic head H, equal to the difference between vacuum and atmospheric pressure. H is the height of the water column. In this manner the upper surface of the liquid in the barometric leg is at a pressure corresponding to the vacuum and the liquid at the bottom of the tail pipe is at atmospheric pressure. The condensed vapours and cooling water continuously discharged to maintain the water column of height H. The water leaves from the bottom of the tail pipe by way of the hot well at the bottom. Atmospheric pressure corresponds to a hydrostatic head of 10.35 meters of water and absolute vacuum corresponds to zero hydrostatic head. To maintain absolute vacuum requires a leg of 10.35 meter of water be maintained between the barometric condenser and the hot well.

If a pump is used to remove the tail pipe liquid instead of a complete barometric leg, the head supplied by the pump can be deducted from the total barometric height to determine the reduced barometric leg. This assembly is known as a low-level condenser.

Estimation of Cooling Water Requirement

In a condenser the cooling water removes heat from the vapour to be condensed causing temperature of the cooling water to rise. In direct condenser the cooling water and vapour come in direct contact, hence condensate and cooling water leave at same outlet temperature.

Let, m = quantity of cooling water required to condense one kg. of water vapour,

C = Specific heat of $C_p$water kJ/kg K

t_1 and t_2 = Inlet and outlet temperature of cooling water

$H = (h + \chi L)$ = Enthalpy of saturated vapors at evaporator temperature and pressure (from saturated steam tables.)

(h=Specific Enthalpy of liquid water; χ = Dryness fraction of vapour; L= Latent heat of vaporization of water)

$h_1 = S \times t_2$ = Heat content of one kg of condensed vapour

where, S is the specific heat of condensed vapour

Then,

Heat received by circulating water from one kg of vapour

$= m \times C_p \times (t_2 - t_1)$ kJ (Eq.15.1)

Heat given out by one kg of vapour to circulating water

$= H - h_1$ kJ (Eq.15.2)

From energy balance we get $m \times S \times (t_2 - t_1) = H - h_1$

Therefore, $m = (H - h_1)/[C_p \times (t_2 - t_1)]$ kg.

Example

Determine the amount of cooling water required in a jet condenser to condense one kg. water vapour, if cooling water inlet and outlet temperatures are 30°C and 40°C respectively. Take heat content of water vapour (H) as 2574 kJ/kg at 40°C and specific heat of cooling water as 4.187 kJ/kg.

Solution

$m = (2574 - 167.57)/[4.187 \times (40 - 30)] = 57.5$ kg

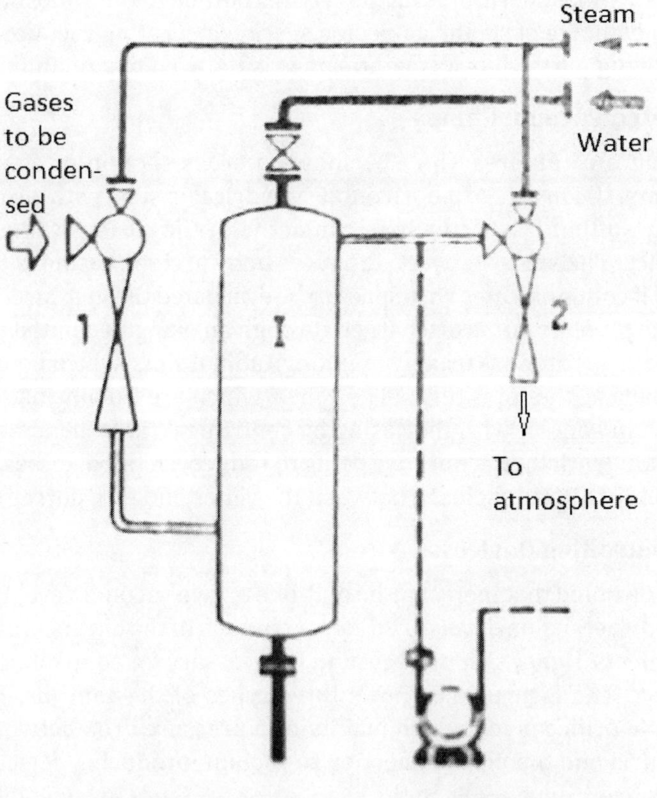

Figure 15.6. Two Stage Vacuum Producing System

Vacuum Producing Equipment

There are number of devices available to produce low pressure in the evaporator, whether it is a single or multiple of effect.

Working Principles of Steam Jet Ejector

Positive displacement pump of reciprocating type and steam jet ejectors are commonly used to produce vacuum. A pump is normally used for producing 24" Hg vacuum or less. The single stage steam jet ejector may be used for developing 25" Hg vacuum. The two stage team jet ejector may be used for developing 28.8" Hg vacuum or three stages to produce 29.8" Hg vacuum. The steam used is at 7.0 bar.

High pressure steam is admitted to a nozzle "1". It is converted to a jet of very high velocity steam into the throat of a venturi-shaped tube. The non-condensable gases to be removed enter through suction chamber as shown in Figure 15.6. By proper proportioning of the throat of the venturi, volumetric flow rate and velocity of steam, the steam can be made to draw the non-condensable gases from the space. For a very high vacuum, the steam air mixture from these jets goes to a condenser "I", where the water vapour is condensed by a jet of cold water and the residual air passes to a second nozzle "II". The discharge from second nozzle can usually be made to reach atmospheric pressure and is discharged to the atmosphere. Another important advantage of steam jet ejector over reciprocating vacuum pump is that it has no moving parts, thus its maintenance is reduced to a minimum.

Oil Lubricated Vacuum Pumps

The diagram (Figures 15.7a,b) shows a cross-section of a typical rotary vacuum pump. It consists of a horizontal cylindrical casing, with a rotor mounted eccentrically so that it is virtually in contact with the casing at one point of the circumference. The space between the rotor body and the casing is thus crescent shaped, and it communicates with space to be evacuated through an elongated inlet port with the pipeline. Air is discharged through an elongated outlet port with the exhaust pipe. The rotor has usually four longitudinal slots, which house vanes free to slide radially as the rotor turns. The vanes, which are usually made of asbestos fiber, remain in contact with the casing by centrifugal force. In some designs the vanes slide tangentially, the purpose being to reduce frictional losses. As the rotor turns, pockets of air are enclosed between the vanes and transferred to the outlet.

Vacuum Controlling Devices:

It must be noted that there may be only one or two vacuum developing systems actually used, the vacuum levels are different in each of the effects, which may be as many as 6 effects. Hence, a control system is necessary for controlling the vacuum in each effect. The Figure 15.8 shows three types of the controls. The first one (a) is a simple orifice place, which maintains a pressure drop between the jacket of the calandria and the pipe connection to vacuum producing device. The size of the orifice is fixed, depending on the evaporator pressure to be maintained in the

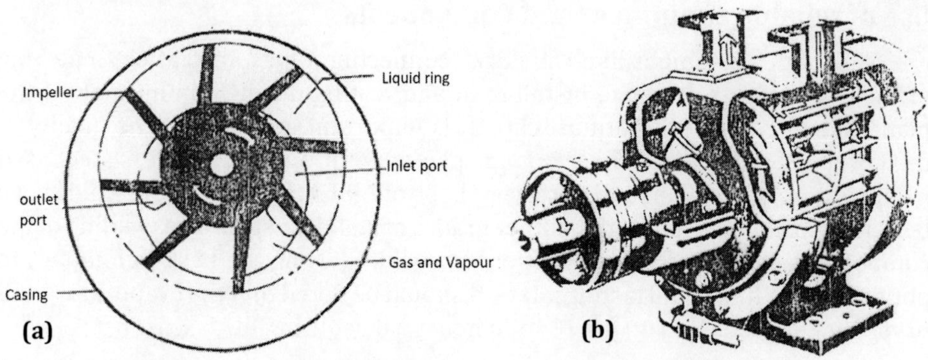

Figure 15.7. Working Principle of Water-Ring Vacuum Pump

effect. The other is a spring-loaded ball valve, depending on the spring pressure, pressure difference is maintained. This too is more or less fixed in its operation range. The third one is a more flexible in its operational range. It is a control valve directly attached to the evaporator. By opening or closing of the control valve the vacuum is maintained.

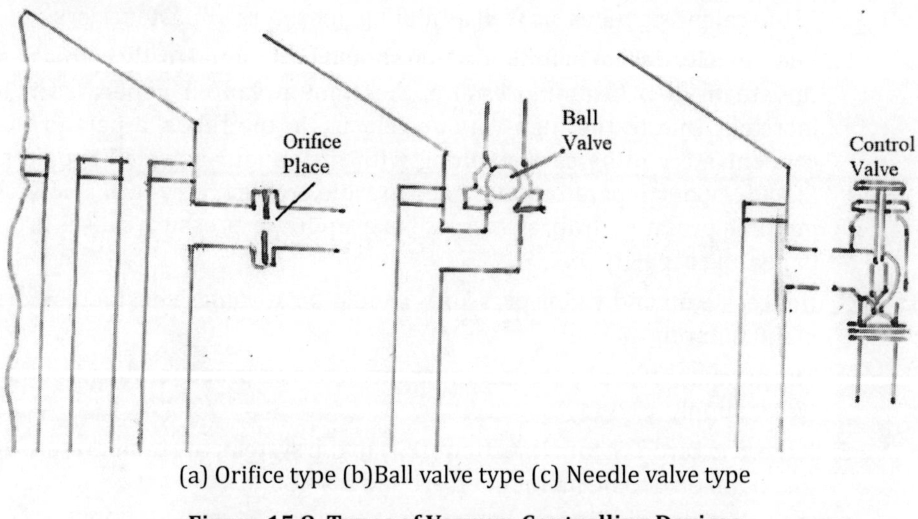

(a) Orifice type (b)Ball valve type (c) Needle valve type

Figure 15.8. Types of Vacuum Controlling Devices

Care and Maintenance of Evaporators

The performance of any process equipment depends on 1) Reliability of equipment 2) Care taken during operation 3) Regular maintenance of the wearable parts.

Use of reliable equipment and Components

The main equipment like calandria, connecting pipes, insulation, matching of the components selected, installation and testing during commission of the plant, must be done with utmost care. It is important not only for the quality of final product, but also for the economic plant operation. The stainless steel used in fabrication of the plant and parts should be of AISI 316 or AISI316L, which have better corrosion resistance than other grades of stainless steel. The welding of the components should be done by argon arc welding using the best techniques, to obtain smooth finish and lasting joints. It should be noted that the evaporators and dryers are expected to run for 18 to20 hours a day during flush season.

The evaporator installation should be such that the milk is evenly distributed in the calandria pipes, in a range of concentrate inflow. The nozzle has to be optimally sized depending upon the pressure drop determined by the quantity of the liquid concentrate to be handled and the degree of flashing when entering the space above the tubes.

Care During Operation

1. The quality of milk being fed should be of good quality, to prevent clogging of the calandria tubes, as well as fouling in heaters or pasteurizers.
2. The capacity designed for operation should be adhered to. By increasing the steam flow rate or pressure, the amount vapour generated will increase. Due to the high vapour velocity in the tubes, a part of the concentrate will be carried along with the vapour as small droplets. Liquid-vapour Separator is designed to avoid product carry-over at lowest possible pressure drop, as increase in evaporator pressure causes drop in heating capacity too.
3. Hence, steam and water pressures should be at values specified by the manufacturer.
4. The pumps that are operating under vacuum need proper sealing/cooling water supply.
5. If there is Direct Steam Injection (DSI) in the system, the temperature at this point also to be maintained around 75 – 80 °C.
6. When any power failure occurs, the steam valves to TVR and DSI have to be closed immediately to avoid burning of product.
7. Follow cleaning procedures of calandria regularly, after each operation of the plant.
8. Follow manufacturer's instructions.

Operational Efficiency

The term capacity of evaporating plant gives the output in terms of water evaporated per hour. It depends on the heat transfer surface area, temperature difference and the overall heat transfer coefficient. The surface area may be used on the basis of external or inner diameter of the tube. The number of tubes and the length of tube will decide the area of heat transfer with fixed diameter of tubes. Thus, the capacity of plant increases in direct proportion to the number of tubes for same diameter and length. With larger number of effects, the total surface area may be distributed among different effects keeping in mind the product flow rate. Thus, the first effect has always larger number of tubes while the subsequent effects will have reduced number of tubes.

The other important parameter is the temperature difference which is the difference in temperature of steam condensing in the first effect jacket and the temperature of vapour condensing in the last effect. With first effect evaporating temperature of 70 °C, the steam temperature may be assumed to be around 80 °C and the last effect vapour may be assumed to be condensing at 45 °C. Thus, the available temperature difference (80-45=35) is about 35 °C. With the increase in number of effects the available temperature difference will decrease and hence the capacity will remain constant or may even decrease because of increased loss of heat from surface of calandria jacket.

The last and one of the important parameter is the overall heat transfer coefficient expressed in $W/m^2.K$. The U-value or the overall heat transfer coefficient is affected by several factors, such as velocity of flow of product, thermal conductivity of tube material, fouling and viscosity of the liquid product flowing down the tube. The change in specific gravity with change in concentration also affects the U-value. Thus the U-value is much higher for the first effect but goes on decreasing with increasing product concentration in subsequent effects. The surface tension and viscosity of the product indirectly affect the velocity of flow and thickness of product film causing reduced heat transfer rate. The fouling of milk contact surface with hard scale is the main cause for reduction in U-value with time. The scale formation occurs basically due to inverse solubility of calcium and phosphorus salts present in milk at the tube wall temperature and protein denaturation at temperature above 70 °C at heat transfer wall. In addition to this low flow rate or inadequate wetting of tube surface causes burning of the thin film at the tube wall even at slightly low temperature. Therefore, care is required in the operation of the plant to avoid conditions for hard and tough scale formation in the evaporator tubes. Over hard scale deposit, softer deposits are also formed causing further reduction in heat transfer rate.

The other factors which affect the plant operation are steam pressure, flow rate of product, even distribution of feed in the tubes, maintaining constant vacuum, preventing leakage of air and operation of condenser for rapid condensation of water vapours and removal of condensate and non-condensables. If thermo-compressor is used, the working of plant is affected by variation of steam pressure at inlet to the thermo compressor. The provision of separate steam main from boiler with regulating valve will solve the problem. The product flow rate can be easily be controlled by a feed pump of adequate size and needle valve for flow regulation. Automatic flow controller and flow measuring device may be used. The uniform distribution of product in all the tubes is another major problem in the design and operation of falling film evaporator. The term wetting rate indicates the relationship between product feed rate and heating surface area. Every tube requires a certain minimum quantity of product to cover the entire inside surface of the tube. Therefore, in a single pass evaporator, the number of tubes per calandria will decrease from an effect to next effect, because the milk volume is constantly reduced with increasing concentration. The first effect is always the largest. In large evaporators, the calandria will be divided in two sections in order to overcome the problem of distribution of the product over a very large number of tubes, and to obtain correct wetting rate. When the first effect is in two sections, the milk flow will be in series, but the steam flow will be parallel.

Care and Maintenance

1. Preventive maintenance guidelines from the manufacturer must be followed.
2. Air leaks which may develop around valves, joints, cover and observation posts must be checked as leaks cause fluctuations to the operating vacuum and temperature.
3. All gaskets must be changed periodically.
4. Use of high pressure steam then the required should be avoided.
5. Cooling water flow to the water cooled condenser should be checked.
6. At least 3°C temperature difference between condenser discharge water and cooling water should be maintained for better economy.
7. It should be ensured that the heating surfaces are clean and free from the deposits.
8. Scaling of heat transfer surfaces constitutes a serious problem and tubes should be cleaned after an optimum operation time. The plant should be descaled at least once a year with a suitable acid solution.
9. Water supply for vacuum system should be checked (working of vacuum pumps).
10. Periodical detection of steam coil leakage should be done by hydraulic pressure test.
11. Condensate must be removed rapidly from the heating surface.
12. Vacuum must be immediately released in case of sudden power failure. Also steam valves and cooling water supply must be shut off at once.

REFERENCE

"Milk-powder technology", GEA process engineering, (2010), GEA Niro, Copenhagen

Chapter 16

Principles of Drying

Introduction

In dairy industry, the concentration of milk to remove moisture can reduce the bulk, and preserve the milk constituents to some extent. To remove moisture to the extent of 'bone dry', evaporators are unsuitable, because of high viscosity of concentrated product. The purpose of further drying milk and milk products is to allow longer periods of storage with minimized packaging requirements, reduced transport and refrigeration costs. The quality of the dried product is greatly influenced by the drying operation. The quality of a dried product is judged by the amount of physical and biochemical degradation occurred during the dehydration process. The drying time, temperature and water activity influence the quality of final product. The structure, density and particle size of the dried milk powder play an important role in its reconstitution.

Drying Fundamentals

Drying is defined as a process of moisture removal by simultaneous heat and mass transfer. Heat transfer from the surrounding environment to the product evaporates the surface moisture. The internal moisture in the product then can be either transported to the surface of the product and then evaporated or evaporated internally at a liquid vapor interface and then transported as vapor to the surface.

The mechanisms of water transfer in the product during the drying process can be summarized as follows: 1) water movement due to capillary forces, 2) diffusion of liquid due to concentration gradients, 3) surface diffusion,4) water vapor diffusion in pores filled with air, 5) flow due to pressure gradients, and 6) flow due to water vaporization–condensation. In the pores of solids with rigid structure, capillary forces are responsible for the retention of water, whereas in solids formed by aggregates of fine powders, the osmotic pressure is responsible for water retention within the solids as well as on the surface. The type of material to be dried is an important factor to consider in all drying processes, since its physical and chemical properties play a significant role during drying due to possible changes that may occur and because of the effect that such changes may have in the removal of water from the product.

The transfer of heat energy in a drying process using hot air depends on the air temperature, air humidity, air flow rate, exposed area of food material, and pressure. The dehydration equipment generally utilizes conduction, convection, or radiation to transfer energy from a heat source to the food material. The heat is transferred directly from a hot air or indirectly through a metal surface.

In convectional drying the heating medium, generally air, comes into direct contact with the product being dried. Various types of ovens, rotary, fluidized bed, spray, and flash dryers are typical examples. In conduction drying, the heating medium is separated from the product by a hot conducting surface. Examples are drum, cone and through dryers. In radiation dryers, the heat is transmitted as radiant energy. Some dryers also use microwave energy to dry food materials at atmospheric pressure or under vacuum.

Free and Bound Water

Water is a basic element in foods which contains positively charged two hydrogen atoms and one negatively charged oxygen atom hence water is polar solvent. The bonds between oxygen and each hydrogen atom are polar bonds, having 40 per cent partial ionic character. Water that can be easily removed from foods is known as free water, whereas water that cannot be removed as easily is termed as bound water.

Several properties of foods significantly depend on water content. Solute molecules and ions, or chemical groups and surfaces, can be hydrated. There are several relations between water content and a macroscopic property because the part of water is present in the constituents. This water is called bound water. Hydration of water in food components involves small quantities of water. Only polar groups, and to a lesser extent dipole, can "bind" water molecules.

Free Moisture

Free moisture is the water in excess of the equilibrium moisture content at a given temperature and humidity. Free moisture can be easily removed by drying which uses the principle of water vapour pressure difference between the moisture in the product and surrounding air.

Bound Moisture

Bound moisture in a food material is that water which is retained in a manner that exerts vapour pressure less than that of the free water at the given temperature.

Water may become bound to the solid by retention in small capillaries, by solution in cells or fibre walls, by homogeneous solution throughout the solid or by chemical or physical adsorption on solid surfaces. Many food constituents can hold onto water molecules which cannot be removed simply and they do not behave like free water molecules.

The bound water exists in food products in the form of 1) water molecules which are bound to ionic groups such as carboxyl and amino groups 2) water molecules which are hydrogen bonded to hydroxyl and amide groups and 3) unbound free water found in interstitial pores in which capillary forces and soluble constituents causing lowering of vapour pressure.

Some properties of bound water are:

1. Bound water cannot act as a solvent for salt, minerals and sugars present in food material
2. Freezing point of bound water is lower than of the free water.
3. Bound water exhibits essentially no vapor pressure.
4. The density of bound water is greater than that of free water.

Unbound Moisture

Unbound moisture exerts a vapor pressure equivalent to that of the free water at given temperature. Thus, in a non-hygroscopic material, all the water is unbound and in a hygroscopic material, the unbound moisture is the water in excess of the Equilibrium Moisture Content (EMC), corresponding to saturation humidity.

Water molecules attach to ionic sites in molecules like starches, pectin and protein. Water adjoining to these molecules has firm bonds between them; the following water layers are attached less firmly, finally the structure of free water exists. Water may be entrapped in foods such as gels, fruits and vegetables. Entrapped water is not easily mobilized in cells, but if released during physical damage, it flows freely.

To compare the difficulty in removal of water in a food product by application of heat energy, the free water would be evaporated and removed initially, followed by water molecules that are hydrogen bounded and finally, the water which is bound to ionic groups. Hence, the energy requirements for moisture removal will have to take these into consideration while designing the equipment for dehydration.

Evaporation occurs when the vapour pressure of the moisture on the solid surface is equal to the atmospheric pressure (around). This is done by raising the temperature of the moisture to its saturation temperature at atmospheric pressure (Example: Roller Dryers). If the material to be dried is heat sensitive; then, the method that should be adopted is evaporation under vacuum. When drying is carried out by a convective process such as passing warm air over the product the, moisture removal takes place by vaporisation. The air is cooled by the product and the moisture is transferred to the air by difference in vapour pressure exerted by moisture in air and that at the product surface and is carried away by air flow. In this case, the saturation vapour pressure of the moisture over the solid is less than the partial pressure of water vapour in the atmospheric air.

Role of Humidity in the Air

The relative humidity of the air to which the moisture is transferred from the product being dried is an important factor to be considered.

All products containing moisture have certain equilibrium moisture content with surrounding air. The equilibrium moisture content depends upon the temperature and relative humidity of air. The material will lose or gain moisture from the surrounding air over a period of time to attain this equilibrium value. Figure 16.1 shows various regions of moisture content for bound, unbound and free moisture in a food product.

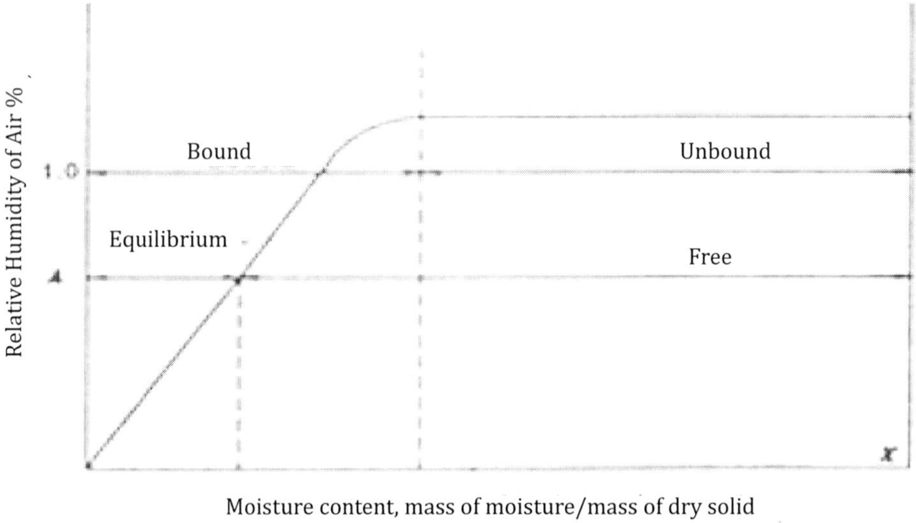

Figure 16.1. Equilibrium Moisture Isotherm for a Food Product

When hot air is blown over a moist food product, heat is transferred from air to the surface, the temperature of water, thus water vapour pressure increases to increase vaporization. Water vapour diffuses through boundary layer of air and is carried away by moving air. This creates a region of lower water vapour pressure at the surface of the food and a water vapour pressure gradient is established from the interior moist region of the food to the dry air. This gradient provides the 'Driving force' for transfer of water removal from the food to the drying air.

Foods are characterized as hygroscopic and non-hygroscopic. Hygroscopic foods are those in which the partial pressure of water vapour varies with the moisture content. The difference is found by sorption isotherms. When a food product is placed in a dryer, there is short initial settling down period as the product surface heats-up to the wet bulb temperature of air. Drying then commences and water moves from the interior of the food at the same rate as it evaporates keeping the surface to remain wet. This drying regime is known as constant rate period of drying and it continues until certain moisture content is reached. This moisture content is called critical moisture content. In practice different areas of food surfaces dry out at different rates, and overall, the rate of drying declines gradually. Thus the critical point is not fixed for a given food and it depends on the amount of food in the dryer and the rate of drying. The three characteristics of air that influence drying in the constant rate period are:

1. A moderately high dry bulb temperature
2. Low RH
3. High air velocity

The boundary layer of air surrounding the food acts as a barrier for the transfer of both heat and water vapour during drying. The thickness of the boundary layer is determined primarily by the air velocity. If air velocity is too low, water vapour leaving the surface of the food increases the humidity of the surrounding air and causes reduction in water vapour pressure gradient and rate of drying.

Stages of Drying

The typical drying process consists of three stages: heating the food to the drying temperature, evaporation of moisture from the product surface occurring at a rate proportional to the moisture content (Figures 16.2a,b). After the surface moisture is dried, a critical moisture point is reached from where the falling drying rate stage starts. The critical moisture point depends greatly on the drying rate since high drying rates will raise the critical point and low drying rates will decrease it.

When the moisture content of the food falls below the critical moisture content, the rate of drying slowly decreases to zero at equilibrium moisture content. This is known as falling rate period. Non-hygroscopic foods have single falling rate period where as hygroscopic foods have two periods. In the first period the plane of evaporation moves inwards and water in the food diffuses through dry solid to the drying air. It ends when plane of evaporation reaches the center of the food and the partial pressure of water vapour falls below the saturated water vapour pressure. The second period occurs when the partial pressure of water vapour is below the saturated vapour pressure as drying proceeds by desorption.

During falling rate period the rate of water movement from the interior of the food to the surface falls below the rate at which water evaporates to the surrounding air. The surface therefore becomes dry. This is usually the longest period of drying operation.

The surface temperature of the food remains close to the wet bulb temperature of the drying air until the end of the constant drying rate period, due to cooling effects of evaporating water. During the falling rate period the rate of water evaporating from the surface gradually decreases, but as same amount of heat is being supplied by the air, the surface temperature of the product rises until it reaches the dry bulb temperature of the drying air. Most heat damage to food occurs during the falling rate period. Figure 16.2 (a) and (b) shows general drying curve for solid food product.

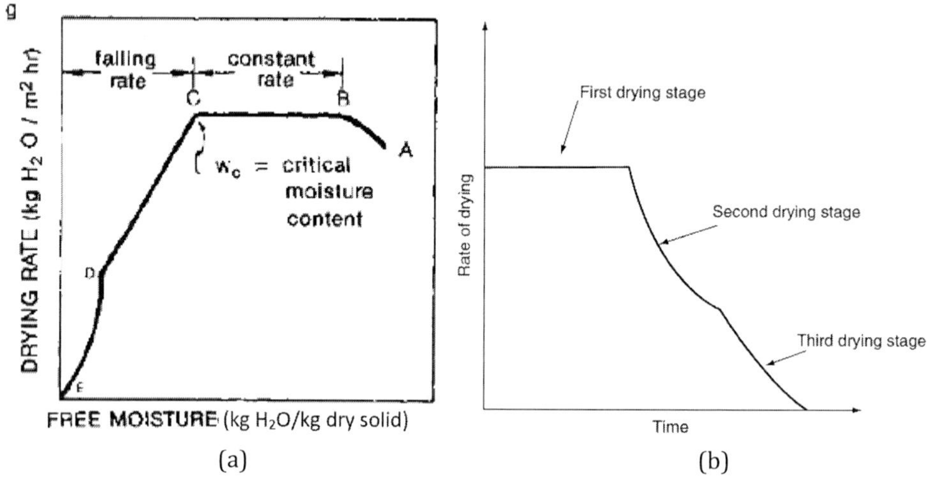

Figure 16.2. Rate of Drying Curve

Constant Drying Rate Period

During constant drying rate period, water in the product dries from the surface in a manner similar to a body of water open to atmosphere. Diffusion of moisture from within the product maintains saturated surface conditions and as long as these last, evaporation takes place at constant rate. When a solid is dried under constant drying conditions, the moisture content and the drying rate typically fall. The Time vs Rate of drying plot (Figure 16.2 b) is linear initially, then curves and eventually levels off. Constant rate drying period proceeds as long as free moisture appears from the surface, the moisture removal rate subsequently drops progressively. The moisture content at which the drying rate ceases to be constant is known as the critical moisture content. During the constant rate period, the moisture from interior of the product migrates to the surface by various means and is vaporized. As the moisture content decreases, the rate of migration to the surface drops. If drying occurs at too high temperatures, the surface forms the layer of closely packed shrunken cells which are sealed together. This presents a barrier to moisture migration and tends to keep the moisture sealed within. This condition is known as 'case hardening'.

In the constant drying rate period, the moisture being evaporated at the same rate as the moisture flow from interior of the product that is being dried. As long as the rate of surface evaporation continues at constant rate, the latent heat supplied for evaporation by conduction (in certain types of driers) is equal to the convective rate of heat transfer at surface by the drying air.

The constant rate period is characterized by a rate of drying that is independent of moisture content. During this period, the solid is so wet that a continuous film of water exists over the entire product surface, and this water acts as if solids are not present. The temperature of the wetted surface attains the wet bulb temperature of drying air.

Wet bulb temperature (WBT) is the steady state temperature shown by the thermometer whose bulb is covered with a wet wick and from which water is evaporating into a surrounding air stream at high velocity. The quantity of water evaporated is not significant to alter the temperature and humidity of the air stream.

Falling Rate Period

This period starts after the drying stage reaches the Critical Moisture Content. From this stage, the drying rate gradually decreases. The decrease may be linear initially and at a decreasing rate subsequently. In the first falling rate period the saturated surface area decreases as the moisture movement within the solid is reduced. When the entire surface area reaches the unsaturated state the internal moisture movement becomes the limiting factor. At this stage, the moisture movement within the product takes place by a) water movement by capillary forces b) diffusion of water c) surface diffusion and d) water-vapour diffusion.

Estimation of Drying Time

In order to determine the time required to achieve the desired reduction in moisture content of the product, the rate of moisture removal or drying rate must be predicted. The rate of drying depends on properties of drying air (the dry bulb temperature, RH, and velocity of air and the surface convective heat transfer coefficient), the properties of food (moisture content, surface to volume ratio and the surface temperature) and rate of moisture loss. The size of product pieces has an important effect on the drying rate in both the constant and falling rate periods. In the constant rate period, smaller pieces have a larger surface area available for water evaporation where as in the falling rate period smaller pieces have a shorter distance for the moisture to travel from interior of food to its surface. Other factors which influence the rate of drying are:

1. The fat content of the food (higher fat contents generally results in slower drying, as water is trapped within the food).
2. The method of preparation of food (cut pieces lose moisture more quickly than through the skin.
3. The amount of food placed in a dryer in relation to its size (in a given dryer faster drying is achieved with smaller quantities of food).

For constant rate drying period the following general expression would apply:

$$R_c = \frac{d_w}{d_t} = (w_o - w_{c)}/t_c \quad \text{(Eq. 16.1)}$$

where,

w_o = Initial moisture content (kg water/kg dry solid); w_c = Critical moisture content (kg water/kg dry solid) and t_c = Time for constant rate drying

During falling rate drying, the following analysis would apply.

$$\frac{-d_w}{dt} = R_c/w_c\,(w) \text{ or}$$

$$w_c/R_c \int - d_{w/dt}\, w = \int dt \quad \text{(Eq.16.2)}$$

The limits of integration on the left side of equation are critical moisture content w_c and final moisture content, w and on the right side of equation are t_c and t, where t is time taken to reach the final moisture content, w.

On integration:

$t - t_c = w_c/R_c \ln(w_c/w)$ or time for falling rate becomes

$$t_f = (w_c/R_c) \times \ln(w_c/w) \quad \text{(Eq.16.3)}$$

The total drying times becomes

$$t = (w_o - w_c)/R_c + w_c/R_c \times \ln(w_c/w) \quad \text{(Eq.16.4)}$$

The above equation indicates that the time for complete drying from some initial moisture content w_o to a desirable final moisture content w depends on knowledge of critical moisture content w_c, the time for constant rate drying t_c, and constant drying rate R_c.

Example 1

A tunnel dryer is being designed for drying apple halves from initial moisture content of 70 per cent (wet basis) to final moisture content of 5 per cent (wet basis). An experimental drying curve for the product indicates that the critical moisture content is 25 per cent (wet basis) and the time for constant drying is 5 min. Based on the information provided, estimate the total drying time for product.

Solution

Initial product moisture content, w_o = 0.7/0.3 = 2.33 kg H_2O/kg solids

Critical moisture content, w_c = 0.25/0.75 = 0.333 kg H_2O/kg solids

Final moisture content,w = 0.05/0.95 = 0.0526 kg H_2O/kg solids

Time for constant rate drying, t_c = 5 min

Total drying time:

$R_c = (w_o - w_c)/t_c = (2.33 - 0.33)/5$ min = 0.4 kg H_2O/kg solids min

Falling rate drying time $t_F = w_c/R_c \times \ln(w_c/w)$

$= 0.333/0.4 \times \ln(0.333/0.0526) = 1.54$ min

Total drying time becomes, t = 5 + 1.54 = 6.54 min.

Example 2

Estimate the drying rate and time needed to reduce the moisture content of a 100 μm diameter spherically shaped droplet, falling in a spray dryer from 60 to 35 per cent. The initial density of the droplet is 900 kg/m³. The droplet is in an air stream such that dry bulb temperature T_α = 200 °C, pressure, P = 101.3 k Pa, convective heat transfer coefficient, h = 200 w/m² °C and T_{wb} = 60 °C. Assume that constant rate drying applies over the total drying process and droplet doesn't change in size.

Solution

The surface area of the droplet is $\pi D^2 = \pi (0.0001)^2 = 3.14 \times 10^{-8}$ m³

The volume of the droplet is $1/6\, \pi D^3 = \pi/6 \times (0.0001)^3 = 5.24 \times 10^{-13}$ m³

The initial mass of the droplet is $\rho \times V = 900 \times 5.24 \times 10^{-13}$

$= 4.71 \times 10^{-10}$ kg

Applying solids balance: $m_p \times 0.4 = m_f \times 0.65$

So, final mass of the product $m_f = m_i \times 0.4/0.65 = 2.9 \times 10^{-10}$ kg.

$$\text{Drying rate, } R = \frac{h\,A\,(T_\alpha - T_s)}{h_{fg} \times 1000}$$

where,

h_{fg} = latent heat of evaporation corresponding to given pressure P.

$$= \frac{200(3.14 \times 10^{-8})\,(200 - 60)}{2370.7 \times 1000}$$

$= 3.71 \times 10^{-10}$ kg s

The drying time is given by $t = (m_i - m_f)/R = (4.71 - 2.9)\, 10^{-10}/3.71\, 10^{-10}$

= 0.49 seconds.

REFERENCES

"Food Process Engineering" by Heldman D.R. and Singh R.P

Further reading: "The Handbook of Industrial drying" 3rd Edn. (2006) Edited by A.S. Majumdar

Chapter 17

Classification of Dryers

Drying is a method in which the shelf life of food products, including dairy products, is increased, by reducing the moisture content to minimum possible. As the physical condition of the initial product, the method of drying process and condition of final product vary with foods, there are number of methods by which dehydration process is carried out. The drying methods can be classified as per the source of energy used and scale of operation, as shown in Chart 17.1.

Dehydration systems can be classified on the basis of scale of operation as shown below:

1. Domestic: a) Open sun drying b) Solar drying
2. Small scale: a) Roller drying b) Tunnel drying c) Cabinet/Tray d) Trough drying e) Microwave drying f) Dielectric
3. Medium and Large scale: a) Spray b) Fluidized bed c) Freeze d) Puff e) Vacuum f) Bin g) Pneumatic h) Rotary i) Tower

Roller Drying

For small scale drying operation for concentrated milk, the roller drying (or drum drying) was found suitable during the initial stage of development of drying in the dairy industry. Later on, the spray drying for pilot and small scale operations was developed. However, for products like infant milk powder Roller drying was initially found suitable.

Advantages and Disadvantages

Drum drying requires less space and is more economical than spray drying for small product volumes. The ratio of steam economy for roller drying varies from

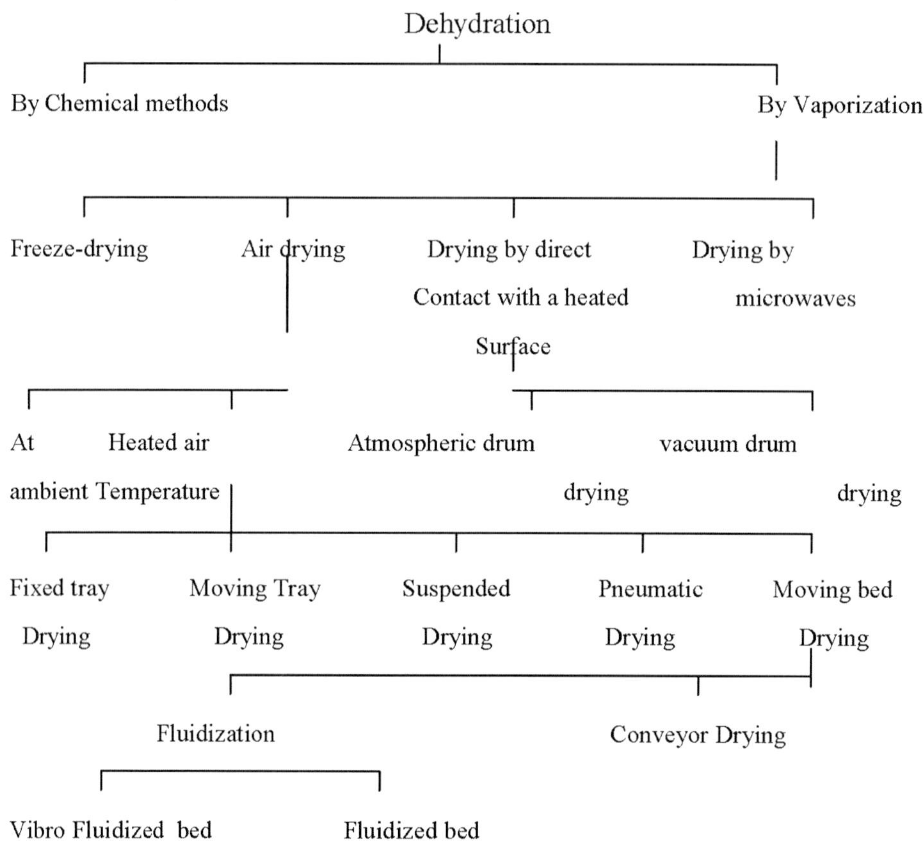

Chart 17.1. Classification of Drying Methods.

1.2 to 1.6 kg of steam used per kg of water evaporation. The major disadvantages of roller drying are the scorched flavour, and low solubility below 85 per cent of the powder. The high temperature gradient formed between the milk film and the steam condensing inside the drum causes protein denaturation and adversely affects the product solubility. To overcome this disadvantage, the entire drum arrangement is placed under vacuum allowing drying at lower temperatures.

There are several types of Roller dryers which are classified as shown below (Figure 17.1):

1. Methods of feeding the milk: a) trough above for top feed b) spray or splash feed c) Sump below for dip feeding d) trough below for pan feeding
2. Number of drums: a) single drum b) twin drum c) twin drum with individual feeding
3. Pressure on the product: a) atmospheric b) Vacuum

4. Material of construction: a) steel b) steel alloy c) stainless steel d) cast iron e) chrome, or nickel-plated steel. Though cast iron is usually used, due to issues of hygiene its use is discouraged.

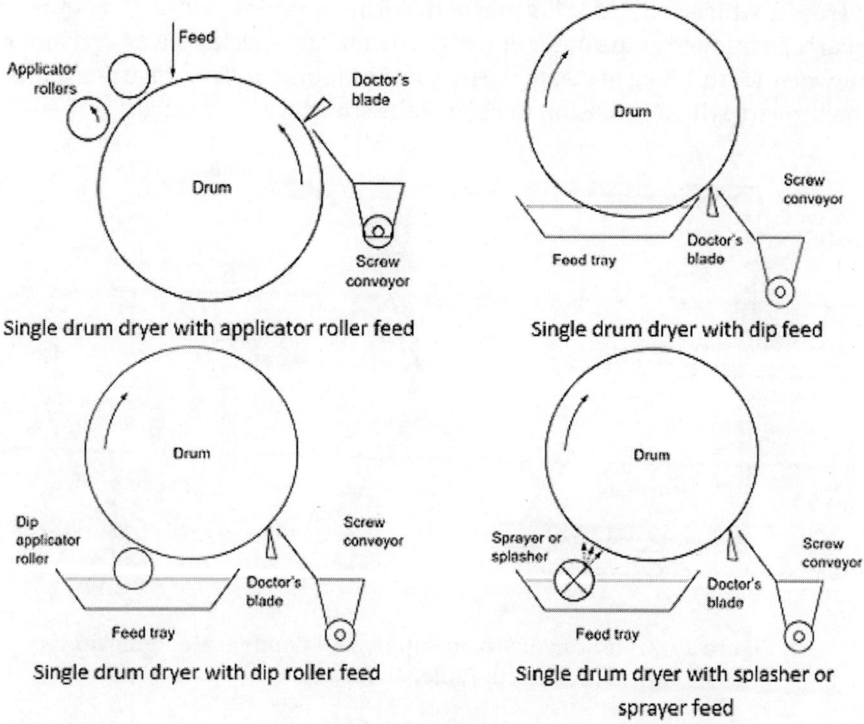

Figure 17.1. Methods of Supplying Feed to Roller Drier

The Roller dryer or drum dryer is an indirect type dryer in which the milk to be dried is formed as a thin film on a steam heated rotating drum. The milk (milk as such or in concentrated form) is spread over the outside surface of the dryer as a thin film. Drying of the product film continues as the hot drum rotates. After completing 3/4 of a revolution, the drum comes into contact with a sharp blade called 'doctor blade' which scrapes the dried film from the drum surface.

Drums are of 60 to 120 mm in diameter and about 360 cm long. The drums are carefully machined inside and outside. The thickness of the drum varies from 20 to 40 mm. Non-uniform drying occurs if thickness of drum is not uniform. Steam at a pressure of 3 – 6 Bar condenses inside the drum so that surface temperature is below 130 °C. Speed of the drum is adjustable. The speed affects thickness of the milk film and time of its contact with the drum surface. In double drum design both the drums rotate at the same speed. The residence time is 3 seconds or less and the product is scraped off from surface after ¾ or 7/8 of the revolution of the drum. The gap maintained between the drums is about 0.5 to 1.0 mm. One of the drums can be adjusted to vary the gap in between them while keeping the drums perfectly parallel.

The steam is supplied through a pipe that is concentric with the roller shaft. The condensate which collects at bottom of the drum is drained through another pipe similarly positioned at the other end of the drum. The specific steam consumption is between 1.3 to 1.5 kg of steam/kg of water evaporated. The drive mechanism is through gears with a provision to change the speed within a range.

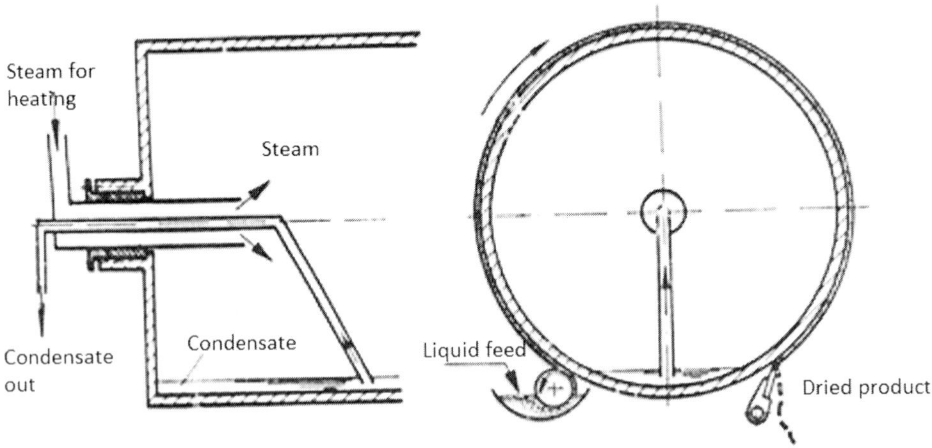

Figure 17.2. Method of Steam Supply and Condensate Removal in Roller Drier

Operating Precautions

1. Drums must be correctly aligned and have identical characteristics of speed, heat transfer, wear *etc.*
2. Knives must be reground regularly and the drums surfaces must be kept smooth.
3. Uniform level of milk must be maintained in the trough.
4. Variations in steam pressure and speed of the drums must be avoided.
5. Dried product must be scraped off as any product adhering to the drums will burn and results a finished product of poor quality.

Rate of Evaporation in Drum Dryers

The rate of moisture removal in a drum dryer essentially proceeds at constant rate drying as the product is continuously fed between the drums and the dried product is removed. The rate of water evaporation in a drum dryer is given by:

$$\frac{dW}{d\theta} = \frac{UA\Delta t_m}{\Delta h_{fg}} \qquad \text{(Eq. 17.1)}$$

where,

$\frac{dW}{d\theta}$ = is rate of water evaporation

t_s = Drum surface temperature

t_a = Room temperature

$\Delta_{tm}\ t_s - t_a$

Δh_{fg} = Latent heat of vaporization.

The overall heat transfer coefficient, U for a drum dryer is in the range of 1.1636 to 2.0945 kW/m ^{2} K (1000 to 1800 kcal/hr m^2 °C) under optimum conditions, although it may drop to 1/10th of these values when conditions are adverse. Since the thickness of drum material is small compared to the diameter of the drum, the area "A" can be regarded as the outer surface area of the drum. The "U", the overall heat transfer coefficient, depends on h_s, k and h_p and other coefficients.

$$U = \frac{1}{(1/h_s)+(x/k)+(1/h_p)} \qquad \text{(Eq. 17.2)}$$

where,

h_s = Convective film coefficient of condensing steam, kcal/hr m^2 °C;

x = Thickness of drum material, m;

k = Thermal conductivity of drum material, kcal/m hr °C;

h_p = Convective film coefficient of product;

The factor having the maximum effect on "U" is the condition of the product film and the drum speed. Drying rates for the drum dryers can be extremely high when a thin film of low viscosity is evaporated, and it is thus permissible to use high temperature difference across drum surface. In addition to assuring adequate heat transfer the drying system must provide for removal of water vapour leaving the product film.(If the speed of a particular drum is increased, then the "U" value and the moisture content of the product will be increased, if the other conditions are unchanged). The overall thermal efficiency of drum dryer is in the range 35 to 80 per cent.

The moisture content "*x*" of a milk product containing m_w amount of water and m_s amount of dry matter including fat is given by:

$$x = \frac{m_w}{(m_w + m_s)} \qquad \text{(Eq. 17.3)}$$

The proportion of milk solids (TS) is given by :

$$T_s = \frac{m_s}{(m_w + m_s)} = 1 - x \qquad \text{(Eq. 17.4)}$$

During concentration of a product if the original moisture content x_0 containing amount of water m_{wo} is reduced to a final moisture content x containing an amount of water m_w, then amount of water removed is:

$$m_{wo} - m_w = [x_0 / 1 - x_0] - [x / 1 - x] \quad \text{(Eq. 17.5)}$$

Expressing as a fraction of the water originally present, Eq. 17.5 modifies to:

$$\frac{m_{wo} - m_w}{m_{wo}} = 1 - [x(1 - x_0) / x_0(1 - x)] \quad \text{(Eq. 17.6)}$$

Example 1

A drum dryer is designed for drying a product from initial TS of 12 per cent to a final moisture content of 4 per cent. An average temperature difference between the roller surface and the product is 65 °C, the overall heat transfer coefficient is 1.745 kW/m ^{2} K (1500 kcal/hr m^2 °C). Determine the surface area required for the roller, required to provide a production rate of 50 kg product/hr.

Solution

Calculations on basis of 1 kg of product P = F – V

F: amount of feed/kg of product; V : amount of vapour/kg of product.

Solid balance: 0.96 x 1 = 0.12 F

Hence, F = 8 kg; V = 7 kg

For 50 kg product, F = 50 x 8 = 400 kg/hr

Water removed as Vapour = 50 x 7 = 350 kg/hr = 0.097 kg/s

Using equation: U x A x ΔT = W x h_{fg} (from Steam Tables)

0.097 = 1.745 x A x 65/2,268

Thus, A = 1.94 m^2

The effective surface area needed for drying is 1.94 m^2. Assuming that about ¼ of the surface area is not used, then the total surface area required would be: 1.94 x 4/3 = 2.583 m^2. If the length of the drum is assumed as 1.5 m, then the diameter of the drum d will be:

π x d x 1.5 = 2.583

Thus, d = 0.55 m

Example 2

Milk having 18.5 per cent TS is dried to 3.8 per cent moisture content on a drum dryer at a rate of 10.5 kg of dried product/hr. The diameter of the drum is 60 cm and length 90 cm and the dried product is scrapped at ½ of a revolution. The steam temperature is 160 °C and evaporation temperature of the moisture is 104 °C. Determine out the overall heat transfer coefficient for the drum dryer.

Solution

Basis: 1 kg of milk

V + P = 1 (material balance)

where,

V = amount of water evaporated P = amount of the dried product

1 X 0.185 = P(1-0.038) = P x 0.962 (material balance)

Thus, V = 0.808 kg and P = 0.192 kg/kg of milk

Given, P = 10.5 kg, then,

dw/dØ = 10.5 x 0.808/0.192 = 44.18 kg/hr =0.01227 kg/s

The heat transfer area, A = 0.9 x 0.6 x Π/2 = 0.847 m^2

Using the equation: $\frac{dW}{d\phi}=\frac{UA\Delta t_m}{\Delta h_{fg}}$

0.01227 = U x 0.847 (160 – 104)/2260

Thus, U = 0.546 kW/m^2 K

Example 3

A drum dryer is being designed for drying a product from initial TS of 14 per cent to a final moisture content of 3 per cent. Overall heat transfer co-efficient, U is 1750 W/m^2 K. An average temperature difference between the roller surface and product 75°C is assumed for the design purpose. Determine surface area of the roller required to provide a production rate of 25 kg/h. The product is scrapped after ¾ of revolution and the latent heat of vaporization is 2420 kJ/kg.

Moisture removal/kg of product = (0.86-0.03) = 0.83 kg

Total moisture removed = 25 x 0.83 = 20.75 kg/h

Rate of heat required to remove moisture = 20.75 x 2420 = 50215 kJ/h

Area of heat transfer for above,

$$= Q/(U \times \Delta T) = 50215/(1750 \times \frac{3600}{1000} \times 75) = 0.106\ m^2$$

Let A = Total Area of drum, then A x0.75= 0.106 m^2. A = 0.0795 m^2

Example 4

Calculate amount of milk powder obtained per hour from double drum dryer. Given:

Initial T.S. in concentrate = 35 per cent, Moisture content in the dried product is 2 per cent, The steam pressure = 7 bar (gauge); Evaporation temperature of milk concentrate = 95°C; Over all heat transfer coefficient = 1600 kCal/h m^2 °C; Drum diameter = 1 m

Length of drum = 200 cm; the product is scraped at 3/4th of the revolution of the drum.

Solution

a) Pressure of steam in absolute = 7.0 + 1.0 = 8 bar; and saturation temperature of steam is 169.62°C

b) Area of the drum available for heat transfer = π d L (3/4) = 3.14 x 1.0 x 2.0 = 4.71 m^2

c) Rate of heat transfer Q = U A ΔT = 1600 x 4.71 x (169.62 – 95) = 562336.32 kCal/h

d) (Vapours generated) x h_{fg} = Rate of heat transfer (neglecting heat losses to surroundings)

   W x h_{fg} = 562336.32;

   W = 562336.32/486 (h_{fg} at 8.0 bar = 486 kCal/kg)

   = 1,157 kg/h

e) Let F = Feed rate and P= Product rate

   0.35 x F = P x (1-0.02)

   P = 0.357 F

f) Vapours generated = Feed – Product;

   1.157 = F – 0.357 F; F= 1,799.4 kg/h

g) Product rate = 0.357 x 1,799.4 = 642.4 kg/h

Chapter 18

Spray Drying

Spray drying operation converts food in liquid form into a dried product in form of solid particles. This is achieved by spraying the liquid food into hot drying medium. Spray drying is one of the best methods for drying of liquid food products. This process is applied to fluids of high solids content and high viscosity. The major advantages of spray drying over drum drying are 1) high production rate 2) gentle drying 3) superior flavour retention, appearance, and solubility 4) uniform product 5) process can be automated 6) less chances for product contamination 7) High thermal efficiency. The disadvantages compared to roller drier are 1) large and complex equipment 2) Requires tall building for housing the spry dryer 3) high initial costs 4) economical only for large volumes of milk, to be dried.

Principle of Spray Drying

Spray drying is dehydration process of a liquid food product containing dissolved and/or dispersed solids. The liquid is transformed into a spray of small droplets which are dispersed into a stream of hot air. The very large surface area presented by the spray droplets causes the evaporation of the water rapidly, leaving dry solid particles.

A conventional spray dryer consists of the following key components (Figure 18.1):

1. Drying chamber
2. Hot air and air distribution system
3. Liquid product feeding system
4. Liquid atomizing device
5. Dried solid particle separation system

6. Pneumatic conveying and cooling system
7. Fluidized bed -drying/cooling
8. Instrumentation and automation

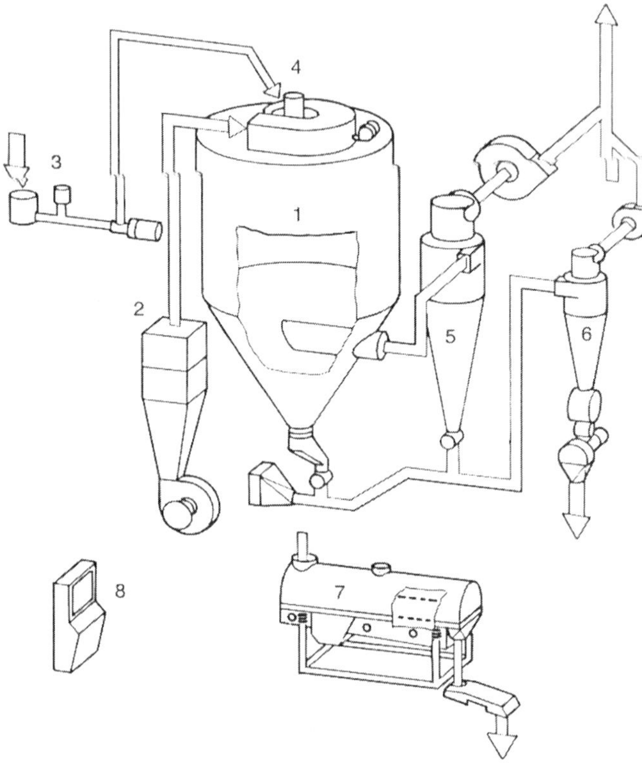

Figure 18.1. Spray Drying Plant

Referring to the above figure, the milk from balance tank (3) pumped to the Drying Chamber (1), through an Atomizer (4) and sprayed as fine droplets into a stream of hot air blown from air heater (2). As the heat transfer surface area is large due to large number of droplets, the transfer of heat to the droplets from the hot air and moisture from droplets to the hot air is almost instantaneous. Depending on the design of spray drying chamber and atomizer, the dried particles and the moisture laden air are drawn to the main Cyclone separator (5) where the separation of dried particles from air occurs. The dried powder is cooled and conveyed to a second cyclone separator (6) where it is separated and collected at the bottom of the cyclone. The purpose of main cyclone (5) is to separate fine powder from the exhaust air. In some of the designs, the product from the drying chamber is allowed to dry only partly and is lead into a Fluidized bed drier (7), where it is dried to the final moisture content and then cooled.

The temperature of hot air supplied to the drying chamber is commonly in the range 180 to 200 °C, while the leaving air is about 90 °C. The temperature of the milk droplets remain at the wet bulb temperature of the drying air and is usually between 49 to 54 °C.

Spray Driers can be classified on the basis of their major components:

1. Heating of the drying air: air heaters with accompanying supply air fans, air filters dampers and ducts.
2. Atomization of feed into a spray: atomizer with feed supply system consisting pumps, tanks and feed pretreatment equipment.
3. Contacting of air with sprays and drying of sprays: drying chamber with air disperser, product and exhaust air outlets.
4. Recovery of dried products and final air cleaning: complete product recovery with product discharge, transport and packing, air exhaust system with fans, wet scrubbers, damper and exhaust duct.

Classification of Spray Dryers

Spray dryers are mainly classified according to:

1. Method of atomization: (a) pressure atomization (b) centrifugal atomization, (c) pneumatic atomization
2. Method of heating air :(a) steam (b) thermic fluid, (c) electricity
3. Orientation of the drying chamber: (a) vertical (b) horizontal
4. Direction of air flow in relation to direction of product flow: (a) counter current (b) co current (c) mixed current
5. Pressure in dryer (a) atmospheric (b) vacuum
6. Geometry of the bottom of the chamber: (a) flat bottom (b) conical bottom

Drying Chamber

The drying chamber construction, shape and size varies according to product handled and matching with the other components like the type of atomizer, air and spray contact *etc.* (Figure 18.2). The shape of the chamber is decided on the basis of type of atomizer used. The ratio of cylindrical height to the diameter is 0.6 to 1.0 for disc atomizer, while that with nozzle atomizer is 3 to 5. The average residence time of the product in chamber is the ratio of chamber volume and the volumetric air flow rate, based on outlet air temperature.

The figures below show the major shapes of the drying chambers with different types of air flow and spray:

A) Chamber shape: a) wide body b) tall form c) horizontal box type
B) Product flow: a) Product leaving the drying chamber with the exhaust air b) partially separated from the exhaust air in drying chamber

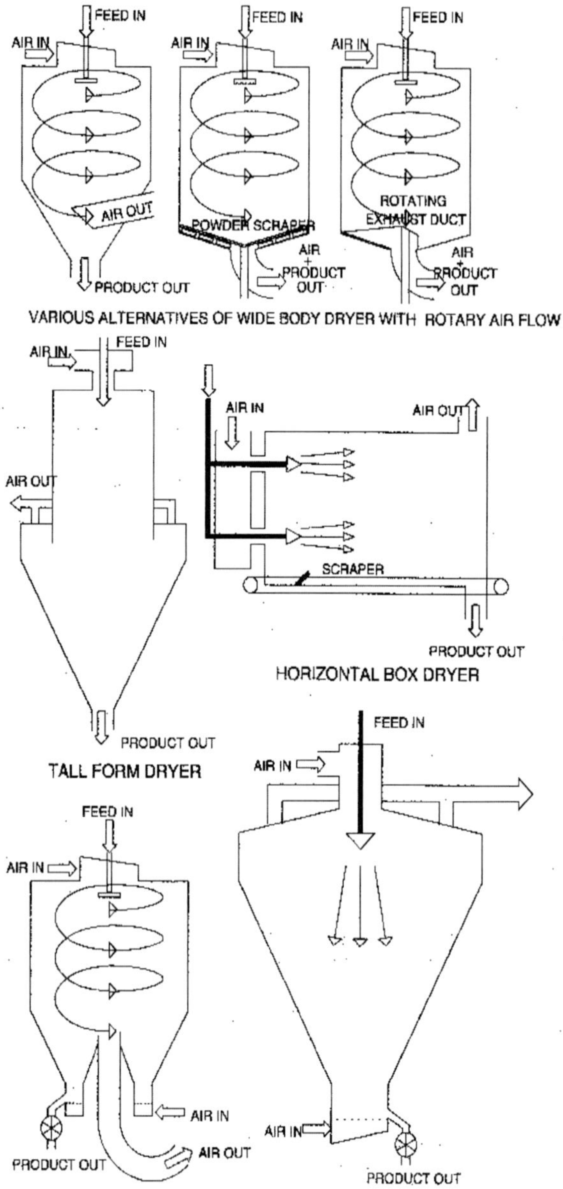

Figure 18.2. Different Types of Spray Drying Chambers

C) Product discharge: a) by gravity (conical bottom) b) mechanically (flat bottom)

D) Air flow pattern: a) rotary downward b) straight downward c) straight horizontal

E) Air/spray mixing: a) co-current b) counter-current

F) After treatment of dried product: a) none b) pneumatic transport system c) external fluidized bed d) integrated fluidized bed

General Observations on Spray Drying

In spray drying, the residence time varies between 25 to 30 s. The initial drying rates (constant rate period) are very high and vary with the velocity of air and spray. At end of this time, a crust of solid phase forms over the droplet which offers resistance to moisture transport causing the drying rate to fall. Hence, the active evaporation occurs in initial 4 to 5 s. Sufficient quantity of heated air must be supplied to furnish energy for heat of vaporization and to provide moisture carrying capacity. Air flows at about 6 to 10 m/s. During drying, the temperature of air-water vapour stream falls rapidly from initial 180 – 200 °C to the outlet air temperature between 66 – 96 °C. Evaporation is rapid and the temperature of milk droplet is kept down to wet-bulb temperature. For exact calculation of air requirement and properties of air-water vapour mixture, reference to high temperature Psychrometric chart is necessary.

Hot Air System, Feed and Atomizing

In a spray dryer, the hot air supply system consists of a) Air Supply Fan b) Air Filters c) Air Heater d) Air Distribution ducts.

Air supply Fan

The process air for the main drying chamber is supplied by a directly driven centrifugal fan. The speed of the motor is controlled by a variable frequency drive.

Air Filters

The drying air is usually drawn into the system from outside the building. It is normally pre-filtered by a coarse filter. However, it must be realized that dust laden air can cause faster clogging of the filters, and contribute possibly to bacteriological problems or even fire hazard.

When using atmospheric air, the intake filters should be placed on the windward side of the factory, reasonably high above the terrain and far enough from known sources of dust (busy roads, chimneys, exhaust stacks from other dryers *etc.*).

Air Heater

The drying air can be heated by: a) Indirect method: Steam/Oil/Gas/Hot oil b) Direct method: Gas/Electricity

Indirect Heater

A steam heater is a heating coil. The temperature of air leaving the heating coil depends on the steam pressure. Under normal conditions it is possible to obtain air temperatures 10°C lower than the saturation temperature corresponding to pressure of the steam supplied. Modern steam heaters are divided in sections, such that the cold air first is heated by the condensate section, then by a section with lower steam pressure (which is usually the largest section in order to utilize as much low-pressure steam as possible). Air is heated to final temperature by the high-pressure steam section. The air heater consists of rows of finned tubes housed in an insulated metal case.

The heat load on the heating coil is calculated from the air flow rate, specific heat and temperature rise of the air. The heater size depends upon the heat transfer properties of the tubes and fins and is usually about 0.058 kW/°C/m^3 (50 kcal/°C/h/m^3)for an air velocity of 5 m/s. Steam-heated air heaters usually have an efficiency of 98-99 per cent. The steam supplied is usually at a pressure in the range of 9.8 to 11.76 bar (10 to 12 kg/cm^2). As the steam boiler is usually located at some distance from the air-heater, 2-3 bar extra pressure at the boiler should be anticipated due to pressure loss in the steam pipe line and regulating valve. The relative humidity of the air leaving the air heater is usually < 3 per cent. To avoid corrosion of the tubes in the air heater it is recommended to use stainless steel tubes.

Direct Heater

Direct gas heaters are used only when the combustion gases can be allowed to come in direct contact with the product. Therefore this type of heaters not used in the food and dairy industries.

Air Dispersers

Uniform mixing of the drying air and product droplets and the control of the air and particle flow is essential for the success of whole process. These factors have a decisive influence on the quality of final product.

Air disperser promotes the spray-air movement during the critical first period of droplet drying. The manner of spray-air combined motion determines the rate and extent of drying. It also determines the residence time. Design of drying-chamber and air disperser must create a flow pattern that prevents deposition of partially dried product at the spray dryer surface near the atomizer. If eddy currents of air are significant, the dried particles back flow to the hottest region and even to the air disperser.

There are two basic types of air dispersers:

a) Air disperser creating rotary air flow and used in vertical wide body drying chambers [Figure 18.3 (a)].

   This type of air disperser operates usually with a rotary atomizer but is also suitable for pressure nozzles.

b) Air disperser creating straight air flow [Figure 18.3(b)] and used exclusively with pressure nozzles for vertical downward air flow (for instance in the tall drying chamber, multi stage dryer and the integrated fluidized bed dryer). Depending on the type of dryer, one chamber can accommodate either just one or several air dispersers of this type (arranged symmetrically in the ceiling). The common goal is to have an air distribution and nozzle assembly that minimizes powder deposits on inside surface of the drying chamber, and that the nozzles are interchangeable during the production to allow for continuous operation of the dryer for weeks without interruption.

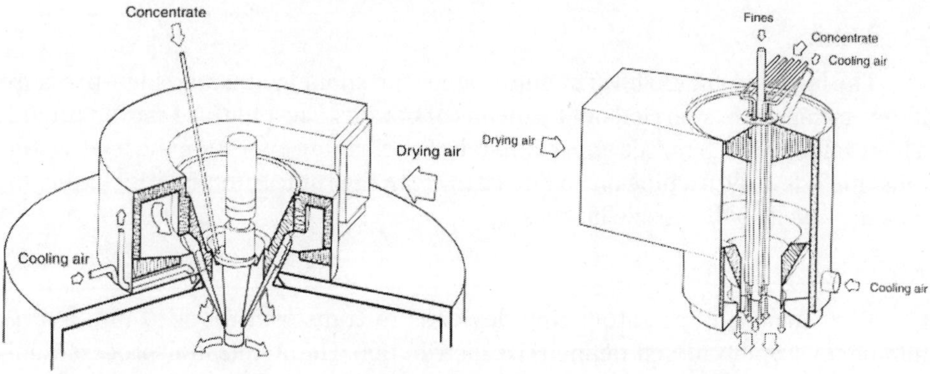

(a) (b)

Figure 18.3. a) Air Disperser with adjustable Guide Vanes, b) Plug Flow Air Disperser (*Source*: GEA Niro).

Feed Supply System

The function of the feed supply system is to deliver feed to the spray dryer via the atomizing device. It is actually a link between the evaporator and the spray dryer, and must compensate also for the capacity fluctuations of both units. The components of the feed supply system are feed tank, feed pump and feed line.

Feed Tank

Feed tanks act as a buffer for capacity variations. Usually, two feed tanks are installed to enable change-over from one tank to the other after several hours of operation for washing of the emptied tank. This is essential for bacteriological reasons as the temperature of 45-50°C at which the concentrate leaves the evaporator is ideal for the growth of thermophile bacteria. The size of a feed tank must be in relation with the plant capacity corresponding to about 15-30 minutes plant operation. Closed feed tanks with inspection covers, maximum and minimum level transmitters, water connection(s) and CIP nozzles are common.

Feed Pump

It supplies the concentrate to the atomizing device. Therefore the type of pump depends on atomizer type. For low pressure systems, as in case of wheel (rotary) atomizers, any type of pump is suitable. A positive displacement pump is usually used. For high pressure nozzles, a high pressure pump has to be used. If the pump has to process whole milk or other fat containing fluids, it can be combined with a homogenizer, preferably with two stages. In such cases, it must be designed for a total pressure equal to that required for homogenizing over both the stages plus nozzle atomizing pressure, including a safety factor. The homogenizer can also be used as a feed pump for the wheel atomizer. It is then advantageous to install a mono-pump as second feed pump for non-fat products.

Feed Line

The feed line should be of stainless steel and should withstand high-pressure, if the atomization is carried out by means of nozzles. The pipe size should provide a feed velocity of 1.5 m/sec approximately. For cleaning operation the feed system must include a return pipe and a device to clean the rotary atomizer, including the wheel, as well as the nozzle lances.

Atomizing Devices

The purpose of the atomizing device is to transform the feed into a large number of droplets of well-defined size distribution. The atomization process highly increases the surface area of the milk concentrate before being exposed to the hot drying air. The rate of evaporation is proportional to the total surface area of the droplets. Thus, fine atomization has positive influence on several properties. One litre of concentrate, when atomized to droplets of average size of 100 μ size, surface area is more than 50,000 m^2. The spray of droplets is characterized by mean droplet size and droplet size distribution. Both depend on the type of atomizing device, operating conditions and the properties of the concentrated feed (concentrate viscosity, surface tension and density).

Droplets travelling too rapidly, will form a deposit on the inside surface of the drying chamber. If the spray is fine, it will follow the air movement. But if it is coarser, the path will be more independent of air flow.

There are three types of atomization methods: a) Rotary Atomization b) Pressure nozzle atomizer c) Pneumatic or two fluid atomization.

Rotary Atomization

In rotary atomizers (Figure 18.4) the feed liquid is continuously accelerated to the wheel edge by the centrifugal force, created by the rotation of the wheel. The liquid is distributed centrally and then extends over the wheel surface as a thin film, and discharged at high velocity at the periphery of the wheel. The degree of atomization depends upon peripheral speed of the wheel, density, viscosity and surface tension of the feed and feed rate. The peripheral speed is widely accepted as the main variable for adjustment of a specified droplet size. The peripheral speed depends on the diameter of the wheel and the wheel speed and is given by:

$$V_P = \pi \times D \times N/(1000 \times 60) \quad \text{(Eq. 18.1)}$$

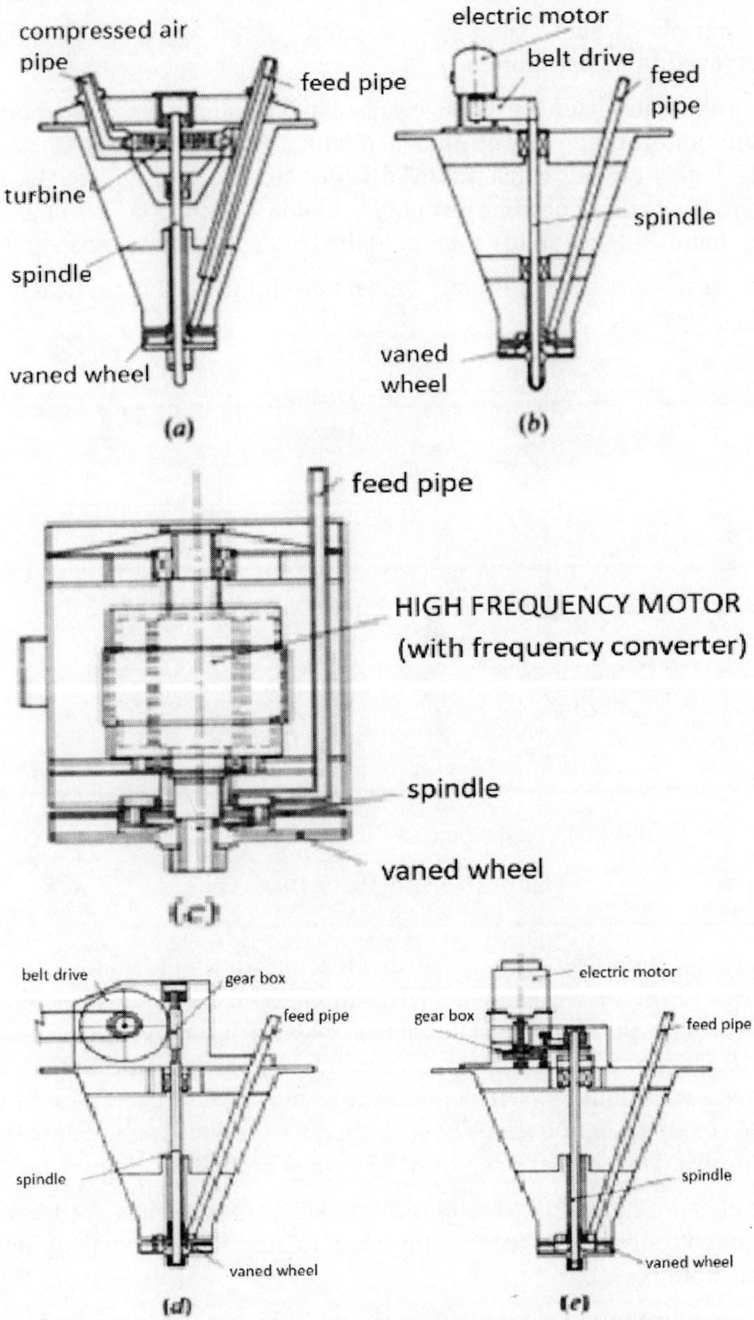

a) Turbine drive; b) Direct belt drive; c) Direct drive with variable frequency drive; d) Gear and belt drive; e) Gear drive

Figure 18.4. Rotary Atomizer with different Drives (from Masters, 1979)

where,

V_P = Peripheral speed (m/s); D = Diameter of the wheel (mm); N = speed of wheel in revolutions per minute.

The rotary atomizer is extensively used in dairy industry. Its advantages are: a) Flexibility in through-put b) Ability to handle large volumes of feed c)Ability to handle highly viscous concentrates d)Different wheel designs to give different characteristics to dried product e)Ability to handle products containing crystals f) Ability to handle higher solids content in the feed giving better economy.

The pressure nozzle and rotary wheel have influence on physical properties of the final product.

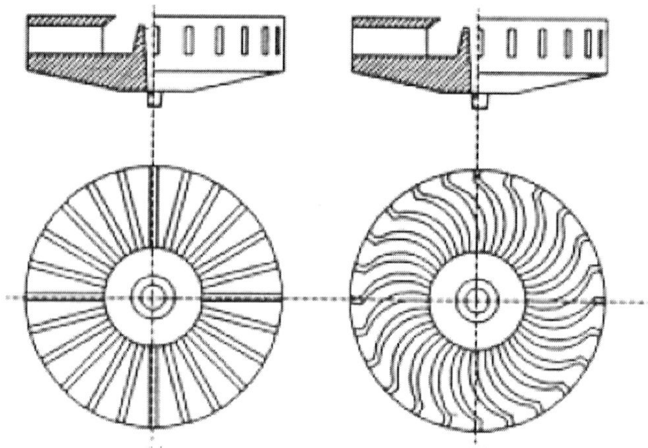

a) With Straight Vanes b) With Curved Vanes

Figure 18.5. Atomizing Disc Types

The wheel should be so designed, that it brings the liquid up to the peripheral speed prior to the disengagement. Therefore, atomizers have vanes of different design to prevent the slippage of liquid feed over the internal surface in the wheel (Figure 18.5). The vanes also allow the liquid feed to concentrate at the disc edge, producing there a liquid film. In general, the speed of rotation of the atomizer disc is between 6,000 to 20,000 rpm. The spray pattern produced is similar to the shape of an umbrella. The droplet size normally ranges from 50 to 250μ.

The effect of individual variables on the disc atomizer has been established over a limited range and there are only a few related to high capacity, high speed industrial atomizers.

The various parameters that affect the size of liquid droplets are given below. D_1 and D_2 respectively are initial and desired values of the disc diameters and N_1 and N_2 respectively are the corresponding disc speeds. This definition of notations is valid for all other parameters listed below (from Masters, 1979).

1. Speed of rotation: $D_1/D_2 = (N_2/N_1)^p$, where P is in the range 0.6 to 1.5 in certain terminal velocity limits.
2. Feed rate: $D_1/D_2 = (M_1/M_2)^q$, where q is in the range 0.1 to 0.12, though fine sprays can still be obtained at higher feed rates, under certain design conditions.
3. Feed viscosity: $D_1/D_2 = (\mu_1/\mu_2)^{0.2}$
4. Surface tension: $D_1/D_2 = (\sigma_1/\sigma_2)^{0.1}$
5. Feed density : $D_1/D_2 = (\rho_2/\rho_1)^{0.5}$

The above relationships show that the droplet size varies directly with feed rate, viscosity, surface tension and inversely with the speed of rotation of the disc atomizer and density of feed.

The average size of the droplet is predicted by many researchers, and the one that is widely accepted is the Sauter Mean Diameter, D in microns. The below approximation is for a co-current design of flow of air and droplets and is applicable under certain flow conditions.

$$D_{vs} = [\ 1.4 \times 10^4\ (M_L^{\ 0.24}\ ]/[\ (N\ d)^{0.83}\ (n\ h)^{0.12}\ ] \qquad \text{(Eq. 18.2)}$$

where,

D_{vs} = Sauter Mean Diameter (μm)

M_L = liquid mass flow rate (kg/h)

N = speed of the disc atomizer, RPM

d = diameter of the atomizer (m)

n = the number of holes

h = Atomizer wheel vane height (m)

Pressure Nozzle Atomizers

The basic function of pressure nozzles is to convert the pressure energy provided to the liquid feed by the high-pressure pump to kinetic energy to form a thin fluid film, the stability of which is determined by the properties of the liquid. Most of the pressure nozzles are designed with a swirl chamber giving the liquid a rotation, so that it will leave the orifice forming a hollow cone.

Pressure nozzles used in the dairy industry are of two types. (Figure 18.6) a) Pressure nozzle with grooved core insert, b) Pressure nozzle with swirl chamber.

Both types operate on similar principle and are often called centrifugal pressure nozzles. It is the grooved core and the swirl chamber which bring the liquid into a rotary motion. The liquid flow rate through the nozzle is directly proportional to the square root of the pressure. Increasing the pressure increases the capacity but decreases slightly the angle of spray. Influence of the physical properties of the liquid on the droplet size is similar to that in wheel atomization.

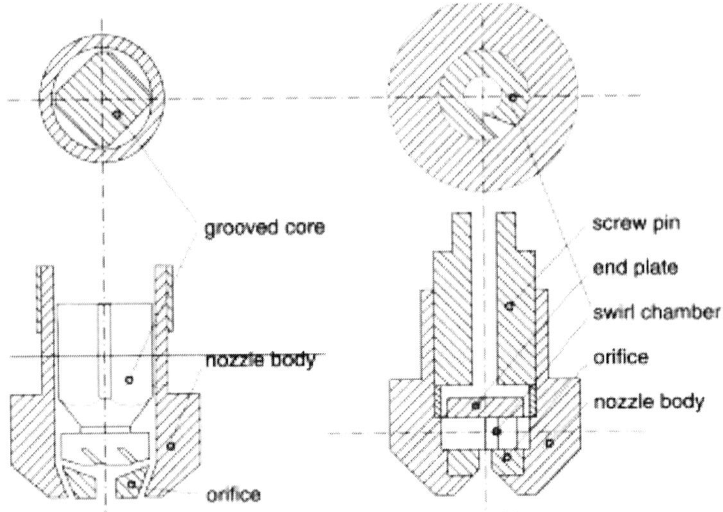

a) With Gloved Core b) With Swirl Chamber

Figure 18.6. Pressure Nozzle Atomizers

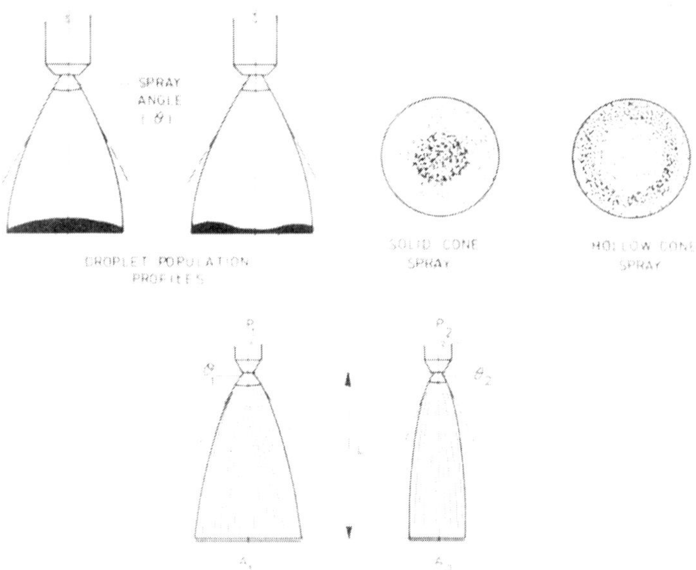

Figure 18.7. Spray Angle Characteristics (from Masters, 1979)

Increasing viscosity decreases the spraying angle. The droplet size depends mainly on the atomization pressure and spray angle. Higher the pressure wider is the spray angle smaller is the mean droplet size (Figure 18.7). However, the effect of spray angle on the droplet size decreases with increasing pressure. Higher the pressure, smaller is the spray cone angle. Due to the influence of the operational variables on pressure nozzles performance it is essential to choose carefully the combination of grooved core/swirl chamber and orifice and to keep the operational variables as constant as possible.

Referring to Figure 18.7 the pressure P_2 is greater than P_1; θ_1 is greater than θ_2 and the projected areas of sprays, A_1 is greater than A_2. Nozzles for milk drying operate usually in the range of 150-250 bar (80-500 bar in exceptional cases). Diameter of the nozzle orifice is about 0.125 mm. Pressure energy is provided by 3 or 5 cylinder positive displacement pump. The nozzles are usually made of tungsten carbide or Stellite coated, to withstand the abrasion.

The power requirement in kW is given by :

$$P_k = 0.98\, Q\, \Delta P \qquad \text{(Eq. 18.3)}$$

where,

Q is the feed flow rate in m^3/h and ΔP is pressure drop across the nozzle in Pascal.

The most important advantage of the pressure nozzle in comparison to wheel atomizers is the formation of almost air-free droplets thus achieving dried product of low content of occluded air and of high bulk density. Another advantage is good free flow of the final powder.

Furthermore, it is easy to direct the spray cloud in any direction. This means that in multi-nozzle dryers, it is possible to combine the individual nozzles close to or apart from each other and let them spray towards or away from each other depending on whether primary agglomeration is required or should be avoided.

Operational difficulties of nozzle type of atomizer include a) choking of nozzle b) increase in diameter of nozzle due to erosion c) unsuitable for feed containing solids in suspension.

Comparison of Pressure and Centrifugal nozzles:

1. Rotating disc atomizer is expensive, but feed pump is less costly when compared to the high pressure pump required for pressure nozzle.
2. Centrifugal force available in rotating disc is much higher than pressure energy of the nozzle atomizers. Hence more viscous products can be handled by the disc atomizer.
3. The control of dryer with nozzle atomization is more difficult. In nozzle type dryer when pressure changes atomization also changes, but with rotary atomizer variation in feed rate will not affect atomization.
4. As earlier stated powder made with nozzle atomizer has less entrapped air and higher bulk density in comparison to powder made with rotating disc atomizer.

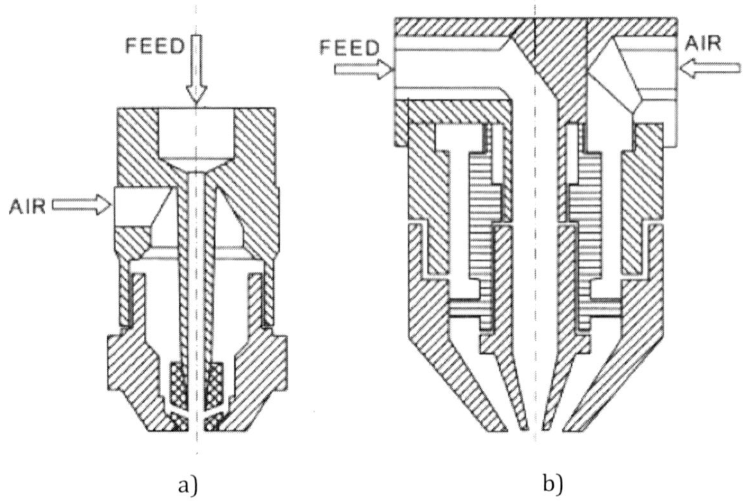

Figure 18.8. Two Fluid Nozzle with a) Internal Mixing; b) External Mixing

Two-fluid Nozzle or Pneumatic Atomization

The energy available for atomization in two-fluid atomizers is independent of liquid flow and pressure. The necessary kinetic energy is supplied by compressed air. The atomization is achieved due to high frictional shear forces between the liquid surface and the high velocity air. Two-fluid atomization is the only successful nozzle method for producing very small droplets, especially from highly viscous liquids. There are two types of two fluid nozzles, *i.e.* with internal or external mixing (Figure 18.8). However, this type of nozzle is seldom used in the dairy industry. Two fluid nozzles with internal mixing are used only in small pilot scale dryers and for special applications, where small amounts of liquid must be atomized into very fine sprays (for example spraying lecithin/oil mixture on powder to produce an instant whole milk powder).Energy requirement is 2 to 3 times more than the pressure nozzle or rotary atomizer.

Spray-Air Contact

The manner how spray-air mixture combines determines the rate and extent of drying. The resulting spray-air movement also determines the residence time. Air disperser is the device that channels the incoming hot air. The drying is introduced into the dryer tangentially for uniform distribution. The design of drying chamber and air disperser must create a flow pattern of air which prevents the deposition of partially dried product on the inside surface of drying chamber and on the atomizer. Droplets travelling too rapidly, result in wall deposits. The eddy currents of air can also carry dried particles to the hottest region, and under certain circumstances lead to burning of the powder particles. Because of range of dryer designs and unknown interdependence of variables, there is no generalized relation for spray-air movement prediction.

Factors affecting Bulk Density, Penumatic Conveying and Controls in Spray Drying

Factors affecting Bulk Density

The bulk density is an economically, commercially and functionally an important property. When transporting dried powders over long distances the high bulk density is desired in order to reduce the volumes to transport. On the other hand, for the purpose of ready reconstitution and agglomeration, low bulk density is necessary.

Bulk density is defined as the mass of a given volume of powder and is expressed in g/cc. Also, it could be loose bulk density, where the container is not shaken or tapped and other is a packed bulk density. The bulk density of milk powders is a very complex property, as it is a result of several other properties. The major factors that affect the bulk density are:

1. Particle density, a function of product composition and the content of occluded air within the particle itself.
2. Interstitial air between the particles, as in agglomerated powder
3. Powder's ability to flow freely

The particle density, (which varies inversely with the occluded air content) is influenced by many factors, as given below:

1. Pasteurization temperature of the milk prior to the evaporation: High pasteurization temperature results in increased amount of denatured whey proteins, which are initially compact, hence increase in particle density.
2. Amount of air in the concentrate: The amount of air in the feed gives a high content of occluded air, especially if surrounding air temperature during the critical stage of drying is high, causing case-hardening. During pumping of the concentrate from evaporator to the feed tank, splashing of the concentrate takes place when air incorporation is likely. However, in a curved vane type atomizer, this air is partly separated from concentrate due to centrifugal force. At the same time, hot air is sucked in and whipping of concentrate takes place in the chamber initially. Even in Nozzle type atomizer some quantity of air is incorporated, especially in high-capacity nozzles.
3. Foaming ability of the concentrated feed: Foaming ability of the feed depends on the composition, temperature and the solids content. Concentrates with low solids content have an extensive foaming ability, which even increases with temperature.

4. Type of atomizer wheel used or size of nozzle: The type of wheel and nozzles is decisive to the amount of air that will be whipped into the concentrate.
5. Solids content in the concentrate: Lower the solids in the concentrate, more air is likely to get whipped in to the concentrate.
6. Drying conditions (single or two-stage): Low surrounding temperature as in two-stage drying results in low occluded air.

Interstitial Air: It is the air in the space between the particles. For efficient drying, the particles must be more uniform in size. But, this results in more interstitial space, and low bulk density. The wide particle size distribution with more small sized particles will have higher bulk density, but at the same time, the powder will be dustier. This, also causes decrease in flowability.

Pneumatic Conveying

Immediately after the powder is separated from the drying chamber, in a single stage drying, the powder is highly hygroscopic, because of higher temperature, about 55 to 60°C. Hence, there is a need for cooling by air, usually to 30°C or lower. The air for cooling has to be at a lower temperature as well as dry. This is done by blowing cool, dry air immediately after the powder is discharged by star wheel, which also conveys the powder to the cyclone separator.

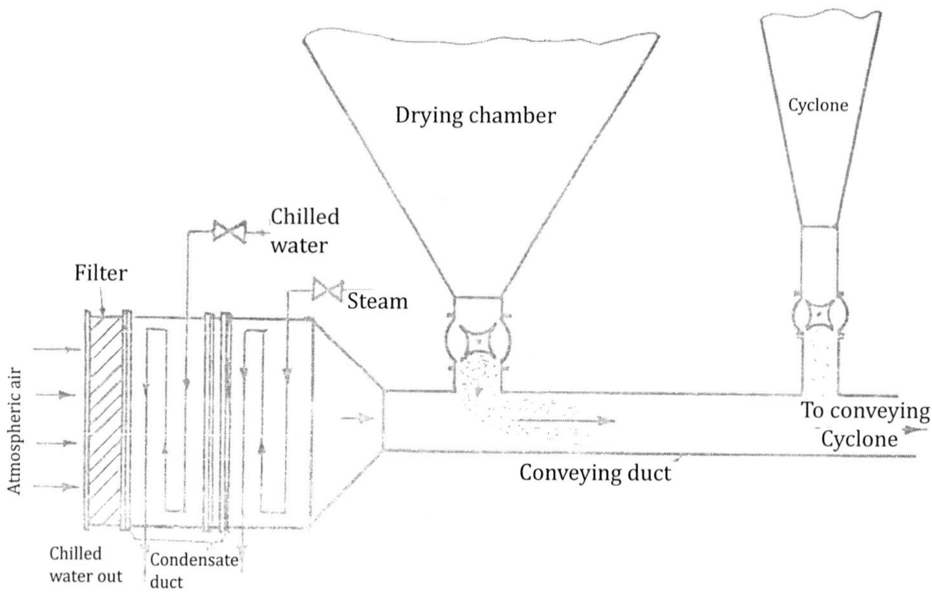

Figure 18.9. Pneumatic Conveying Milk Powder

A crucial parameter in pneumatic conveying is the ratio of volumetric quantity of powder to the air quantity. This depends on temperature of powder leaving the dryer, as well as its composition. For products with high fat content, the ratio is 1: 5 while, for skim milk, the ratio is 1: 4. The air velocity is usually about 20 m/s. Another factor is that due to the discharge of powder is by mechanism of star wheel, the powder loading into the air is not continuous. This can be improved by blowing the conveying air through a perforated plate. The problem of high humidity in the air surrounding the drier is solved by dehumidifying the air, which involves passing the air over cooling coils, with chilled water circulating through them. Care must be taken to prevent the condensing water droplets from being carried over into the conveying duct. This is done, by cooling the air initially below its dew point temperature, and then reheating it by steam heating coils, to reduce the relative humidity

The use of pneumatic conveying is considered for its lower cost, compared to other modes of conveying. However, the agglomerated powder, made for instant reconstitution properties, is likely to break into smaller particles during pneumatic conveying, resulting in slower reconstitution properties of the dried product.

Two Stage Drying

Initially drying in the Spray drier proceeds rapidly but as the moisture inside the droplets is depleted, removing remaining moisture and to bring the product to its final moisture content, takes longer time. This can result in scorching of the particles and less efficient drying. The drying chamber has to be taller too, for this stage, though in a conical shape. Hence, drying is divided into two steps. The first step is undertaken in the spray drying chamber itself, removing most of the moisture.

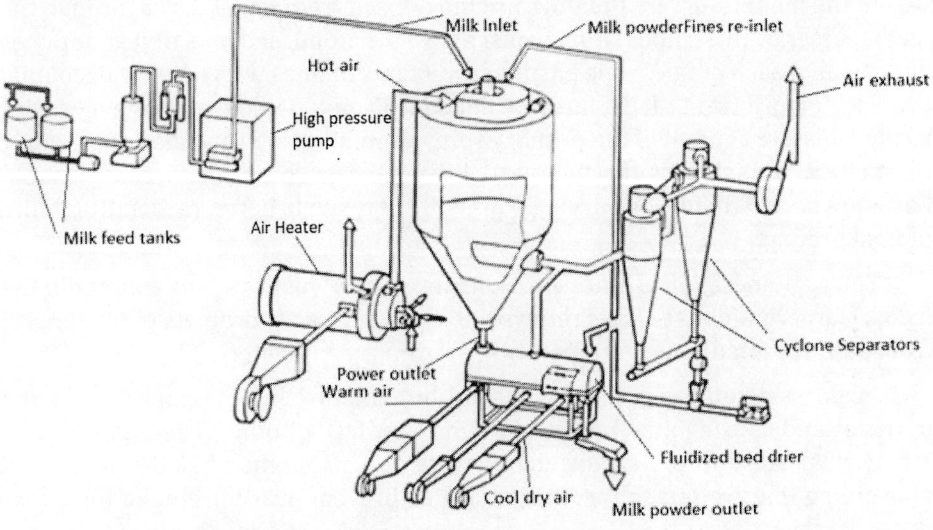

Figure 18.10. Spray Dryer with Fluidized Bed Drier

The second stage of drying is done in a fluidized bed dryer. This stage involves drying with air at a lower temperature than air in first stage of drying. The powder is suspended in air by making it to float over a mesh, with air blowing from below. In this stage of drying moisture in the powder is reduced from about 9 per cent to the final moisture content of below 4 per cent. This process is called Fluidized bed drying. In some of the designs, the fluidized bed forms lower part of drying chamber itself (Figure 18.10). The two - stage drying can be used in the manufacture of agglomerated powder as well as non-agglomerated powder.

The benefits of two-stage drying are many. The outlet temperature of air can be reduced by 15 to 20 °C below that in single stage drying. Consequently, the air temperature at critical stage of drying and particle temperature can be lower. This allows high concentration of solids in feed to be spray dried. In the fluidized bed too, in spite of specific heat consumption being high, the overall heat requirements is reduced by 10 per cent or more of that required for single stage drying. However, the moisture in the powder leaving the spray drying chamber should not be more than 7 to 8 per cent, for skim milk or whole milk powder, to the powder from sticking to metal surfaces.

Agglomerated Powder Production

As an extension of two-stage drying, the fine dried particles blown out of fluidized bed are reintroduced at the atomizer stage, to obtain larger, porous particles. The powder particles so obtained have rapid reconstitution properties. This process is called Agglomeration. The same may also be obtained under special circumstances, without re-entry of fines at the atomizer stage.

In the modern dryers the fines are introduced from top of dryer through the air disperser by fines pipes situated just above the atomizer cloud. Deflector plates at the end of each of fines pipe ensure correct mix of fines with the atomized milk droplets (Figure 18.11). If the fines are introduced too close to the atomizing device as the moisture content of the primary spray droplets is high and also due to their plasticity and stickiness the fine particles penetrate the primary particles and become a part of them while drying. This will not result in better reconstitutability of final product.

The agglomeration is more easily achieved by a) higher solids content in the feed b) larger number of fines returned to the atomizer c) droplets of higher size from atomization d) lower temperature at primary air heater.

Agglomeration can also be achieved by moisturizing the surface of dried particles and causing attachment of them to surface of other dried particles, at a particular stage in the drying chamber itself. This method has the advantage of avoiding fines re-inlet altogether, and thus prevent over burning of fines that occur occasionally.

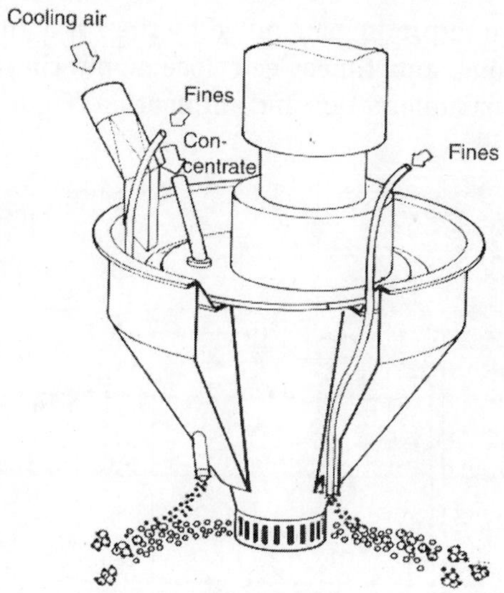

Figure 18.11. Fines Re-inlet at Atomizer

Dryer Efficiency

Though Spray drier is less efficient than Multiple Effect evaporator with respect to energy consumption/moisture removal, at concentration levels of 50 per cent, Drum drier and Spray driers are the only options for removal of the remaining moisture. Among these two, the Spray drier is the more efficient. The dryer efficiency is defined by the ratio of the heat theoretically required to supply the latent heat of vaporization to evaporate the water to the actual heat used in heating the air supplied to the dryer. In actual practice, 2.2 to 3.2 kg of steam (at temperatures suitable to heat air) is required to evaporate one kg of water in the dryer. The thermal efficiency is the ratio of total heat utilized for heating and evaporating water to the heat input.

$$\eta_{thermal} = [(1-R/100)(t_1 - t_2)]/(t_1 - t_0) \quad \text{(Eq. 18.4)}$$

In the above, R=radiation losses, per cent of heat input to air; t_0, t_1 and t_2 are atmospheric temperature, inlet air temperature and outlet air temperature respectively in °C.

The overall thermal efficiency of spray dryer is only 50 to 60 per cent. The thermal efficiency of Spray dryer can be increased by a) high inlet air temperature b) low outlet air temperature c) use of outgoing exhaust air for heating incoming air d) reducing heat losses and air leakage from spray drying chamber.

Spray Drier Controls

The spray dryers range from small laboratory models, pilot models and to industrial dryers. The industrial drying chambers are as high as 3 or 4 stories.

There is also variation in products handled by the spray dryers. Their size and complexity of operations, sometimes need close monitoring of many parameters, necessitating use of instrumentation and automation (Figure 18.12)

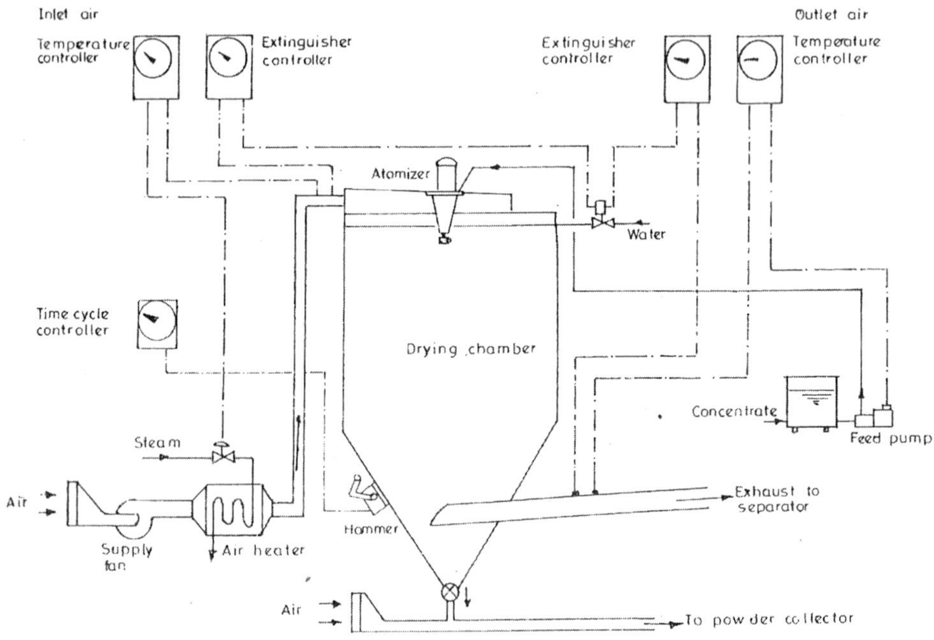

Figure 18.12. Control System Spray Drier

The operations also need certain safety aspects as there are chances of explosion, charring of the product *etc.* This involves provision of safety interlocks, which provide automatic controls and tripping of certain motors, and activation of cooling water spray. The controls also depend on the type of operations, like disc or nozzle atomization, two or three stage drying, fines re-inlet for agglomeration *etc.*

Objective of automatic controls of spray drying plant is to maintain the desired product quality irrespective of disturbances to operation and feed supply. The controls can be manual in case of small units or automatic for large plants. The product parameter is its moisture content, while the operating parameter is outlet air temperature. The two broad types of controls are commonly used. Control system "A" maintains outlet air set point temperature by controlling the feed rate or inlet air temperature to the drying chamber controlled by air heater regulation. Control "B" system, controls spray dryer outlet air temperature by controlling air heater, while feed rate is held constant.

Control system "A": The temperature of exhaust air is measured and transmitted to the temperature indicating controller, which counter acts by varying the feed rate (Figure 18.13). The temperature of air supplied to the dryer is measured and transmitted to temperature controller and corrected by control of fuel and air to the burners (oil and gas fired heaters) or steam pressure in case of steam air heaters.

Control system "B": It is particularly used with nozzle atomization where feed rate cannot be widely changed (Figure 18.14). The outlet air temperature is measured and transmitted to a Programmable Logic Controller (PLC) and is corrected by regulation of steam pressure in case of steam air heaters.

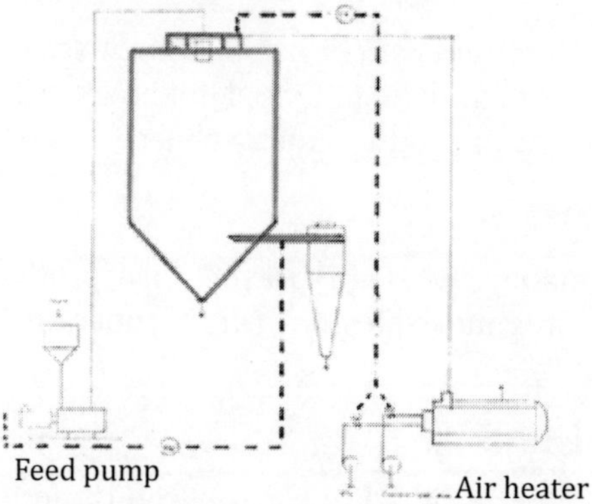

Figure 18.13. Control "A"

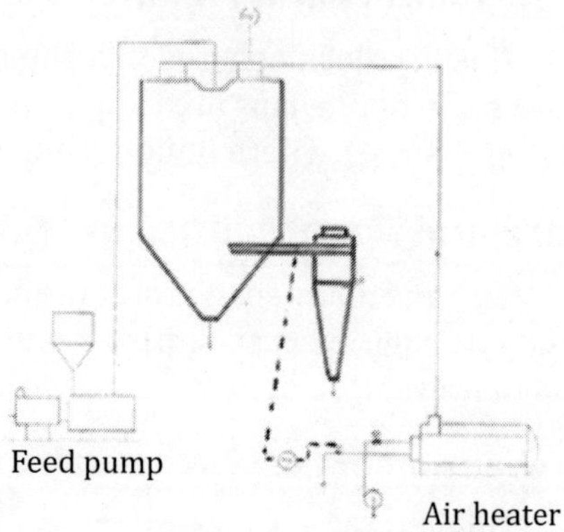

Figure 18.14. Control "B"

Safety Interlocks

Interlocks are closely connected with the control and operation of dryer. Interlocks are installed to ensure safe start up, operation and shut down of the dryer. Some examples of typical interlocks are as follows:

For Prevention of Drying Chamber Damage

An interlock ensures that an exhaust fan cannot be started before the start of supply fan. This is installed to prevent chamber collapse under the negative pressure caused by the exhaust fan operating alone. This interlock is usually over ridden when the chamber inspection door is open. This allows the exhaust fan to operate alone for shut down and cleaning purposes. For example, for an "air sweep" chamber cleaning operation. An alternative system for prevention of negative pressure involves a vacuum switch mounted on the chamber ceiling in a powder free area. The switch is normally set to shut down the exhaust fan if chamber pressure reaches -4 in W.G.(-100 mm W.G.)

Prevention of Drying Chamber Flooding

An interlock ensures that feed is not supplied to the rotary atomizer when it is not in operation. Furthermore, the interlock will cause the feed supply stop in case of atomizer failure.

Prevention of damage to Air Heater

It is an interlock to ensure that the burner cannot be ignited unless the storage oil supply pump, burner fuel pump, combustion fan, and dryer fans are in operation. For indirect air heaters, a differential pressure switch can be installed to check the correct air flow, and link it to the combustion and supply fans.

Prevention of Burnt Product

This interlock ensures that the main dryer fan cannot be started unless the necessary cooling fans are in operation. These fans cool potential hot areas in the drying chamber where build-up and scorching of product can take place.

Care and Maintenance of Spray Driers

Care and maintenance of spray driers mainly involves a) prevention of product losses through exhaust air b) prevention of fires and explosion in drying operations and c) prevention of heat loss at various stages.

Prevention of Losses through Exhaust Air

Though cyclone separators in dryers are designed to recover solids, under certain circumstances, the losses may increase due to poor maintenance and/or operational negligence. The air velocities through the ducts have to be monitored, and whenever the air flow dampers are operated or adjusted, they have to be checked for solids losses as well as power losses due to friction. Excessive losses from cyclones are due to a) over loading of dried product in cooling and conveying duct b) finer size of powder particles entering the cyclone c) malfunctioning of the cyclone powder discharge mechanisms.

The reasons for over loading of cyclone separator could be i) partial or total blockage of one or more in the battery of cyclones, thereby overloading the cyclones with clean outlets ii) incorrect drying chamber operating pressure, leading to a larger cyclone iii) sharp increase or decrease in air flow through cyclone handling powder of fixed amount of flow. The reasons for finer size of powder particles could be i) Too fine atomization ii) feed viscosity very low iii) too low solids in the feed iv) reduced flow of feed at constant atomization conditions v) Excessive friction and breakage of powder particles in conveying duct. Malfunctioning of cyclone could be due to i) blocked or partial blockage of cyclone base ii) Leakage of air into cyclone from base iii) incorrectly adjusted vortex air locks.

When excessive powder loss occurs, the feed, atomization, chamber and cyclone conditions must be investigated. The steps to be taken are: i) Visually inspect the cyclone for cleanliness. If deposits exist in inlet or outlet ducts or on cyclone inside surface, the cyclone must be completely cleaned. ii) Check the air flow through the cyclone for correct flow. If air flow is not as specified, adjust air dampers to achieve the correct flow. iii) Determine the ratio of the flows of air and powder leaving the chamber (through the rotary valve at dryer bottom) and cyclone and determine whether cyclone is overloaded. If overloaded, check chamber pressure and reestablish to correct pressure. iv) Measure particle-size distribution (by sieve analysis), bulk density and moisture content at the point of bagging-off. If power is fine, control the solids in feed, feed rate and atomization conditions. If bulk density or moisture content is low, check for any increase in dryer's inlet and outlet air temperatures.

If no real deviations are found during the above investigation, the fault may be at the cyclone base with excessive air leakage into the cyclone. Excessive losses from bag filters are invariably due to broken or ill-fitting bags and insufficient water supply for the scrubber.

Fires and Explosion in Spray Dryer

The reasons for fire and explosions in spray dryer are due to 1) spontaneous ignition of powder deposits 2) Oxidation 3) Extraneous particles 4) friction between metallic surfaces 5) Static electricity 6) Electrical discharge

1. Spontaneous ignition of powder deposits: Normally temperature inside the dryer is between 85 to 100°C, which is equal to outlet air temperature. The temperature at air disperser is between 160 – 220°C, which is the air inlet temperature. While this is the high temperature zone, the dry powder tends to self-ignite when exposed to 120 – 140°C for a period of about 10 h. Hence, it is hazardous for dry powder to form deposits in the hottest region. The deposit formation can be due to a) imperfect cleaning leaving behind wet deposits b) incorrect startup procedure leading to formation of wet deposits c) Improper stopping of the dryer which may be due to power failure may cause back draught carrying dried particles into the air distributor.
2. There can be ingress into the drying chamber, extraneous particles like dried powder soot floating in the ambient air, which can contaminate supply air and act as a source of ignition. Malfunctioning/damaged filter may lead to entry of such materials into air heater.
3. Oxidation: It is an exothermic process, which proceeds rapidly with increasing temperature. Oxidation itself generates heat which increases reaction rate. When heat generated exceeds heat losses from the surface, the material temperature will raise to the ignition temperature.
4. Sparks at high temperature zones can result from friction between metallic surfaces *e.g.* bearings, atomizer wheel *etc.* Also, incorrect assembly of atomizer wheel, badly cleaned wheel/liquid distributor may cause this type of hazards.
5. Friction between powder particles will generate static electricity. This is especially true in the case of Fluidized bed drier where dry powder particles adjacent to surface collide with each other and metal surfaces.

Following safety precautions will help in prevention of fire explosion occurrence or damage:

1. Dryer Design: i) Correct combination of equipment and their layout ii) explosion proof wiring iii) correct positioning of atomizer iii) Smooth internal surfaces to prevent deposit formation
2. Preventive maintenance: i) Ensuring that safety instrumentation is functioning correctly (interlocks, alarms, 'fail safe' devices) ii) Maintenance of lubrication schedules, and to repair promptly any oil leaks iii) All rotating parts rotating freely (atomizer, supply and exhaust fans and rotary valves)
3. Following recommendations of the manufacturer: i) Special attention to operating temperatures, startup and shutdown procedures ii) Thorough cleaning of dryer components iii) Regular cleaning of supply air filters iii) Regular inspection of all electrical installations, mechanical equipment, bearings, packings.
4. Regular inspection during operation of the areas that are known to be prone to deposit formation.

Other methods are: observing air analysis for any initial burning before actual explosion, any sudden increase in internal pressures *etc.* Avoiding those conditions are more important, than trying to solve the problem after it has started.

REFERENCES

"Milk Powder Technology: Evaporation and Spray Drying" by Vagn Westergaard, 5th Edn, GEA Niro, Copenhagen, Denmark (www.niro.com).

"Spray Drying" by Masters K, 3rd Edn, (1979) John Wiley and Sons, New York.

Chapter 19

Theory of Solid-Gas Separation

Introduction

The separation of solids from gas, in dairy industry can be done by centrifugal action, or flow of air through a fixed bed and making it fluidized by use of flow of air, using an obstruction like bag or filters of suitable size. The requirement of removal of solids from air, is usually in Spray drying operations, removal of dust from air to prevent pollution in milk powder plant or even boiler exhausts. The procedures also depend on whether the solids to be removed are important or the gas that is to be purified, as in air filters that supply cooling air or air for combustion process. The following table can summarize the application of solid-gas separation methods:

Table 19.1. Application of Solid-gas Separation Methods

Principle	*Description*	*Application*
Centrifugal force	The gas is made to enter chamber tangentially at certain velocity and vortex created	Cyclone separator in Spray drying plants;
Filter material	Gas is made to pass through porous material, and hold back the solids depending on size of pores	Dust removal by Bag filters, Removal of contaminants in lubricating oils circulated
Gravity settling	Gas is made jump a gap at low velocity so that solids drop by gravity	Oil separator in refrigeration units, though along with oil, contaminants like abrasion material from compressors also drop out with oil
Membrane technology	The gas is made to pass through membranes at a suitable transmembrane pressure	Selectively separate different sizes of solids in the gas

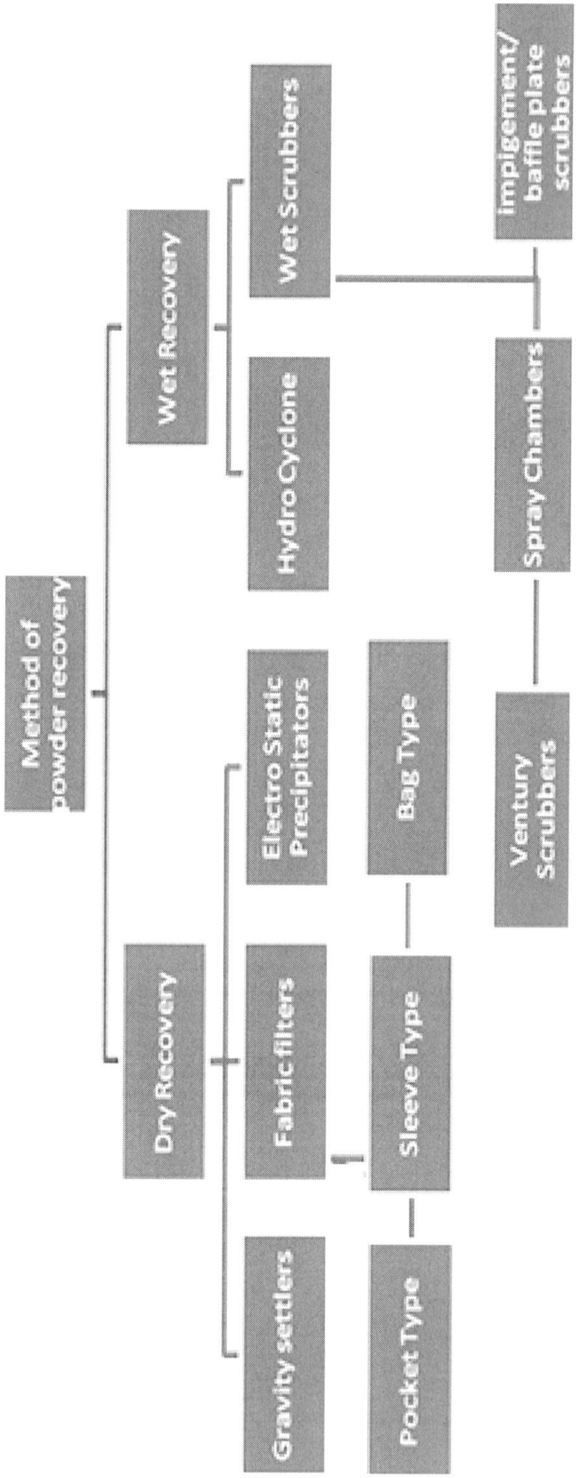

Figure 19.1. Classification of Methods of Separation

In the Spray drying operations, after the cooling and conveying of the powder coming out of the drying chamber, there is a need to separate the powder from the air that is transporting it. It is done initially by large Cyclone separators, and further the smaller dust through bag filters or wet scrubbers. The later operation is done more with an objective to control the pollution of the exhausting air. The broad classification of separation operations is given in Figure 19.1.

Cyclone Separators

The cyclone separators employ centrifugal force to remove the particles. Air with particles at high velocity is introduced tangentially into a cylindrical vessel. As the air enters the larger diameter cyclone chamber, the velocity decreases, permitting the settling of the solids. The air flow forms a vortex and flows down before start moving up in the center. The air outlet is through an extended duct, whose inlet is below the air inlet to the cyclone, so that the incoming air cannot short circuit to outlet.

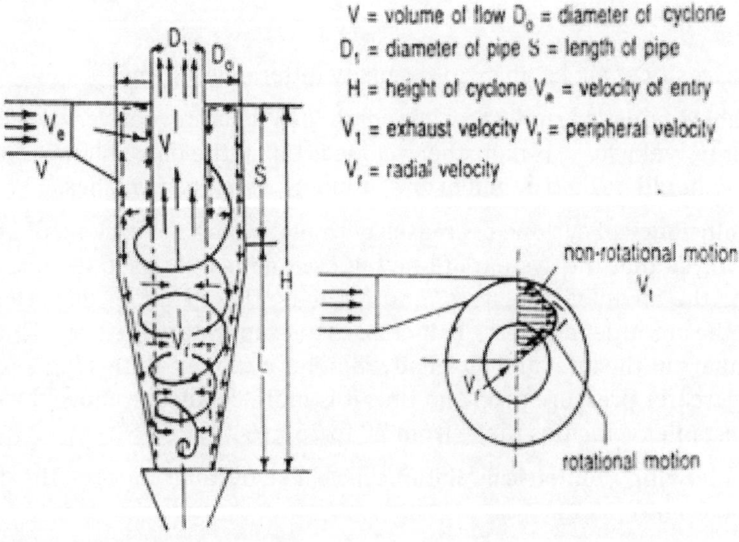

Figure 19.2. Air Flow Pattern in Cyclone Separator

The centrifugal force acting on each of the particle is represented by mV_t^2/r, where m, is the mass of the particle, V_t, the tangential air velocity and r, is radial distance from the cyclone axis. From this equation it can be concluded that the higher particle mass, the better efficiency. The shorter way the particle has to travel the better efficiency, and the closer the particle is to the wall the better efficiency, because the velocity is highest and the radial distance is short. However, time is required for the particles to travel to the cyclone wall, so that sufficient air residence time is taken into consideration when designing a cyclone. From above equation it is evident that small cyclones (diameter less than 1m) will have the highest efficiency.

The cyclone separators will remove about 98 to 99 per cent of all particles above 20μ but only 50 per cent of the particles below 5μ. In a properly designed separator, a powder recovery of more than 99.5 per cent is achieved. Depending on the size and density difference ($\Delta\rho$) between the particle and the air, the particle to be separated reaches the wall of the cyclone under the influence of the centrifugal force quickly. Very small particles, however, leave the cyclone along with the exhaust air. They are carried along, with the air entering the cyclone, whose radial component of velocity V_r, because for such small particles resistance due to the V_r is greater than the centrifugal force due to the peripheral velocity V_t.

The diameter of particles for which the drag equals the centrifugal force on the imagined extended cylindrical surface of the protruding pipe ($\pi D_1 L$) is called limiting diameter, d_l of the particle.

The limiting diameter is found by the equation:

$$d_l = \sqrt{(18 \times \mu \times (V/v_t^2) \times \Delta\rho \times 2\pi L)} \qquad \text{(Eq.19.1)}$$

where,

μ = viscosity of air, kg/m s; $\Delta\rho$ = density difference, kg/m^2

The above equation shows that it is possible to separate even very small particle if, a)the entry velocity v_e is high, the cyclone is tall. If the flow volume is large, the flow of air should be distributed over a number of smaller cyclones.

The efficiency of cyclone decreases with increasing the cyclone diameter D_o, which involves that the system of one big cyclone is only used in special cases. Increasing the air inlet velocity, v_e and reducing the air outlet diameter, v_e and reducing the air outlet diameter, D_l increases the separator efficiency. This is done with dampers at the inlet and by small diameter cyclones. With a higher velocity, there is a greater pressure drop and thus it is more costly to remove the product. The highest inlet velocities range from 20 to 25 m/s.

The following geometrical relations have a favourable effect on the degree of separation (Figure 19.2):

D_o/D_l = 2 to 2.5 (prefer 3); H/D_l = 7 to 8 (prefer 10); S/D_l = 0.8 to 1.3

In general, the air velocity can be up to 30 m/s. The pressure drop across the cyclone separator can be between 5 to 15 mmHg.

Bag Filters

Average powder loss from a high-efficient cyclone should not exceed 0.5 per cent when spray drying skim milk. However, pollution control authorities may conclude that 0.5 per cent is too high due to pollution, thus requiring a final cleaning of the air. This is usually done in bag filters consisting of numerous bags or filters arranged so that each bag receives almost equal quantities of air. The direction of the air is from outside in through the filter material to the inner part of the bag from where the cleaned air enters an exhaust manifold. With a correct selection of filter material high efficiencies can be obtained and collection of 1 micron particles is possible. The collected powder is automatically shaken off by blowing compressed air through the filter bags from the inner side. The powder is collected at the bottom via a rotary valve. (Figure 19.3).

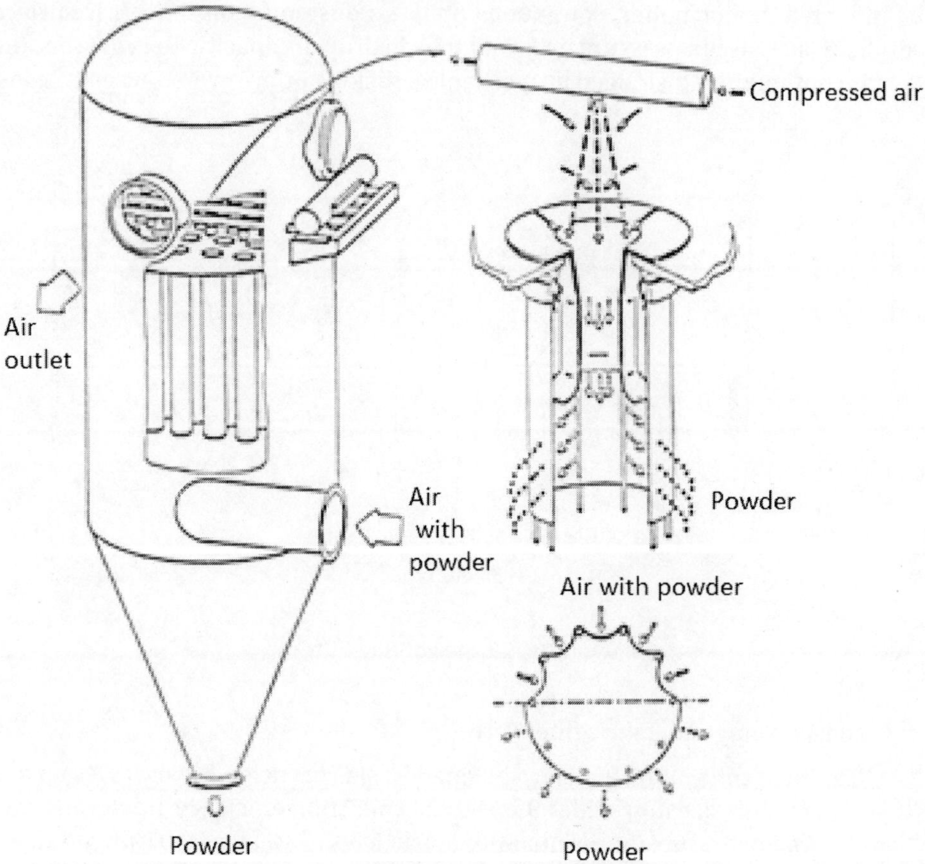

Figure 19.3. Bag Filter Arrangement

The bag filter may also replace the cyclones, a solution often chosen in dryers for whey protein powder or egg white. To prevent condensation, especially on the conical part of the filter housing, warm air, water circulation or heat tracing is established. Powder loss of 15-20 mg/m^3 is reported.

Fabrics filters consist of woven fabrics of tubular shape which are held over certain frames. Among the various designs available the commonest is bag filters shown in Figure 19.4 It consists of freely suspended tubes which are open at both the ends and mounted inclined or vertical. Dust laden air enters from the top and escapes from the pores, while solid particles leave from the bottom. The sleeve type needs internal wire frame for supports. It is mounted inverted. Air enters from the bottom opening and is filtered on the inner side of the sleeve. Pocket type is mounted with opening on the top. Air, enters from the sides and gets filtered on outer surface and leaves through the inside of the tube while the dust is collected on outer surface. For large capacity plants large numbers of sleeves/pockets might be mounted in a chamber. Continuous build up dust layer on the fabric surface results in increase in pressure drop and thus in drop of capacity. To avoid this, the dust is continuously dislodged by mechanical sacking or by reverse air pulsations.

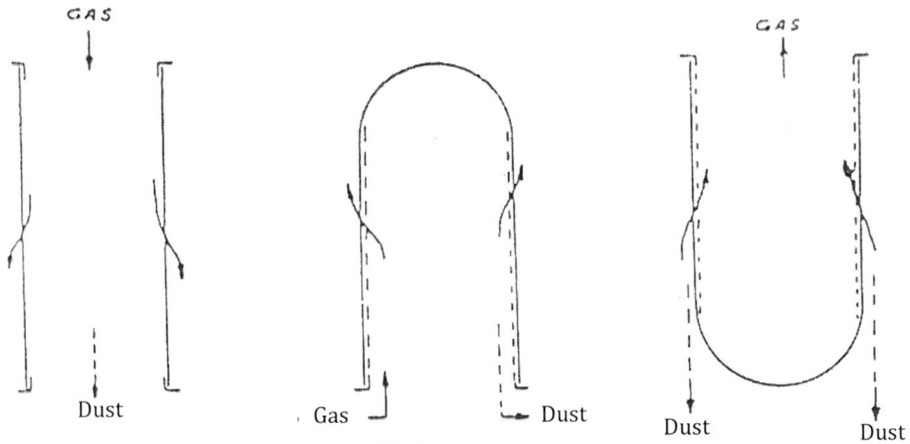

(a) Bag type (b) Sleeve type (c) Pocket type

Figure 19.4 Types of Bag Filters

The advantages of fabric filters are :

a. High collection efficiency for smallest size particles. For particles up to 1 micron size it provides 99 per cent collection efficiency. However, it can remove a substantial quantity of particles as small as 0.01micron size
b. Moderate pressure drop of the order of 7-15 cm of water gauge.
c. Dry recover, and
d. Absences of corrosion and rusting problem.

However, the disadvantages of the system are: a) Large size of the unit b) Difficult maintenance and need for frequent replacement c) Plugging of fabric pores by hygroscopic materials.

Wet Scrubbers

The wet scrubber is based on the venturi scrubber principle. It has an efficiency very close to 100 per cent. The outlet separator is designed according to the well-known cyclone principles, however with a modified outlet, resulting in a minimum liquid level, thereby minimizing bacteria growth, and a design ensuring de-aerating thus avoiding from building (Figure 19.5).

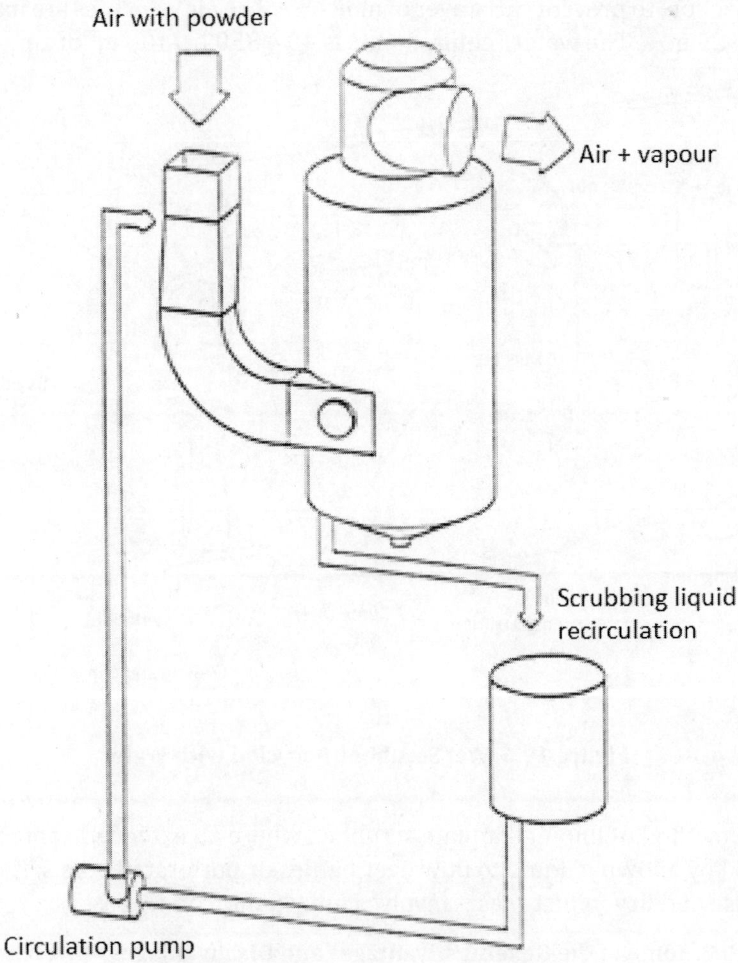

Figure 19.5. Sanitary Wet Scrubber

The outlet air from the spray dryer containing powder particles is accelerated to a high velocity (45 – 120 m/s) in the venturi inlet, where also the liquid is injected through full-cone nozzles. Due to different velocities between the air/particles and the liquid droplets, they will collide, and the powder will dissolve in the liquid droplets. Passing through the subsequent diffuser this process will continue simultaneously with a certain pressure recovery of the air/droplet mix. (Figure 19.5). Passing through the separator, air and liquid are separated again. The air leaves through the centre duct and the liquid through the bottom outlet for further processing or recycling depending on what system is selected.

The other method is providing a Spray chamber, where, the air with powder dust is made to flow through a chamber, in which water is showered, and drained at the bottom. The gas after forcing its way through liquid spray passes through the demist section, to prevent carry over of moisture. The air velocities are maintained at 0.6 to 1.5 m/s. The water requirements is 40 – 350 lit/100 m^3 of air.

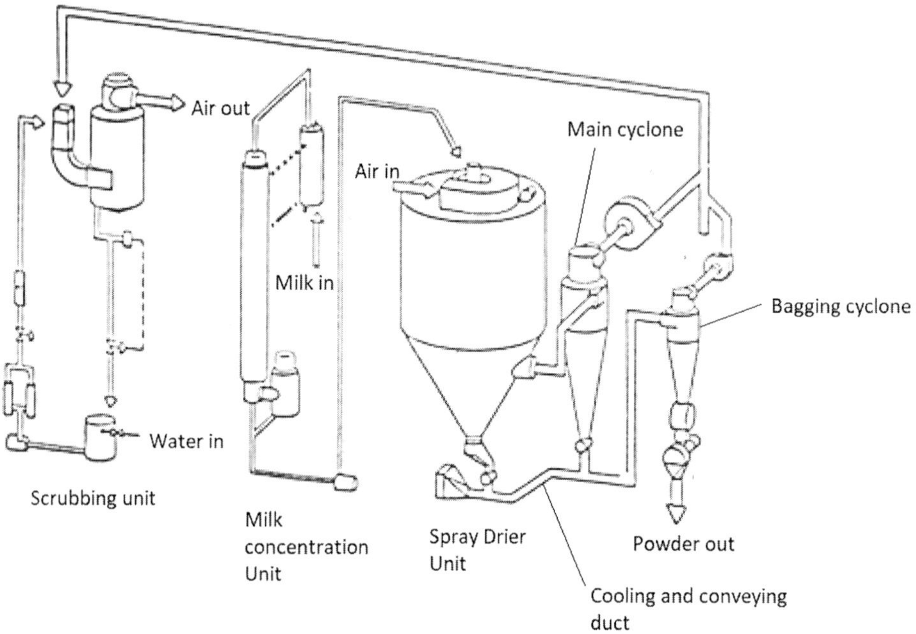

Figure 19.6. Wet Scrubber Recycled with Water

The method of impingent plate scrubber, where an extended contact surface is created by allowing liquid to flow over baffles or perforated plates, is ruled out in the case of dairy industry as it involves unhygienic conditions.

Table 19.2. Some of the General Advantages and Disadvantages of Wet Scrubbers

Advantages	*Disadvantages*
Ability to recover solids as well as liquids	May create liquid disposal problem if the same is not reused
Ability to handle high temperature, high humidity gas streams.	Product is collected wet which subsequent drying
Initial cost is usually low	Pressure drop is higher. Thus, operating cost is more
Small space requirement	Severe corrosion and maintenance problem

Electrostatic (ES) Precipitators

The device works on the principle that if one or more of the materials in a mixture can be imparted surface charge, on or before entering the electrostatic field, those charged grains will be repelled from one electrode and attracted towards other depending upon the polarity of the electrode. Thus, in ES precipitator, strong electric field is created with the help of rectifier and transformer unit and dust laden air flows, in that chamber. Powder particles develop negative polarity by combination with air ions. These particles move towards positively charged collecting plates where from they fall in separate chute after getting neutralized. High D.C voltage of the order of 50,000 volt is used.

The advantages of ES precipitator are: a) High collection efficiency above 99 per cent for particles upto 0.01 micron size, b) Low pressure drop, c) Low maintenance cost.

The disadvantages of the unit are: a) High initial cost b) Higher energy consumption and difficult operation, c) Ionization of air and production of ozone.

Because of these disadvantages ES precipitators are not frequently used in dairy plants. They are preferred in chemical industries, especially those dealing with hazardous chemicals where near complete trapping of even the finest particles is essential.

Table 19.3. Advantages and disadvantages of various recovery devices

Devices	*Advantages*	*Disadvantages*
Fabric/Bag filters	☆ 99 per cent efficiency for particles up to 1 μ ☆ Moderate pressure drop ☆ No corrosion/resting	☆ Large size of unit ☆ Clogging of filters ☆ Maintenance difficult
Electro Static precipitators	☆ 99 per cent efficiency for particles upto 0.01μ ☆ Low pressure drop ☆ Low maintenance	☆ High cost ☆ High energy consumption
Spray Chambers	☆ Inexpensive ☆ Lesser pressure drop ☆ Easy maintenances and cleaning	☆ Low separation efficiency, only 70 per cent for particles above 10μ ☆ High space requirement
Venturi scrubber	☆ 95 per cent efficiency for 1μ particle ☆ Low cost ☆ Easy cleaning	☆ High energy consumption.

Fluidization

It refers to those gas-solids and liquid solids systems in which the solid phase is subjected to behave more or less like a fluid by the upwelling current of gas or liquid stream moving up through the bed of solid particles. Both the forces of gravity and fluid friction act simultaneously to effect fluidization of the bed of solid particles.

Fluid stream transfer the momentum from the fluid to the solid particles. If the fluid flowing is of low velocities, the momentum transfer does not cause particles to move. When stream velocities increase, the rate of momentum transfer and hence the pressure drop for the flow through bed increases gradually. A point is reached where the momentum transferred from the fluid to solid particles balances the opposing gravitational force on the particles.

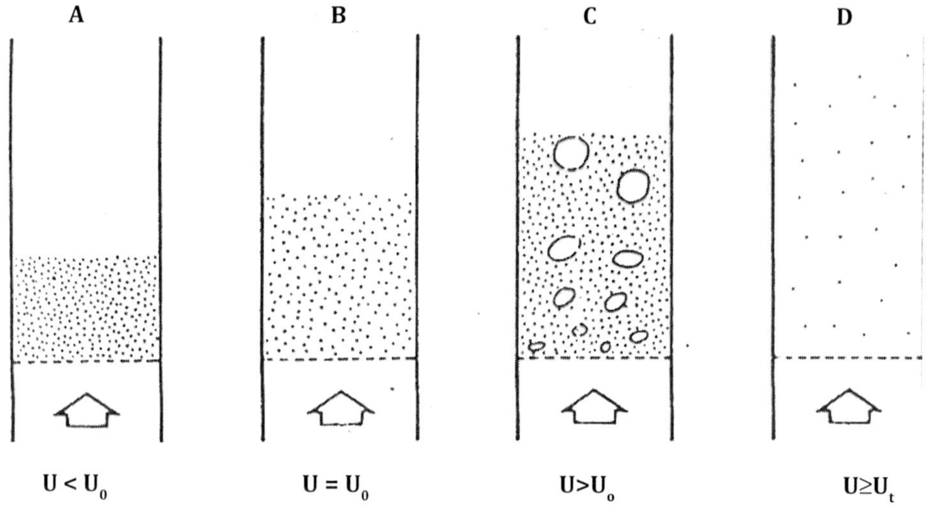

Figure 19.7. Fluidized Bed Flow Regimes

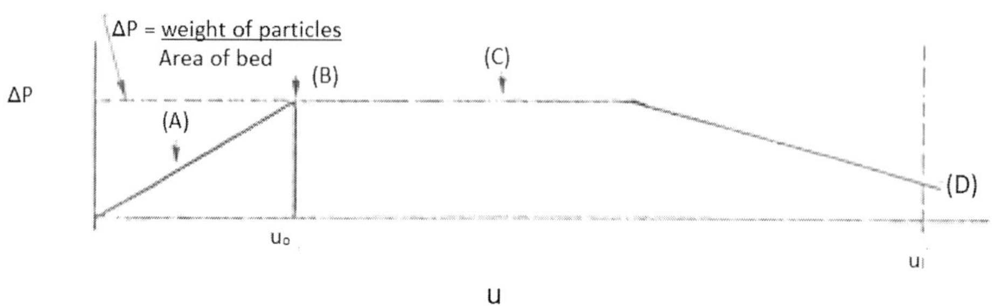

Figure 19.8. Velocity of Air Flow and Pressure Drop across the Bed

As the velocity of air increases, the pressure across the bed ΔP also increases. At some point the ΔP balances the weight of the bed. Further increase in velocity, the bed expands slightly, but increases the voids and offsets the velocity and ΔP is constant. Finally, as the velocity increases further, the particles separate and true fluidization begins.

Porosity of a Static Bed

It is the fraction of total volume of the static bed occupied by voids.

$$\varepsilon_0 = (V_{bed} - V)/V_{bed} \quad \text{(Eq. 19. 2)}$$

where,

V_{bed} = volume of static bed, m^3

V = Volume of particles, m^3

As, density of medium (voids) being small in comparison to that of particles, it can be ignored and hence,

$$\varepsilon_0 = 1 - (\rho_{bed}/\rho_s) \quad \text{(Eq. 19.3)}$$

where,

ρ_{bed} = Bulk density of bed, kg/m^3

ρ_s = Particle density, kg/m^3

For randomly packed static bed, the value of ε_0 of spherical particle bed ranges from 0.38 to 0.42

Porosity of Fluidized Bed

The porosity of a fluidized bed by particles is defined likewise by the formula,

$$\varepsilon = (V_{fbed} - V)/V_{fbed} \quad \text{(Eq. 19.4)}$$

where,

V_{fbed} = Volume of the fluidized bed, m^3

Minimum Porosity

It is the porosity of the bed at which true fluidization just sets in.

Bed Pressure Drop

For a fluidized bed, the bed pressure drop $(-\Delta P_b)$ is equal to the weight of the bed per unit of its cross sectional area

$$(-\Delta P_b) = W_b/A \quad \text{(Eq.19.5)}$$

where,

W_b = Weight of the material in the bed, kg

A = Cross sectional area of the bed, m^2

This should be constant to satisfy the basic hydrodynamic characteristics of a fluidized bed.

The pressure drop of the fluid medium across the height (L) of the fluidized bed can be determined from the relation,

$$(-\Delta P_b) = (1-\varepsilon)(L)(\rho_p - \rho_f)\,g \quad \text{(Eq. 19.6)}$$

$$= (1-\varepsilon_0)(L_0)(\rho_p - \rho_f)\,g$$

where,

L = height of the fluidized bed, m

L_0 = height of the static bed, m

ρ_p = density of solid particles, kg/m^3

ρ_f = density of the fluid medium, kg/m^3

If the fluidizing medium is air or gas, then ρp>> ρ_f

Therefore, $(-\Delta P_b) = (1-\varepsilon)\, L\, \rho_p\, g$

cross section area is $= (1-\varepsilon_0)\, L_0\, \rho_p\, g$

Bed Height

Porosity is a direction function of bed height, if cross section are is uniform.

$(-\Delta P_b)/L_0 = (1-\varepsilon_0)\,(\rho_p - \rho_f)\, g$

and $(-\Delta P_b)/L = (1-\varepsilon)\,(\rho_p - \rho_f)\, g$

Hence, $L_0/L = (1-\varepsilon)/(1-\varepsilon_0)$;

However, the porosity initially is, $\varepsilon_0 = 0$; and hence, $L = L_0/1-\varepsilon)$

Minimum Fluidization Velocity

For any fluidization process the velocity U_0, at which fluidization starts and the velocity at which entrainment occurs are the two limits. Minimum fluidization velocity can be calculated by a) Taking rate of sedimentation as a base b) Equating force acting on a particle is exactly balanced by drag force.

Gravitational force acting on the particle is given by

$$\frac{\Pi d^3}{6}\,(\rho_p - \rho_f)\, g\,(1-\phi_0) \qquad \text{(Eq. 19.7)}$$

where,

ϕ_0 is the porosity of the particle

Drag force acting on the particle due to upward flow of the fluid is given by,

$$C_d \times \Pi d^2\,(\rho_f/2)\, U_0^{\,2} \qquad \text{(Eq. 19.8)}$$

where,

C_d is drag coefficient which is equal to 24/Re for laminar flow.

By equating the above two equations, and substituting for C_d, we arrive at the minimum fluidization velocity U_0 as,

$$U_0 = d^2\,(\rho_p - \rho_f)\, g\,(1-\phi_0)/18\,\mu \qquad \text{(Eq. 19.9)}$$

Batch Fluidized Bed

Though application of Batch fluidized bed is limited in dairy industry, it is used, in Ion-Exchange water softener, where, a granular resin bed is made to expand a little by hard water, and the surface of resin is used for exchange of calcium and magnesium ions in water with sodium ions adsorbed on its surface. The resin is kept on a porous bed support and also an upper protective barrier, so that the resin is not washed away if flow of raw water is rapid.

The other application of Batch fluidized bed can be for biochemical reactions with catalytic actions, or for heat transfer operations involving thermal storage of energy. In many such cases, the fluidization may not happen in strictest sense, but a separation of particles does take place, to make better use of the particle surface for the process intended. It may also be applied for solids size enlargement or solids size reduction. It may be used for coating of solid particles, drying of solid particles *etc.*

Fluidized Bed Driers

The fluidized bed driers are used for gentle drying of heat sensitive products. Especially in milk powder manufacture, the outlet temperature of spray drier can be reduced, by removing the last 8 per cent of moisture in a secondary drier, by way of fluidized bed. Further, to obtain agglomeration of milk powder, the fines blown away in fluidized bed are used for re-inletting them in the atomization zone in the spray drying chamber.

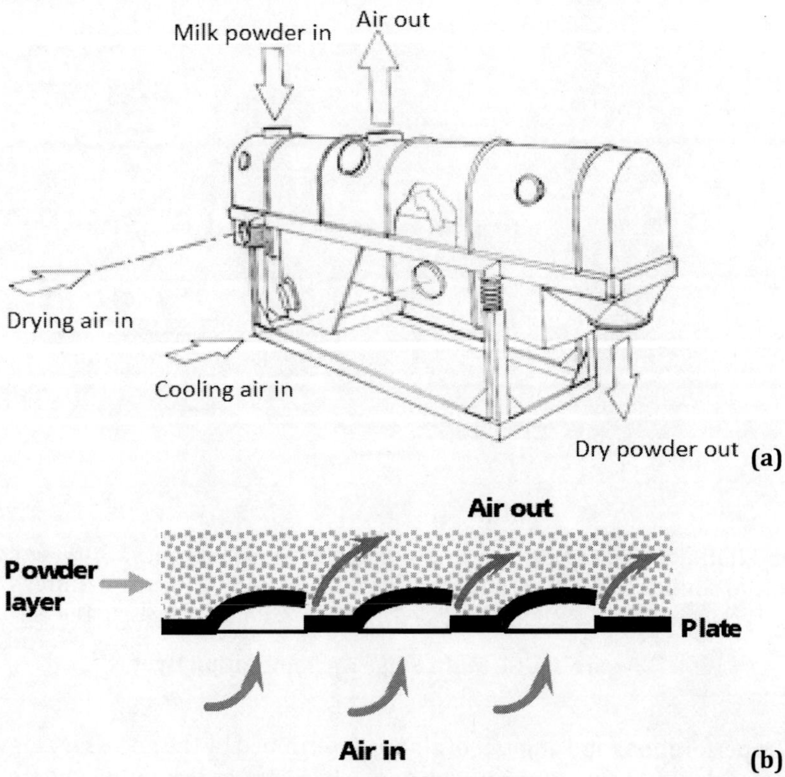

Figure 19.9. Vibro Fluidized Bed Drier

In the application of milk powder manufacture, the fluidized bed drier is also provided with mechanisms to vibrate the bed, to achieve non-sticky operation and with lower drying air velocity.

A typical Fluidized bed drier is shown in Figure 19.9 (a). It is a big horizontal chamber divided in an upper and lower section by a perforated plate. For drying the partly wet powder as well as to cool, air is passed from the bottom of the perforation, taking care to evenly distribute the same. In the initial stages, the air supplied is warm to dry the powder, and later segments, cooler and dry air is blown into the powder. The perforated plate in fact has a surface in the shape of bubble, so that the air entry is at an angle, so that it also aids in moving forward the fluidized milk powder, so that it goes to the cooling section and then to out (Figure 19.9 b).

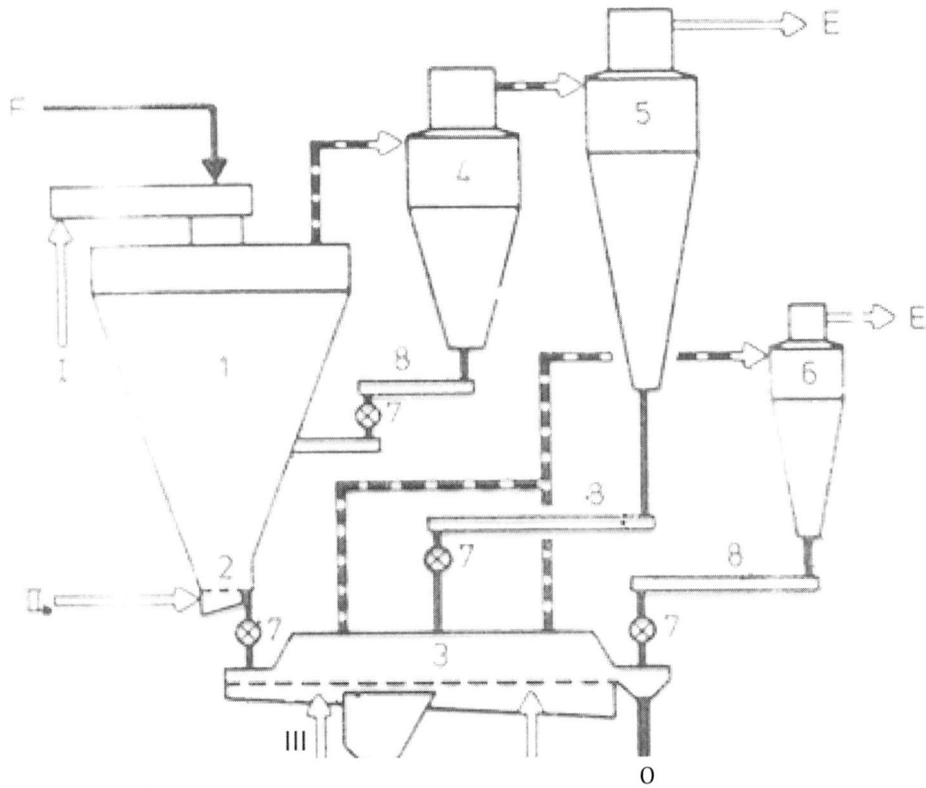

F: Feed; I,II,III: hot air; C: Cool air; O: Product out; E: Exhaust air; 1: Drying chamber; 2: Internal Fluidized Bed; 3: External Fluidized Bed; 4: Fines re-inlet cyclone; 5: Fines cyclone; 6: Fines from fluidized bed cyclone; 7: Rotary Valves; 8: Ducts.

Figure 19.10. Multi Stage Agglomeration Dyer

The perforations and amount of air are determined by the necessary air velocity needed for the fluidizing of the powder, which in turn is determined by the nature of the powder such as the moisture content and thermoplasticity.

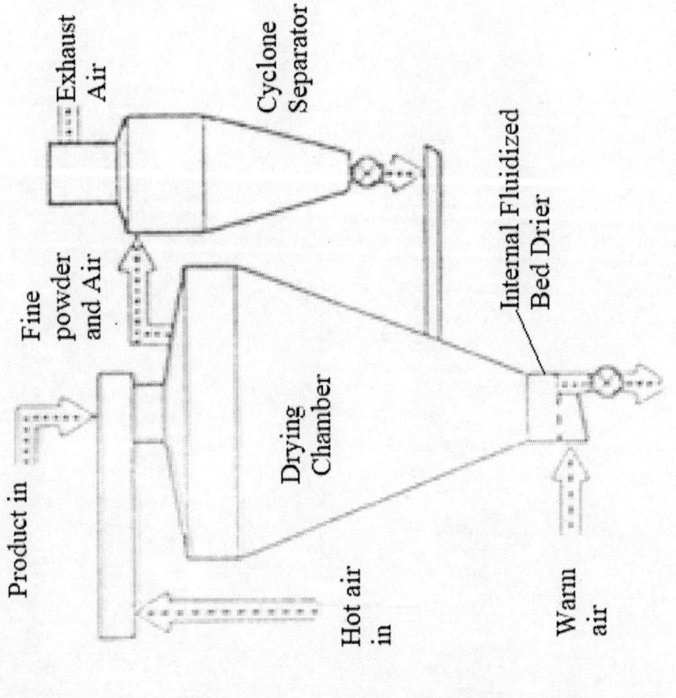

Figure 19.11. Internal Fluidized Bed (IFB)

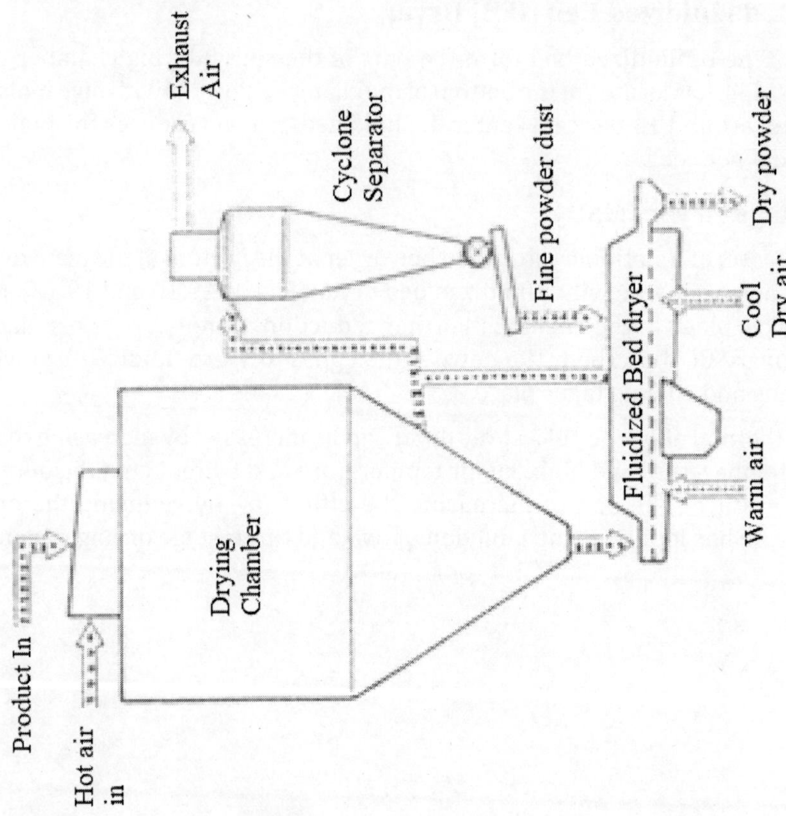

Figure 19.12. Two-Stage Dryer

Integrated Fluidized Bed (IFB) Dryer

This type of fluidized bed form the part of the spray drying chamber itself (Figure 19.11). It is located at the bottom of the chamber. Powder with high moisture content (of around 15 per cent) enter the fluidized bed and reduces the moisture to around 5 per cent.

Multi Stage Dryer (MSD)

It consists of a stationary fluidized bed dryer at the bottom of the main drying chamber and a separate vibro fluidized bed dryer (Figure 19.10 and 19.12). Main drying takes place in the chamber. Further reduction of moisture takes place in the stationary fluidized bed. The powder then goes to Vibro fluidized bed where final drying and cooling takes place.

The thermal efficiency of a spray dryer can be increased by increasing the hot air inlet temperature and ambient air temperature and by reducing the outgoing air temperature. The aim of increasing the efficiency, by reducing the outlet temperature has led to the introduction of two and three-stage drying systems.

Chapter 20

Butter Making Machines

Introduction

For preservation of fat in milk for long storage, it is necessary to convert first into cream by centrifugation and then to Butter by churning. The moisture content in butter is about 16 per cent. In the manufacture of table butter salt and color are added. The sequence of operations for converting Cream to Butter is indicated below:

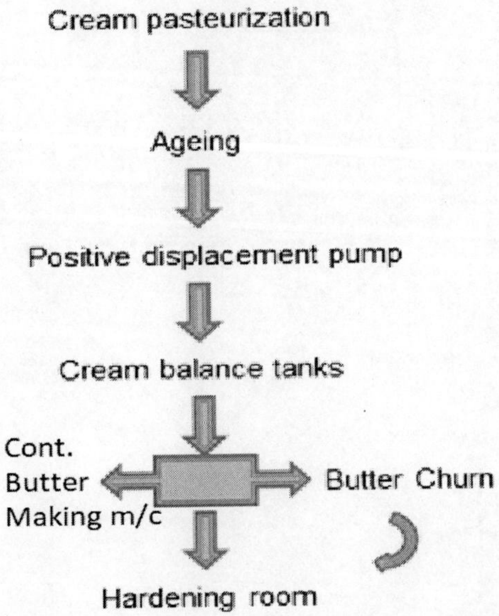

Figure 20.1. Sequence of operations to convert Cream to Butter

Batch Butter Churn

Butter is made industrially in two forms: 1) Cooking butter 2) Table butter. The cooking butter is made for long storage and then reconstituted into milk at times of scarcity. The table butter is for consumption as such along with other food items.

Batch Method using Rotating Churns

The rotating churns consist mainly of a barrel rotated on an axis with internal baffles of various kinds. These are provided to increase the agitation. The intake capacity of batch churns varies from 100 to 3000 Kg of cream per batch. The shape is mostly cylindrical with front opening. The other shapes include cone with cylinder, single cone and double cone *etc.* The churn is short in length and large in diameter. In some designs of churns, ribbed rollers are fitted internally through which butter grains pass. Air vent, sight glass, butter milk outlet, cream inlet, and outlet for butter are mounted on barrel. As butter does not adhere to the wood inside surface of the wooden churns does not require special treatment. For the metal churns, to prevent the butter from sticking to the inside surface it is roughened by sand or lead shot blasting to allow a film of moisture to form on the surface. The water film remains between the metal and the butter and prevents butter from sticking to the inside surface.

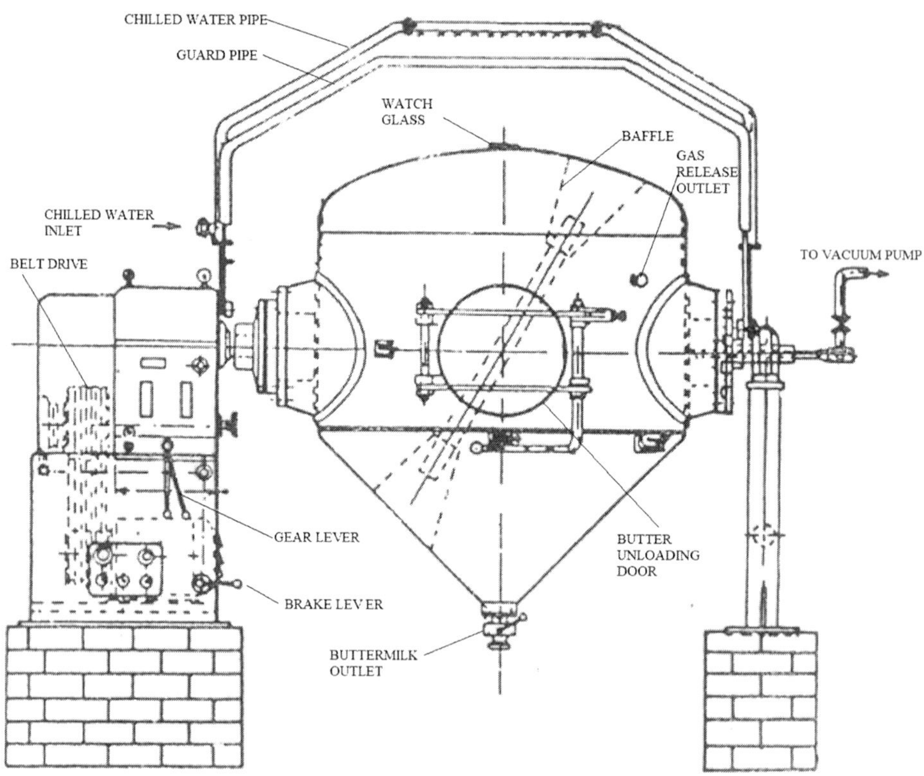

Figure 20.2. Top Churn with Gear Drive

An improvement in the design is the Top Churn, which has a conical bottom, which promotes better agitation of the cream. The churn also has provision for the creation of vacuum for processing fermented cream churning. A perforated pipe placed above, provides for spraying chilled water during churning, for maintaining the temperature of the churn around 11°C.

The drive for the Top Butter churn is through a motor and a speed reducing gear, and usually in three steps of high speed and slow speed forward, slow reverse speed. The high speed is for churning, slower speed, and reverse speed are for working. A brake lever is provided to stop the churn slowly. A safety bar is provided to prevent the keep the operator at safe distance from the churn, while it is rotating.

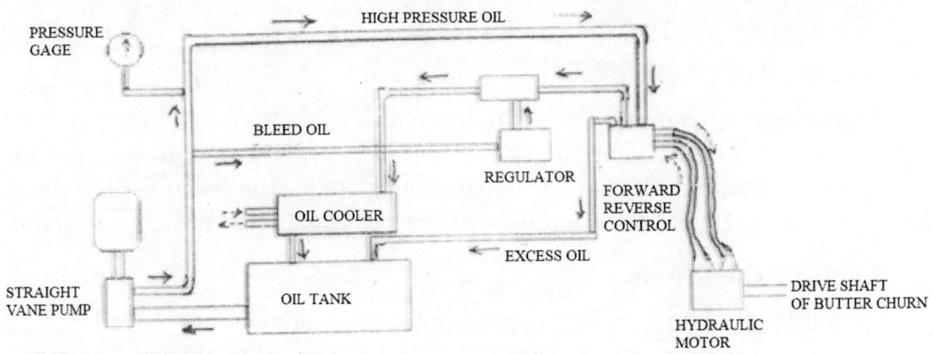

Figure 20.3. Hydraulic Drive Mechanism of Butter Churn

A drive has been developed as hydraulic to replace the gear mechanism and it offers smooth speed control suitable for different capacities and loadings. The Hydraulic power train has a vane pump for developing high oil pressure in a flexible pipe line that transports oil to a vane motor. The Vane motor drives a shaft at varying speeds, depending on the pressure of oil. The shaft is connected to the churn through a speed reduction gear. The oil after turning the vanes in the vane motor returns to the supply oil tank, where, it is also filtered, cooled by chilled water in a heat exchanger. The direction of flow of oil can be controlled, so as to reverse the direction of rotation during working stage. A typical hydraulically driven butter churn is operated at 35 to 40 RPM. The maximum working pressure of oil is about 135 kg/cm ^{2} (2000 psi), and its temperature is controlled and kept below 65° C., The oil tank has 300 liters capacity.

The degree of mixing depends on the amount of cream in the churn and on the speed of revolution. Too low speeds will not give sufficient turbulence. At too high speeds the centrifugal force ($m\,\omega^2 R$) exceeds the gravitational force (mg) and the cream sticks to the periphery and rotates with the drum. The best condition for churning is obtaining maximum turbulence in the cream. It is achieved when the force of gravity just exceeds the centrifugal force.

i.e. $m\omega^2 R < mg$

where,

m = mass of butter (kg)

ω = Angular velocity (Radian)

R = Radius of butter churn (m)

n = Number of revolutions (1/s)

g =Acceleration due to gravity, (9.81 m/s^2)

$(2\pi n)^2 R < g$

Or $n < [g/R]^{1/2}. 1/2\pi$ (Eq. 20.1)

$n < (1/2\pi) \times [g/R]^{1/2}$

The energy consumption of the butter churn is about 7 - 11 kWh per 1000 kg of butter produced. About 90 per cent of this energy is used in churning and 10 per cent is used in working. The lower values are for the cream with a higher fat content. The sequence of operations of the butter manufacture is given below:

1. Loading the Cream

Pasteurized cream with 35 - 40 percent fat, properly aged is pumped into the churn. Cream is filled to 40 - 45 per cent of the volume of the churn. The cream may be ripened or unripened.

2. Churning

The churn can be operated at different speeds. The range of speed depends on the size and shape of the churn. The cream is well whipped by the corners, edges and other irregularities in the churn. Chilled water is sprayed over the churn during the churning operation. It takes about 35 – 40 minutes for the formation of butter granules of peanut size.

3. Buttermilk Draining

The churn is stopped and buttermilk is drained off, when it is observed from watch glass that butter grains are formed. Equal quantity of pasteurized wash water is added at this stage.

4. Washing

The churn is started again. The wash water is drained off after some time. Two or three washings are generally given.

5. Working

After the wash water is drained off and salt is added. The churn is then operated at a lower speed for working. Compared to the churning speed the speed for working is lower. After 3 – 5 min., sample is taken and moisture is adjusted by adding required quantity of water. The working is continued till desired body and texture are obtained. Creating vacuum of 5 mm of water gauge during working gives close texture by reducing the content of air.

6. Unloading and Packing

The butter is unloaded into the trolleys and then packed for sale. Different types of packing machines are employed depending on the size of packages.

The unloading of the butter into trolley, may be 1) for packing for long term storage 2) for immediate making of Ghee. The trolleys may be simple stainless steel rectangular vessel mounted on wheels and provided with handle, or it may have arrangement for hot water circulation. This is because, the butter will be unloaded at lower temperature, and would need to be melted and pumped into Ghee boiler, and would need to be in molten state or at least in a condition suitable for pumping.

Care of Churns

1. Driving gear should be filled with lubricating oil. It should be replaced once in two years.
2. The speed should not be changed while the churn is running.
3. Solid foundation for the churn is necessary. If the butter churn is tilted on one side due to weight, the moisture distribution in the butter will not be uniform. There should not be any vibrations, while churning is taking place.
4. The Gaskets should be maintained leak proof.
5. Proper roughness inside of the churn should be maintained.
6. The churn should be thoroughly cleaned after the operation

Continuous Butter Making

Development of continuous butter making machines resulted in 1) increased operating efficiency 2) better sanitary features 3) no requirement of a sturdy foundation.

The continuous butter making machines are used in dairy industry can be classified into three categories according to the churning principle involved as shown in the Table 20.1.

The theories of continuous butter making can be summarized as described below:

1. Machines operating on Fritz process or floatation churning in which accelerated churning and working takes place. Few equipment of this type are known as Westfalia Separator (West Germany), Contimab (France), HMT (India) and Masek (Czechoslovakia).

Table 20.1. Different Principles of Churning

	Feed	*Process*	*Company*
Using normal cream and agitated at high speed	Cream having 30 – 50 per cent fat	High speed whipping at 1000-2500 rpm for 3 – 5 sec, working at 30 – 65 rpm, draining the butter milk.	Fritz process: Equipment based on this principle are made by Contimab, HMT (India)
Using high fat cream, chilling and agitation	Cream having 80 per cent fat	Cooling to 8 – 15 °C, agitation; no draining of butter milk	Alfa Laval
	Ripened cream; 80 per cent fat and vacuum pasteurization	Chilled and extruded through series of openings	Cherry Burrel
	Cream having 80 per cent fat	Using Emulsion breaker (like homogenization)	Creamery Package

2. Equipment operating on concentration of normal cream followed by phase inversion, cooling and finally mechanical treatment. Machines based on this principle are known with commercial names as Alfa (West 'Germany) and Alfa-Laval (Sweden).
3. Machines operating on concentration of normal cream de-emulsification and re-emulsification into butter. Equipment working on this principle have commercial names of Gold'n Flow Process (USA), New Way Process (Australia) and Creamery Package Process (USA).

The important steps in butter making by these three processes are outlined below. One machine based on each of the three processes is described briefly.

The Fritz Process

Machine based on this principle contains three main parts, *viz.* Churning cylinder, draining and washing cylinder and worker. Capacity can be up to 10000 kg/h.

1. Preparation of Cream

For smooth operation of the churn, the cream must be of uniform quality. The properties like fat content, temperature, age, pH, are controlled. The temperature is kept around 4°C.

2. Churning the Cream

A pump, delivers cream to churning cylinder. The churn shaft is driven by a variable speed drive. A four-armed beater runs inside the cylinder at 1250 – 2800 rpm (some designs operate at 800 rpm) with a wall clearance of 2-3mm. The speed can be varied. The pockets in the churning cylinder impart extra turbulence to the cream film thereby enhancing the butter making action and improving the yield. Butter granules are formed within 3-5 seconds. During the operation the cylinder is cooled by cold water in circulation. A second churning cylinder rotates at 10-25 rpm or alternately a stationary cylinder with paddle-bearing shafts rotating at 34 rpm. This cylinder is also cooled. This cylinder is to expel moisture still in the butter.

3. Butter Milk Separation and Draining

The mixture of buttermilk and butter granules proceed to enter into the adjoining section known as separating cylinder with welded ribs. The next is the buttermilk draining cylinder. This cylinder is perforated and provided with beater studs. Buttermilk gets separated and drained through holes.

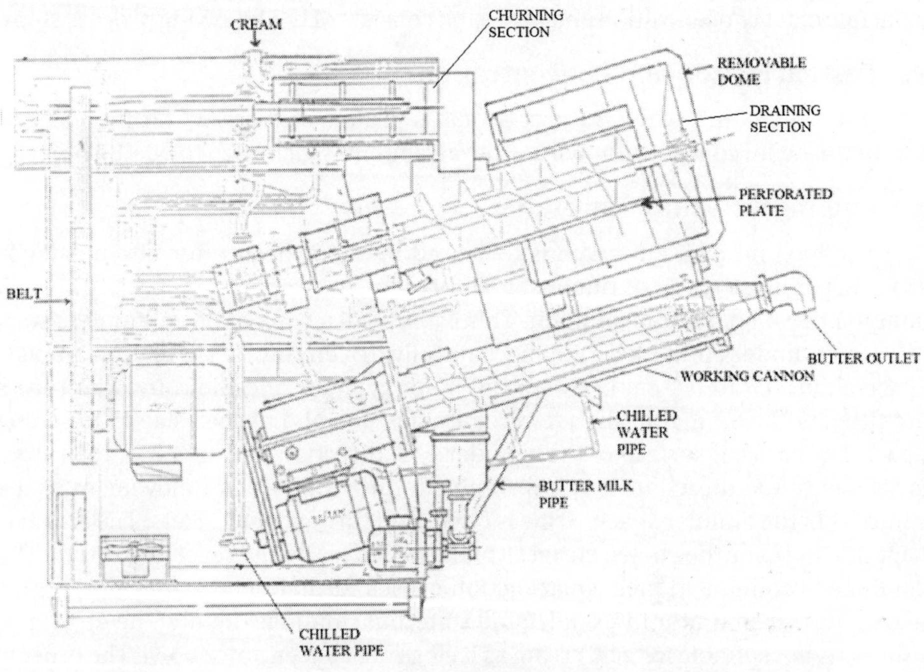

Figure 20.4. The Fritz Process – Major Components of Continuous Butter Churn (HMT, India)

4. Washing the Granules

The butter granules enter the washing section which is also perforated and has water spray for washing. The grains which now reached finishing stage are washed here.

5. Working and Texturizing

The washed butter grains fall into twin worm butter worker (two contra rotating screws). The screws which are inclined force the butter through number of perforated plates arranged in series. This operation gives the fine dispersion of water. The process is assisted by mixing vanes, situated between plates. The butter worker can be operated at either of two fixed speeds of approximately 65 and 30 rpm. In this section, there is provision for salt and colour addition, moisture corrections *etc.* The butter is discharged as a continuous stream. The shape of discharging butter depends upon the geometry of the outlet spout.

Concentration and Phase Inversion Process

Principle of concentration and phase inversion explains that normal cream can be converted into fat rich cream with fat content as high as the minimum permissible fat content in butter. Butter formation during cooling of concentrated cream was observed. The Alfa-Laval (Sweden) butter making process which is based on this principle of fat concentration and inversion consists of the following process steps.

1. Pasteurization and Concentration

Normal cream of 30 – 35 percent fat is pasteurized at 90°C. It then passes through a cream concentrator where the fat content is increased to 80 – 84 percent.

2. Cooling and Phase Inversion

The high fat content cream is delivered by the concentrator into a balance tank, which is stirred continuously. The cream so prepared is forced through equipment called the transmutator. This consists of a bank of three stainless steel jacketed cylinders provided with mechanically driven rotors. There is an annular space about 0.5 to 0.6 cm between the inner cylinder and the rotor. The rotors are fitted with soft metal ribs set in a spiral fashion and of such size as to scrape lightly the internal surface of the cylinder. The direction of rotation of the rotors is related to the direction of spiral ribbing to provide forward movement of the contents in the annular space. Brine is circulated through the jackets. The cream is kept moving over the cooled surfaces of the cylinders by the revolving rotors. The combined cooling and mechanical action causes formation of butter. The butter leaves the transmutator in a semi liquid state, but solidifies rapidly. The Alfa-Laval process is not suitable for acid cream as it clogs the concentrator bowl. The process presents some difficulties in control of moisture content of butter, as per some of the research reports.

3. Concentration, De-emulsification and Re-emulsification Process

The main steps of this method of butter making are described below.

1. Destabilization: Air is incorporated into freshly separated cream of 30 – 40 percent fat. After this step, the cream flows through a destabilizing pump containing perforated rotor blades turning at about 3000 rpm. Better destabilization is obtained by using two pumps in series. The cream is then heated to 60°C in a vertical centrifugal heater. The heating is employed to bring the cream to adequate temperature for completing destabilization. Fat globule membrane is disrupted and weakened by

the mechanical action of pumps and the destabilization is completed by subsequent heating. Fat then moves as a continuous mass which is free of globular fat.

2. Concentration: After destabilization, the cream is concentrated to 85 – 90 percent fat content by a special separator which discharges skim milk, sludge and concentrated cream.
3. Pasteurization: The concentrated cream is pasteurized in a vacreator at 88 - 90°C and subsequently cooled to 38 – 43°C.
4. Standardization: The liquid butter fat in which small droplets are dispersed is constantly stirred and standardized for moisture content, salt and fat.
5. De-emulsification and Texturization: The standardized product is cooled to 4 - 6°C in a chiller. The chiller consists of two horizontal cylinders, installed side by side and cooled by direct expansion of NH_3. Each cylinder is equipped with an agitator provided with scrapper blades. The moisture is finely disturbed by means of vigorous mechanical.

The required consistency of butter is obtained by treatment in the texturator. While passing through the texturator, temperature of butter increases by about 2°C, which initiates further crystallization of butterfat. Butter comes out at 4 - 6°C.

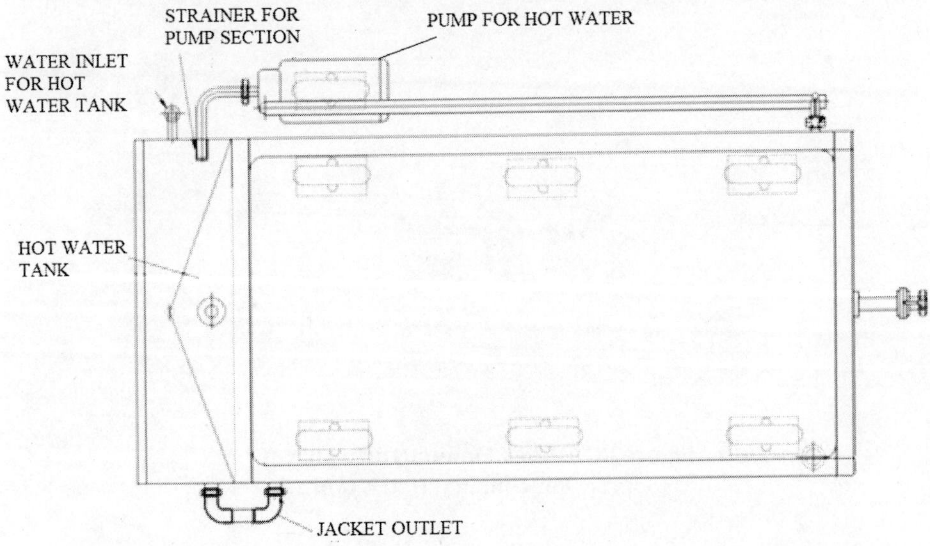

Figure 20.5. Butter Trolley with Hot-water System.

Butter unloading from Churn

The unloading of butter is carried out at low temperature. Depending on the immediate utilization or to store it for long period, the type of Butter trolley is chosen. The butter trolley may have arrangement for hot-water circulation to melt the butter and prepare it for pumping into Ghee kettle for ghee manufacture. The other option is to have Butter trolley without hot-water arrangement, and lead it directly to packing for long term storage.

Ghee making Equipment

In the dairy plant, the butter is usually stored in cold store maintained at -20°C. To manufacture Ghee from stored butter, it is first transferred to 4°C cold store so that energy required to bring it to higher temperature is saved to some extent. The other procedure is to prepare Ghee from fresh or sour cream stored in cream silos after its pasteurization. The steps involved in Ghee manufacture is shown in the flow chart below:

CST =Cream Storage Tank

CBM= Continuous Butter Making m/c

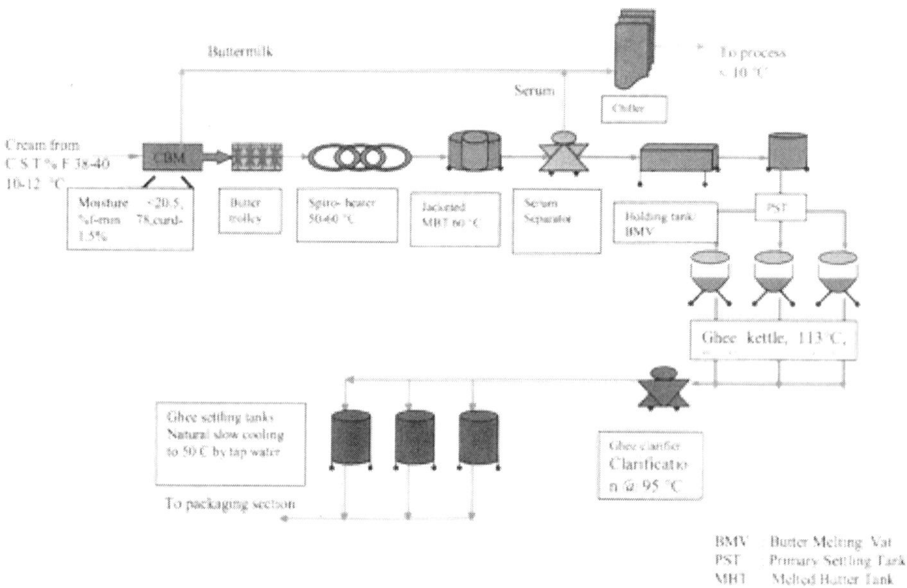

Figure 20.6. Ghee Manufacturing Layout (*Source*: Panchmahal Dairy, Godra).

Butter Melting Vat

The butter in shape of blocks is wrapped in butter paper and in cardboard boxes and kept in deep cold store. It is brought out, unpacked and placed on the support structure of the butter melting vat. The steam is supplied to a vessel, with temperature control. Butter blocks are loaded at a rate that melting takes place. Melting is carried slowly. The butter melting vat can also be used for pre-stratification process and draining of water, to reduce the heating load further. Some dairy plants use specialized separators for this purpose.

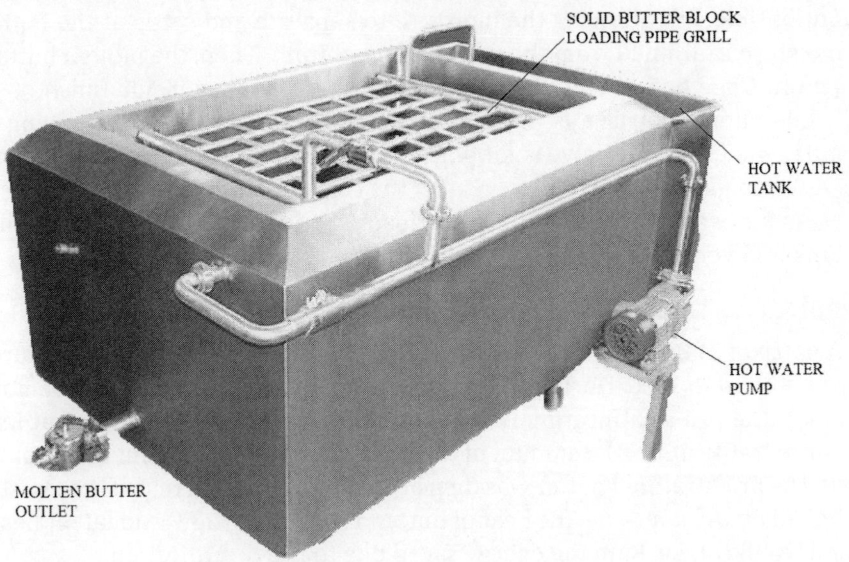

Figure 20.7. Butter Melting Vat with Hot Water generation compartment

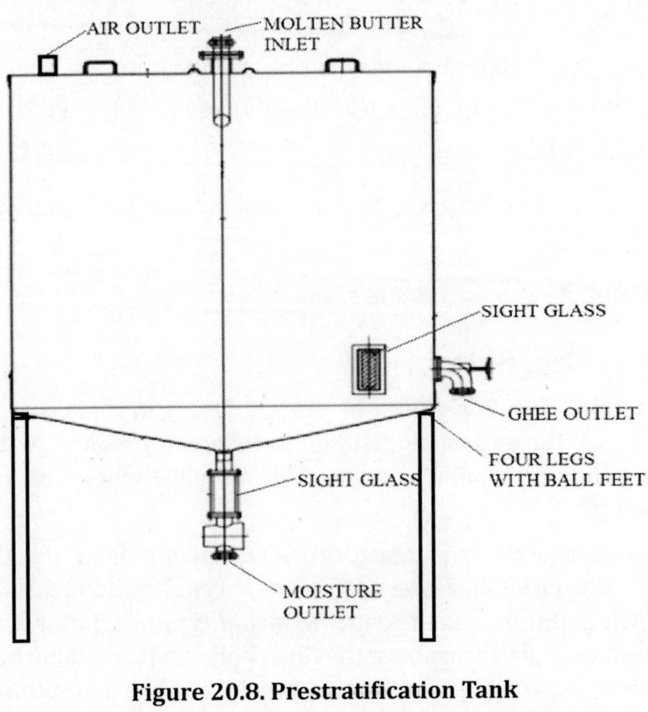

Figure 20.8. Prestratification Tank

The Prestratification reduces the heat energy required in Ghee manufacture from molten butter, by removing most of the moisture present in Butter. This is done by making the molten butter to stand at warm temperature for certain amount of time and allowing the moisture to separate and settle at the bottom. The moisture is drained from the outlet at the bottom. Then, the molten butter is pumped to Ghee boiler. The Prestratification tank shell has inside finish of 150 grit, while outside surface is mill finish. The bottom valve is for removing the moisture, while the side valve is for ghee. The molten butter inlet is non foaming type. A watch glass is provided to inspect the condition of the product inside the vessel. An air vent is provided for ease in loading and unloading of molten butter. The tank is covered with a lid.

Example

A batch of 2000 kg of butter having 16 per cent moisture at a temperature of 4°C was melted to 75°C. During pre-stratification, the quantity of serum removed is 180 kg. The pre-stratification was done in horizontal butter melting vat itself. After pre-stratification, the amount of moisture (theoretical) left out in butter was 125kg. The pre-stratified butter was then heated to 112 °C, and rest of the moisture was boiled off. Assume specific heat of butter as 0.65 kcal/kg°C and latent heat of water as 540kcal/kg. Find the energy saved due to pre-stratification.

a) Energy required for boiling off moisture without pre-stratification:
   = [2000 kg x (112-4) °C x 0.65 kCal/kg°C] + [2000kg x 0.16 kg of moisture/ kg of butter x 540 kCal/kg] = 313,200 kCal.

b) Energy required for boiling with pre-stratification and removal of serum:
   = [2000 x (75-4) x 0.65] + [(2000-180) x (112-75) x 0.65] + [125 x 540]
   = 203,571 kCal

c) The energy saved due to pre-stratification: (313,200 – 203,571) = 109,629 kCal

d) Amount of saturated steam saved = 109,629/540 = 203 kg of steam.

Ghee from Molten Butter

This method is commonly used in dairy industry. In this process the excess fat is stored in winter and is used either to reconstitute along with skim milk powder into recombined milk, or converted into ghee when demand increases due to festivals *etc.*

The equipment used for manufacture of ghee under industrial scale is the Ghee boiler. It is a double jacketed vessel with hemispherical bottom. Steam is supplied to the jacket. The agitation is kept gentle by anchor type of agitator which is driven by a motor, fixed on a platform above the Ghee boiler. A retractable hood is placed above the ghee boiler, in some cases as an improvement, to exhaust the vapours. The steam heating system is provided with usual fittings like, steam flow control valve, safety valve, pressure gauge (along with syphon), and condensate outlet through a steam trap, and also a bypass line with a valve for steam trap replacement. An outlet pipe with control valve is also provided, to the steam jacket, to expel air, at the time of starting the process. An inclined thermowell is provided into the inner vessel, to measure the temperature of ghee. The hemispherical bottom has an outlet with a valve. A pump connected to outlet is used for pumping out the ghee after the process is completed. A rotary pump is used for loading the ghee boiler with molten butter. Sometimes the vessel is used for melting the butter and for pre-stratification. An additional outlet is provided at appropriate level for draining the moisture separated during Prestratification. The hemispherical portion of the vessel is insulated outside with fiber glass wool (about 100 mm thick) to reduce heat loss. The ghee boiler is operated at 110 °C or more depending on flavour requirement of final product. The inner shell of SS is 6 mm thick, the MS jacket is 6 mm thick and outer SS shell is 2 mm thick. The steam jacket is tested by hydraulic test at 5 kg/cm ^{2} pressure. This is done before insulation is applied and outer shell is welded. The diagram of a typical ghee boiler is shown below:

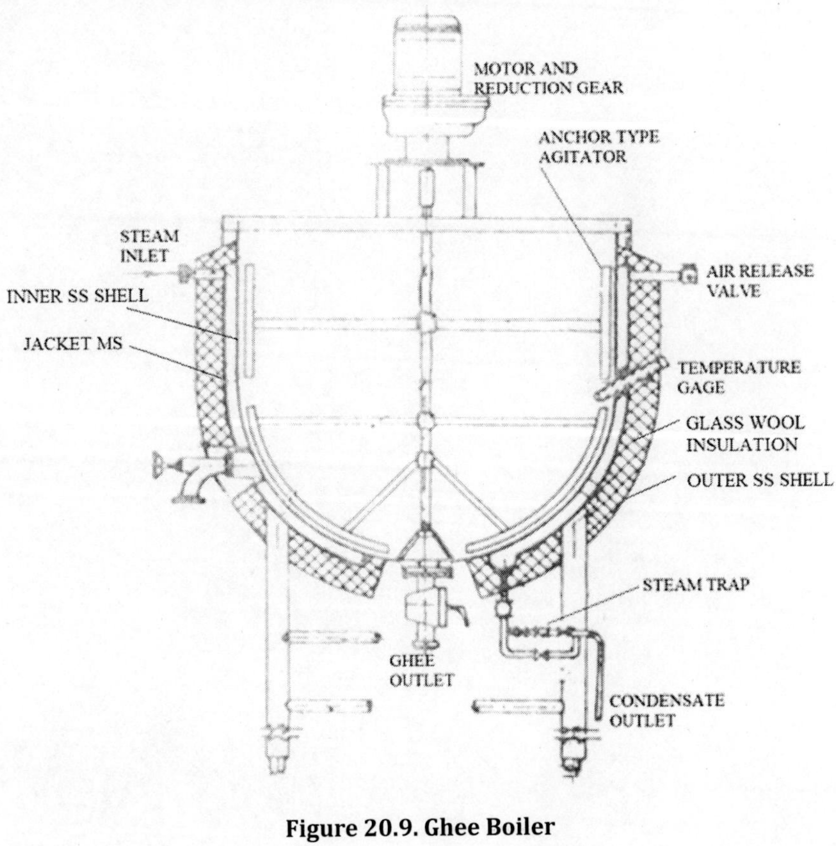

Figure 20.9. Ghee Boiler

Care and Maintenance

1. The loading of butter should be done only to an appropriate level.
2. High pressure steam should not be supplied to the ghee boiler. Also steam should not be supplied when butter is not in the boiler.
3. The condensate removal system should be inspected regularly.
4. The pressure gage should be tested regularly.
5. The safety valve should be tested for satisfactory operation periodically.
6. The agitator must be in proper contact with the inner surface profile of the boiler. If the Teflon edge should be replaced when worn out.
7. The agitator motor must be cleaned of accumulation of dirt
8. The ghee boiler must be cleaned after a day's operation

Chapter 21

Mechanization of Indian Dairy Products Manufacture

Introduction

The term Indian dairy products refer to those milk products which originated in undivided India, as defined by Prof. Sukumar De (1991). The Indian dairy products can be classified as shown in Table 21.1.

Table 21.1. Classification of Indian Dairy Products

Principle of Manufacture	*Products*
Heat desiccation	*Khoa, Rabri Basundi*
Heat- and acid coagulation	*Chhana, Paneer*
Fermentation	*Dahi, Misti Dahi, Chakka, Shrikhand*
Fat concentration	*Makkhan, Ghee*
Frozen	*Kulfi*
Addition of cereals and desiccation	*Kheer, Payasam*
Khoa-based milk confections	*Burfi, Peda, Gulabjamun, Kalakand, Khunda, Milk cake etc.*
Chhana-based milk confections	*Rasogolla, Sandesh, Rasomalai, Chhana Podo, Chhana Murki, etc.*

The manufacture of dairy products in India was initially done on a small scale and for local consumption. There was no attempt for a long time to manufacture the products on a large scale and under hygienic condition for extended shelf life. Due to urbanization and public awareness to hygiene, it necessitated the development of equipment for large scale production, packing and distribution.

Figure 21.1. Kharahi

The traditional methods of khoa manufacture employed Kharahi, which was a shallow, round bottomed pan with handles. As the manufacture of product required continuous stirring and scrapping, it was open to atmosphere making the process unhygienic. Also, the quality of product varied from batch to batch. This not only limited the scale of operation, but also the standardization of manufacture.

Manufacturers and research institutions attempted to solve the problems associated with hygienic manufacture of products continuously on larger scale in enclosed equipment. Many types of equipment of varying capacities and for application of different energy sources have been developed.

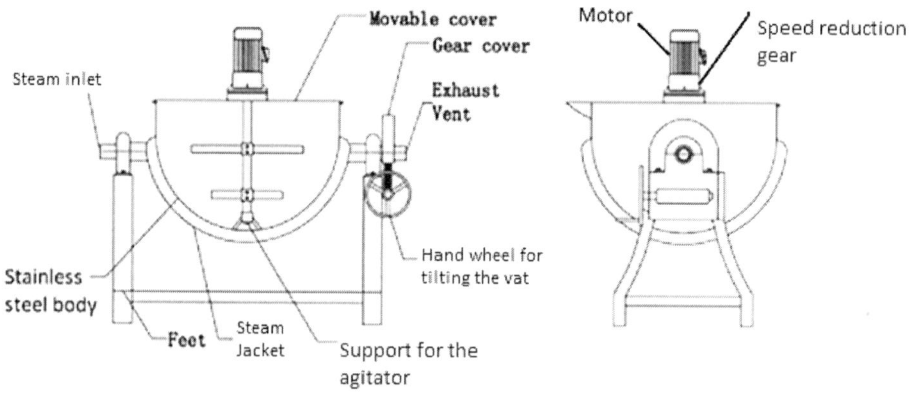

Figure 21.2 Tiltable Khoa Pan, Steam Heated

The improvements in the batch and small-scale equipment can be categorized on the basis of 1) Mechanization of agitation in open pan 2) Heat source that can be better controlled 3) Improvement in the material of construction 4) Method of product loading and unloading.

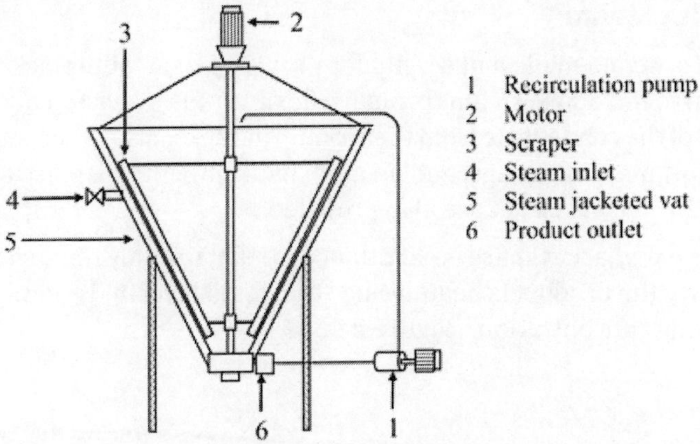

Figure 21.3. Conical Vat with Partitioned Jacket

Mechanization of Agitation

This required a slow speed agitator suitable to the rate of heat transfer, viscosity of the product which increases with progress of process and to match the profile of the vessel surface. The agitator most suitable is anchor type, that has edge matching the profile of the vessel to scrape surface being heated. The scrapping edge was further improved by providing adjustable and replaceable Teflon material (*Polytetrafluoroethylene, PTFE*). It does not damage the material of the vessel, being softer and also it is a reasonably good conductor of heat, for conducting away the heat of friction from edge of the agitator. Certain amount of flexibility is provided by keeping the agitator blade spring loaded (Figure 21.4).

The slow speed requirement necessitated a motor with reduction gear mechanism for the agitator. A support system which is aligned with the agitator and concentric to the vessel has to be provided. The supports should be stable enough in lateral movement to handle most of the highly viscous products. A coupling is required between the shafts of the geared motor assembly stainless steel agitator.

Heat Source

The manufacture of Indian dairy products, involved heating the Karahi with wood or coal. The control of the heat source is not efficient and waste of energy is considerable (according to one estimate the loss is 64 per cent). Hence, use of gas or steam is adopted. They could be controlled better. This became possible with advent of easy availability of gas cylinders for commercial use in rural areas, as well as small scale steam generators that were not controlled by Indian Boiler Regulations (IBR) (IBR boiler has at least 22.75 liters water holding and develops pressure of at least 3.5 kg/cm^2). As mentioned earlier, the production was usually small scale and in widely distributed rural areas the above adoption suited well.

Heating with Steam

The use of steam made a necessity for provision of an outer jacket to the pan, a mechanism to introduce steam through a flexible pipe provision of control valve and removal of the condensate from the jacket. The jacket also is insulated to reduce the heat loss. A pressure gauge, and an air exhaust pipe (to displace air and allow entry of steam) for the jacket are also provided.

In larger pans, jacket space is partitioned, so that the only the surface of the vat in contact with the product is heated since the liquid level in the vessel is reduced with increasing concentration. (Figure 21.3).

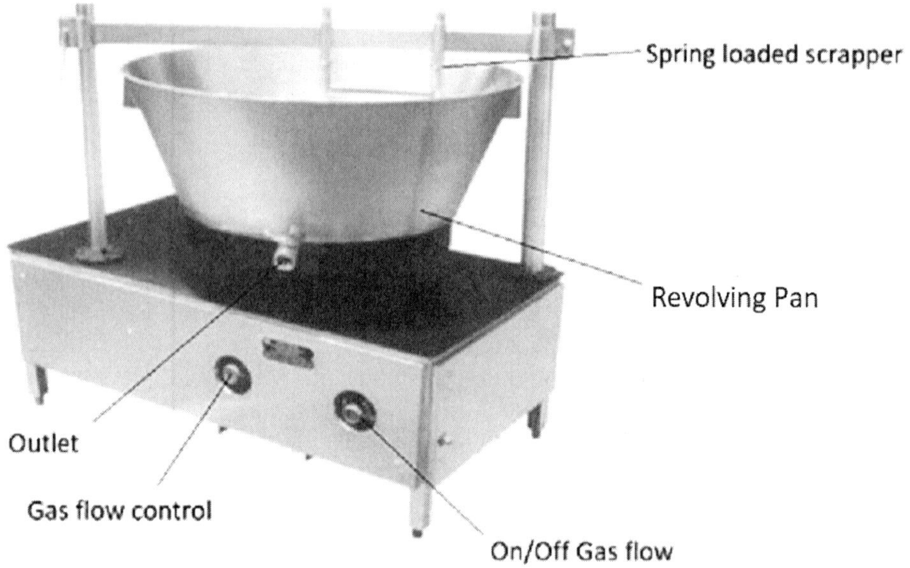

Figure 21.4. Gas Operated Khoa making Equipment

Heating with Gas

The use of gas burner is another method of controlled heating during concentration of the product. But this method required an innovation, similar to the wet grinder system, where the vessel is rotated and the agitator is kept stationary. The vessel is rotated by a motor and belt mechanism fitted to the bottom.

Method of Loading and Unloading

The loading is done manually transferring milk in to the pan. For unloading, for steam heated vessel, a tilting mechanism is provided. For the gas heated vessel an outlet is provided for drawing the product out, while the product is hot and less viscous.

The other features are also provided such as adjustable hood over the pan exhausting the vapours. This will prevent vapors condensing on the walls around and making the surrounding unhygienic.

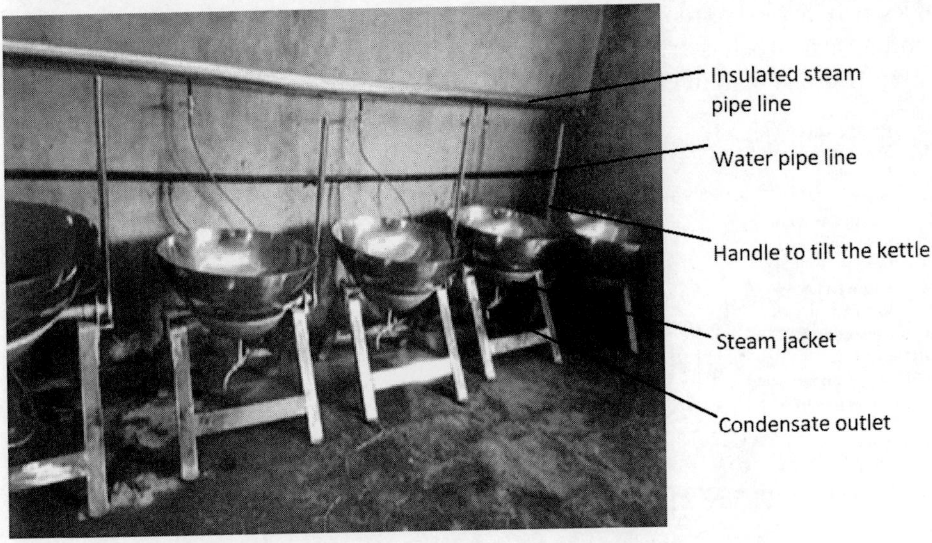

Figure 21.5. Battery of Khoa Pans for Higher Production

For manufacture of large quantities, two or more such pans are arranged in a row, so that steam connections and condensate drain are connected with minimum pipe lengths and supervision. Modeling of the heat transfer regime for steam heated kettle as well as using flue gases as heat source has been done by Das, H. (2005)

Equipment for continuous manufacture of Khoa

In comparison to kettle or karahi, the heat loss is much less for Scraped Surface Heat Exchanger (SSHE). It can also handle viscous and semisolid products. Most of the Indian dairy products come under this category. These units have been used in the food and chemical industries and are available in vertical and horizontal configurations.

Among the various types of designs available for continuous manufacture of Indian dairy products, following two designs have been established for commercial use.

1. Inclined Scraped Surface Heat Exchanger (ISSHE)
2. Thin film Scraped Surface Heat Exchanger (TSSHE)

ISSHE unit (developed by the National Dairy Development Board, Anand) comprises of a scraped surface heat exchanger in inclined position (about 30°). The concentrated milk (about 40 to 45 per cent TS) is pumped from feed tank to ISSHE at a flow rate that can be controlled. The heat exchanger comprises of an inner cylinder, rotor and drive, and outer steam jacket. The inner cylinder is S.S. and forms the main body. It is provided with steam jacket and split into three separate compartments. The steam jacket is insulated and cladded. The design of rotor is critical to the effective operation as it has to scrap as well as move the product forward. It is driven by a variable speed drive. The constructional features of ISSHE are shown in Figure 21.6.

A. Feed balance tank
B. Feed pump
C. Feed inlet
D. Drain
E. Scraper assembly
F. Steam inlet
G. Condensate outlet
H. Steam jacket
J. Vapour outlet
K. Khoa outlet
DR = Drive
ISSHE = Inclined scraped surface heat exchanger
V = Valve

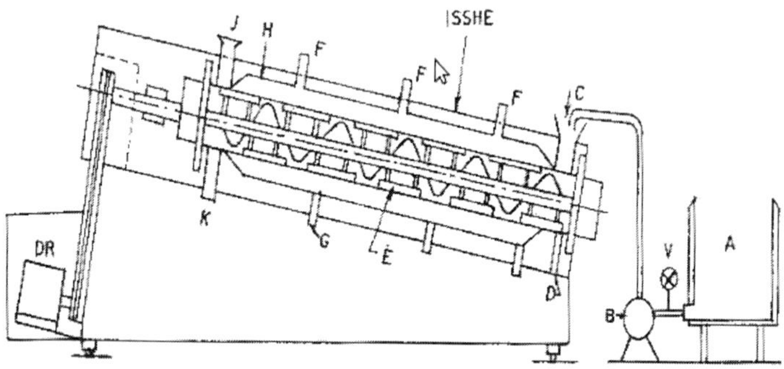

Figure 21.6. Inclined SSHE for Khoa making machine

The inclination of ISSHE permits formation of a pool of boiling milk, critical for development of desired heated or slightly cooked flavor in khoa. Subsequently, fresh concentrated feed enters the pool of boiling concentrates, while an equivalent mass continuously leaves the pool as semi-solid (khoa). The inclination of the equipment enhances the heat coagulation of proteins in a unique manner simulating desired flavour and texture in khoa similar to that made in a karahi. The unit was designed to produce about 40 kg khoa per hour.

Thin film SSHE consists of a jacketed cylinder. Inside the cylinder a rotor is provided to which blades are attached. The blades move over the surface maintaining a very close clearance. The clearance of blades would be such that centrifugal force of the blade is balanced by hydrodynamic force developed in liquid film. The blades besides forming a thin film of a product on the surface, also agitate it violently thereby improving thermal performance of the unit. Heating medium, normally steam condenses in the jacket. If the SSHE is operating as an evaporator, vapour and concentrated product are led to a liquid-vapor separator. Many variations are possible in operation in the SSHEs (A.K. Dodeja *et al.,* 1989; Abichandani and Sarma, 1989).

The design developed by NDRI has two horizontal TSSHEs (Figure 21.7).

The constructional features of this system are shown in Figure below. In this unit the two SSHEs are arranged in a cascade fashion. The rotor of the first SSHE is provided with four variable clearance blades and operated at 200 rpm. Standardized buffalo milk is concentrated in this SSHE in the range of 40-45 per cent T.S. This concentrated milk then enters the second SSHE, which has a different kind of rotor arrangement. It has two variable clearance blades and two helical blades, which operate at a speed of 150 rpm. Both the SSHEs are double jacketed and connected to steam supply and provided with pressure gauge and vapour outlets. Milk is supplied to first SSHE with a centrifugal pump, whereas, the concentrated milk from first SSHE to second SSHE moves by gravity. The unit can produce about 50 kg of khoa per hour.

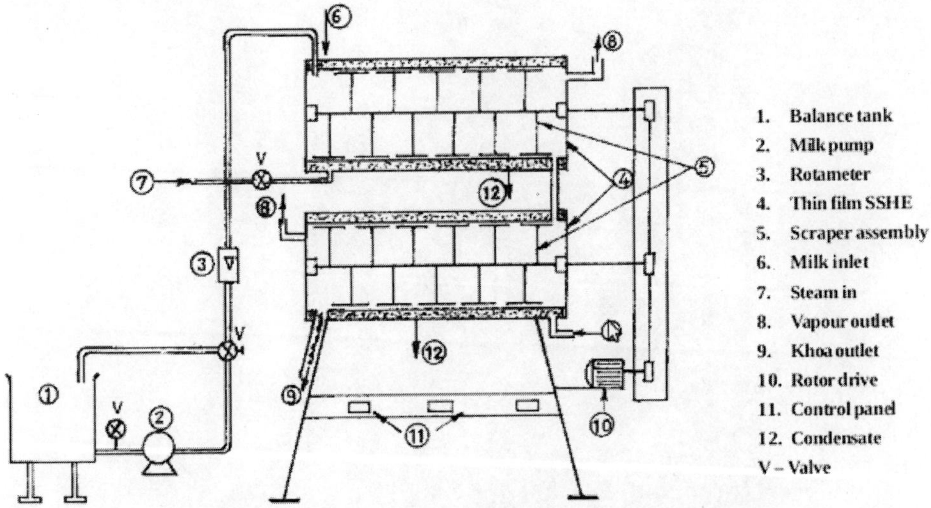

Figure 21.7. Twin Cylinder SSHE Khoa making machine

The equipment needs considerable power to run components like agitators, vacuum or condensation systems *etc.* Researchers have developed equations for predicting the power requirements, based on factors like the percentage of Total solids, the speed of rotation and the number of blades, the overall weight of the agitators and scrappers.

The earlier development of this machine was in two stages where standardized milk was concentrated to 30-35 percent in first stage and remaining concentration was done in second stage. The speed of rotor was slightly higher in first stage compared to that of second stage. Later study revealed that when concentration of milk was increased to 60 percent, Khoa making can be better controlled to simulate the Khoa made in Kharahi. Therefore, a third stage was added in which concentration occurs at Lower rotor speed (Fig 21.8).

Other Indigenous Dairy Products Manufacturing

The basic design of SSHE has also been used for converting cream to ghee and for melting butter. As these products are viscous the SSHE with suitable modifications can be used for continuous manufacture. Moisture is an essential parameter for control in these operations.

A detailed account on the development of equipment for khoa and other indigenous dairy products manufacture can be seen in the review article by Mahesh Kumar (2015); Heartwin *et al.* (2020).

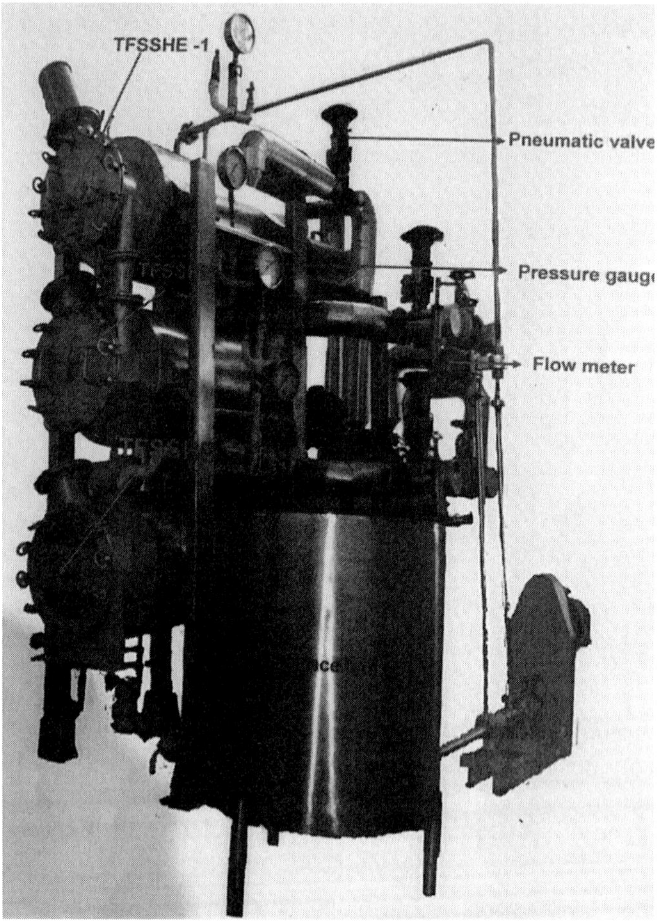

Figure 21.8. Three Stage Khoa making Machine

Use of Extrusion Equipment

Many of the Indigenous dairy products like Sandesh are granular, and need extrusion either during process or in forming the final size and shape of the finished product. The Vented Extruder was found to be capable of functioning as evaporator and to carry many unit operations such as mixing, kneading, concentration, and shaping. Vented extruders are significantly different from non-vented extruders in design and in functional capabilities. This type of extruder is equipped with one or more openings called vent ports in the extruder barrel, through which volatiles can escape. Thus, vented extruder can extract volatiles

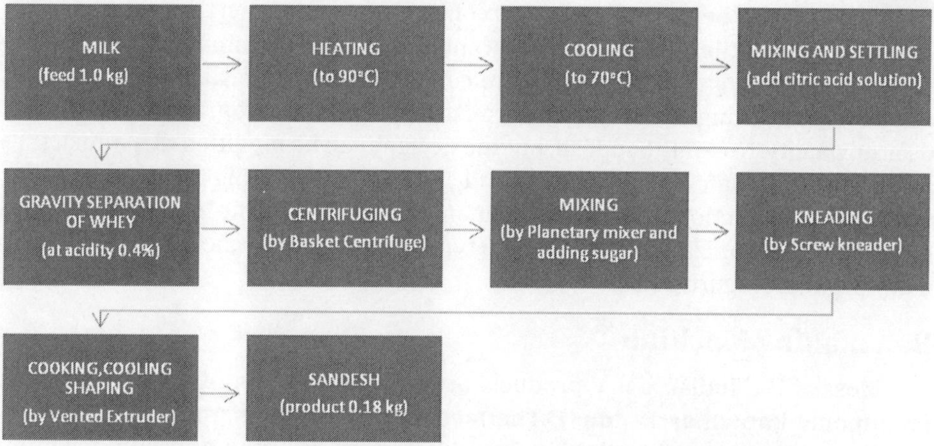

Figure 21.9. Process Steps in Continuous Sandesh making

Figure 21.10. Ball making Machine (*Courtesy*: CDT, SVVU, Tirupati).

from the material being processed in a continuous fashion. Further, instead of the extraction of volatiles, one can use the vent ports to add certain ingredients to the material being processed. Studies were conducted to design single screw vented extruder for cooking of channa-sugar mixture to get the soft grade Sandesh with desired quality. Not only Sandesh, but the designs can be modified to manufacture similar, intermediate moisture foods and thereby mechanizing the process for its continuous production (Rajendra Kumar and H. Das, 2007). The following steps are involved in mechanization of Sandesh, a typical intermediate moisture indigenous dairy product (Figure 21.9).

Ball making Machine

Most of the Indian dairy products are sold in the form of a spherical form (commonly known as Laddus). The mechanization of this process has been made possible by equipment that is mechanically operated and untouched by hand (Figure 21.10). The option to choose size and shape (mostly spherical and cylindrical) is also provided.

The final product can be varied for different consistencies by adjusting the speed, force of pressing (temperature of environment).

REFERENCES

Abhichandani, H and Sarma S.C., "Power requirement during concentration of milk and cream in horizontal thin-film scraped surface heat exchanger", Ind.J. Dairy Sci 42,2,1989)

Das, H. "Food Processing operations analysis" (2005), Asian Book Pvt Ltd, New Delhi, India

Dodeja, A.K., Sarma, S.C. and Abhichandani, H. " Heat transfer during evaporation of milk to high solids in thin film scraped surface heat exchanger, J. of Food Proc. Engg. Vol. 12 (1990) 211-225.

Heartwin A *et al.* "Recent trends in small-scale mechanization of Indian dairy products ", Conference Proc. Pp 69-77, Dairy Industry Conference, Jaipur, (2020)

Kumar, Mahesh "Up-gradation of Khoa production and preservation technologies", SAMRIDDHI: A J. of Physical Sciences, Engg. and Technology, 2013, 4 (1)

Rajendra Kumar R. and Das H. 2007. Performance evaluation of single screw vented extruder for production of Sandesh J. Food Sci. Tech. 44: 100-105

Rajendra Kumar R. and Das H. 2007. Modelling of a single screw vented extruder for concentration cooking of viscous food material. J. Food Sci. Tech. 44: 94-99

Chapter 22

Ice Cream Making Equipment

Ice cream is one of the popular dairy products. It is very profitable for the manufacturer. It is made on a small-scale batch process or on industrial scale by continuous ice cream making machines. There are also varieties of ice creams, with varying processes, ingredient combinations and display for marketing.

The functions of the freezer are:

(a) To freeze a portion (usually 50 per cent) of the water of the mix to get a smooth product,

(b) To incorporate predetermined amount of air uniformly into the mix to obtain proper overrun.

Fast freezing is essential for obtaining a smooth product because ice crystals that are formed rapidly are smaller in size than those formed slowly. Therefore, it is desirable to freeze and draw the product from the freezer as quickly as possible.

Major equipment required for Ice-cream making are:

1. Multipurpose vat for making ice cream mix
2. Homogenizer to homogenize the mix
3. Refrigerated Cooler for bringing the ice cream mix to lower temperature
4. Ice-cream freezer
5. Hardening chamber
6. Deep freezer to store the ice cream
7. Miscellaneous equipment like molds, packaging material, dry store for storing mix ingredients like sugar, stabilizer, emulsifier, flavours *etc.*

The refrigeration system size depends on 1) The freezing point 2) Drawing temperature 3) Load of ice cream 4) Freezing rate.

Freezing Point

The various components in ice-cream mix, especially those going into the solution, decrease the Freezing point. The major role is played by added sucrose and the lactose to the mix. It also has to be noted, that in the Ice cream freezer, about 50 per cent of the moisture in the ice-cream mix only is in frozen condition, with rest of the moisture dissolving the various components of solute. The following methods is used for estimating the freezing point depression, with certain assumptions:

Freezing point depression,

$$\Delta T = [R_g (T_{AO})^2 W_A m]/1000 L \qquad \text{(Eq. 22.1)}$$

where,

Rg = Universal gas constant, (J/mole. K)

T_{AO} = Freezing point of pure water (K)

W_A = Molecular weight of water

m= mol fraction (mol/kg of solvent)

L = Latent heat of fusion of water (kJ/kg)

Example 1

Compute the initial freezing point of Ice Cream mix with the following composition: Fat 12 per cent; SNF 11 per cent; Sugar 15 per cent Stabilizer and Emulsifier 0.3 per cent. Assume that sugars predominately influence the Freezing point and that the mix is sufficiently dilute to use the above equation.

Solution

a) Total solids: 12 + 11 + 15 + 0.3 = 38.3 per cent; and Moisture = 100 – 38.3 = 61.7 per cent

b) On an average Lactose in SNF is 54.5 per cent

Hence, the total sugars + 0.15 + 0.11 (0.545) = 0.2099 g/g of mix

c) Sugars as expressed in water fraction: 0.2099/0.617 = 0.3403 g/g of water

= 340.3 g/1000g of water

d) Mole fraction, m = Molality/Molecular weight =340.3/342 = 0.995

e) Freezing point depression = ΔT= (0.462) $(273)^2$ (18) (0.995)/1000 (333.22)

= 1.85 K

Hence, the freezing point of the mix is (-) 1.85 °C

More accurate results are obtained if the influence of salts is also considered.

For freezing of Ice cream mix, the smallest and simplest of the equipment is Bucket type ice cream maker (Figure 22.1) which can be used at homes. It is basically wooden vessel in the shape of a bucket with a metal cylinder inside. It can be rotated by an arrangement that is supported at bottom and rotating provision at the top. The top dismountable lid is provided with a bevel gear arrangement and support system with a handle. When the handle is rotated, the whole cylinder enclosed with ice cream mix rotates. The central opening at the top lid has a provision to hold a dasher, which is fixed. As the cylinder rotates, the ice cream mix is agitated and inner surface of the cylinder is scraped, the mix is cooled by a mixture of ice and salt. Salt depresses the melting point of the ice thus enabling the ice cream to be cooled to a temperature below 0°C. The Figure below shows the ice cream freezer:

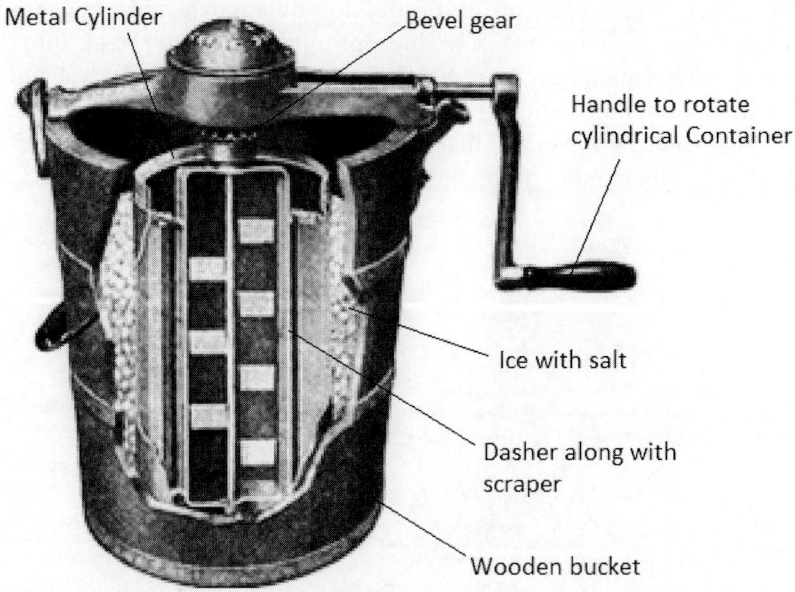

Figure 22.1. Bucket Type Ice Cream Freezer

The larger batch type (about 40 to 100 litres volume) ice cream freezer (Figure 22. 2) has a refrigeration system, with control and mechanized operation of Dasher and scrapper assembly. It basically has a stainless steel or nickel coated Brass cylinder. It is fixed in a horizontal position with a provision for power transmission to Dasher scrapper assembly inside the cylinder from the back end while a bush supports the assembly at inside of the front door closing on to the cylinder. There is an opening at front top of the cylinder to feed the ice cream mix. At bottom of the front an outlet is provided to withdraw the semi frozen ice cream. The cylinder has a slope slightly towards the exit. The cylinder is provided with a jacket inside which the evaporator coils of the refrigeration system are located. Over the jacket, insulation with cladding is provided.

The Dasher having Beater -scrapper assembly is driven by a drive arrangement in such way that the Beater and Scrapper rotate in opposite directions for increased heat transfer and incorporation of air into the ice cream mix. This is achieved by having two co-centric shafts of Beater and scrapper enmeshed in two gears, one driven by motor through belt and pulley, and the other by the driven shaft. For the purpose of incorporation of air, the cylinder is filled to only half the volume. The rotational speeds are in the range 140 to 250 rpm. The scrapper has blunt edge, with toe being 1/3 of the total thickness of scrapper edge, so that the inside of the cylinder does not get grooved over a time.

Controls switch has three positions: 1) Switching Off 2) For switching on Dasher and Refrigeration system 3) For switching on only Dasher

For higher degree of incorporation of air some batch freezers are provided with an air pump with inlet for front door for injection of filtered air into the mix. The air pressure is maintained between 1.5 to 3 kg/cm^2, when the inlet and outlets are closed. An ammeter indicates the current drawn by motor driving the dasher. The current drawn by the motor increases with time as freezing continues. This is due to increase in viscosity of the mix.

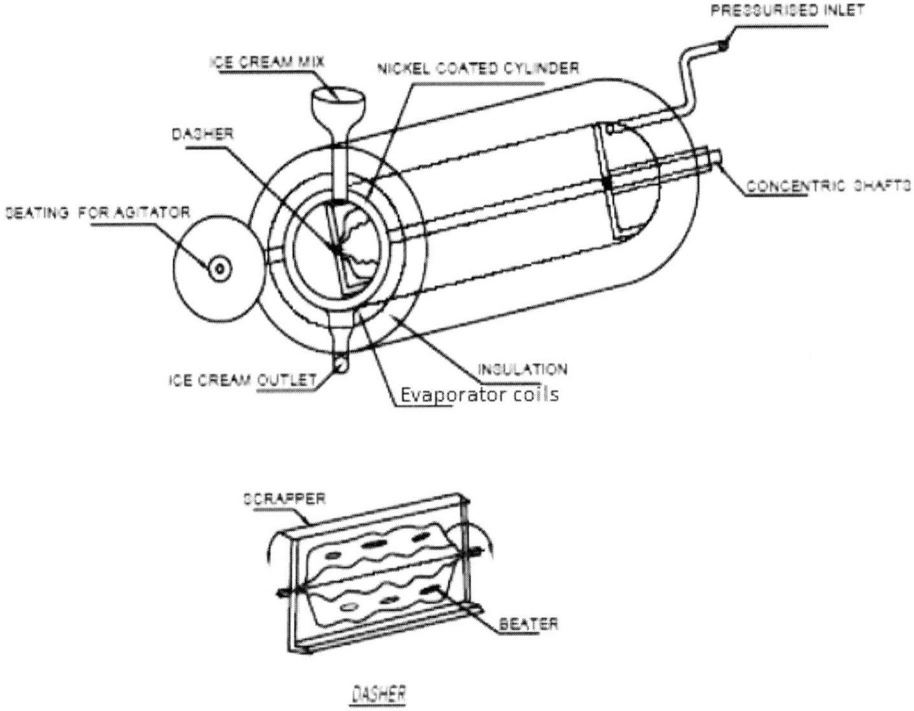

Figure 22.2. Batch Ice Cream Freezer

Example 2

Compute the heat to be extracted from 10 kg of mix of composition Fat 12 per cent; SNF 11 per cent : Sugar 16 per cent. The mix is supplied to freezer at 5.0°C and is frozen to a temperature of -6°C. Consider Specific heat of mix as 0.8 kcal/kg °C and the slush as 0.65 kcal/kg °C. Consider the initial freezing point to be at – 2.7 °C, At the point of drawl, 50 per cent of moisture is frozen.

Solution

a) Total solids: 12+11+16 = 39 per cent; Total moisture = 100 - 39 = 61 per cent

b) Total heat to be removed = (Sensible heat from 5°C to -2.7°C + latent heat at -2.7°C + Sensible heat from -2.7°C to -6°C)

c) = [{10 x 0.8 x (5 – (-2.7)} + {10 x (61 x 0.5) x 80} + {10 x 0.65 x (-2.7 +6)}]
= 327.05 kCal

Continuous Ice Cream Freezer

The continuous ice cream freezer is designed to cool the ice cream mix rapidly and continuously. The continuous operation is made possible by having 1)Narrow and long freezer 2)Continuous supply of mix by a rotary pump 3) Continuous supply of filtered air by a positive displacement pump 4) Refrigeration system for rapid cooling 5) Controls for a smooth operation of the freezer continuously (Figure 22.3).

Construction

The volume of the freezing cylinder is smaller, by having comparatively a solid dasher and by making the mix to flow as a thin layer, compared to one used in batch process. The Dasher rotates at 150 to 200 rpm. It is powered by electronic speed controlled motor or hydraulic pump systems. The scraper blades are made of spring steel. The blades scrape the ice crystals formed continuously. The beater is eccentrically placed, rotated by reaction to Ice cream (Figure 22.4 a). The freezing chamber is supplied with mix from tank by two pumps (Figure 22.4 b). In some designs the first pump is a metering pump, for feeding the mix at a controlled rate. The outlet of this pump leads to a second pump through a pipe, which has provision to allow controlled amount of air. The second pump being larger in capacity (approximately three times) transfers the mixture of air and ice cream mix to the freezing chamber at a pressure of 5 to 6 kg/cm^2. Air is supplied at pressure higher than that in the cylinder. As, the air is compressed it occupies only 1/6th of its original volume and an overrun up to 130 per cent can be achieved with special designs of dasher. The time for freezing is approximately 30 seconds. The process effectively controls Ice crystals within the range of 45 to 55 microns and air voids within the range of 100 to 150 microns. The refrigerant in the jacket evaporates at a temperature in the range of – 20°C to – 30 °C.

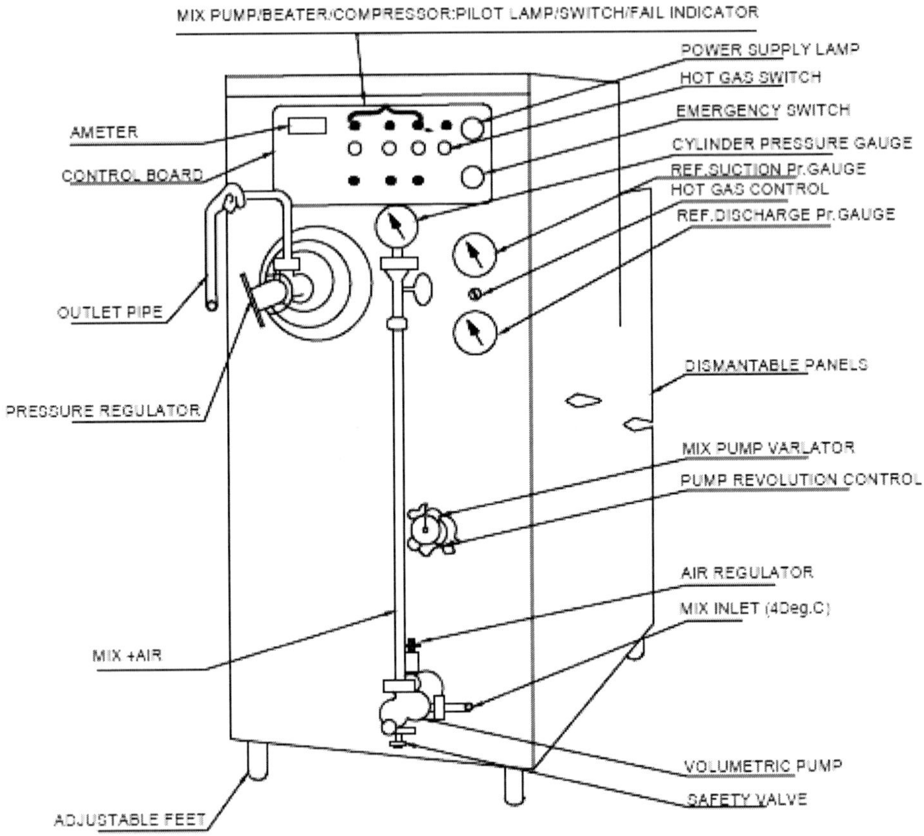

Figure 22.3. Continuous Ice Cream Freezer

In some designs with single large size pump air enters on suction side with the mix. Both horizontal and vertical models are available. Capacities range from 100 to 900 liters per hour.

Advantages of Continuous Ice Cream making Machines

1. Large production capacity
2. Uniform quality
3. More efficient controls
4. Rapid freezing, as it is done under pressure and smaller diameter chamber
5. Labor saving due to automation
6. Increased number of smaller size ice crystals, hence lower is the viscosity and stabilizer requirements.
7. Aging time is reduced because of lower viscosity.

8. Smaller crystals, make melting in mouth faster hence the flavors are pronounced even with a smaller quantity.

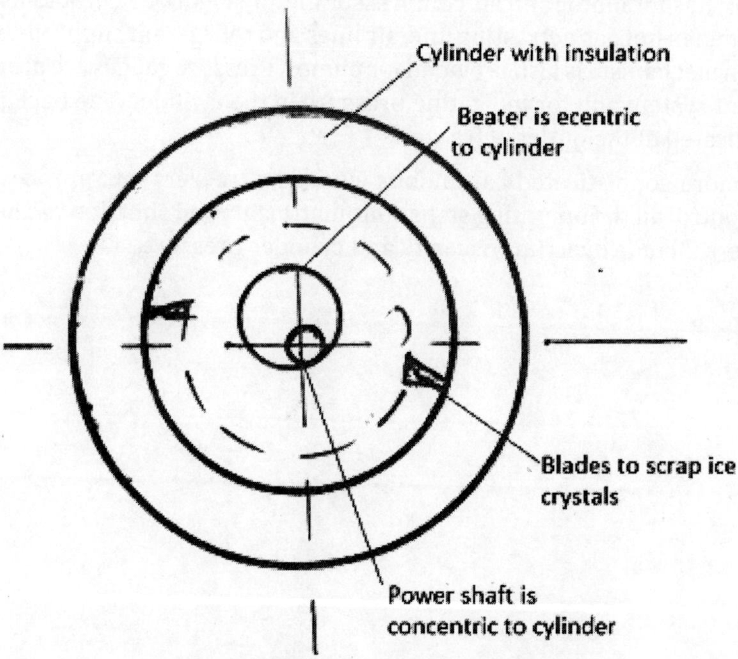

(a) Arrangement of beater and scrapper.

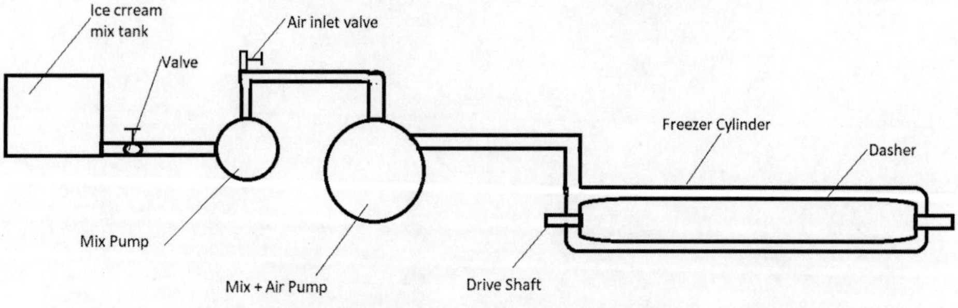

(b) Arrangement of mix and air pumps

Figure 22.4. Continuous Ice Cream Freezer with Mix and Air Pump

Disadvantages

1. High initial cost
2. Economic only at larger capacities

Controls and Automation

As shown in the figure, ON/OFF switches are provided along with pilot lamps for pumps, dasher motor and air compressor motor (Figure 23.3). Solenoid valves are provided in hot gas defrosting line, air lines and refrigerant supply lines. Speed regulation mechanism is also provided for pumps. Pressure gauges are provided for refrigerant system and for indicating pressure in the cylinder. The back pressure valve is located at the outlet of freezer.

For more sophisticated continuous ice cream freezers microprocessors are used to control all the operating steps, automatic start and shut down. The control parameters include overrun, viscosity and cylinder pressure.

Chapter 23

Cheese making Equipment

Cheese is product mostly produced overseas, but finding increased popularity in India. A comparable Indian product to cheese is Paneer, though it is not a fermented product. Paneer is closer to cottage cheese. There are number of varieties of cheese, but the basic raw material is milk coagulum, obtained by coagulating milk with enzyme, called rennin.

Sequence of Operations

The sequence of operations in making cheese making is shown in Figure 23.1.

Brief Explanation of Steps

The process of cheese manufacture has steps that may vary from individual types of cheeses, but in general, the following steps are involved. (For more detailed explanation of steps may be referred from texts of Dairy Technology.)

1. Standardization of milk so that the Fat: SNF ratio is adjusted to give maximum yield and desired texture.
2. Pasteurizer is to eliminate any undesired bacteria and to control bacterial population.
3. Souring is the process of development of acidity due to addition of starter culture.
4. The milk clots due to the development of acidity and enzyme, which is controlled.
5. Cutting of the curd is to hasten the uniform treatment of curd particles.
6. Scalding is the process of heating the curd particles in presence of whey.
7. The process of removal of whey and consolidate the curd particles

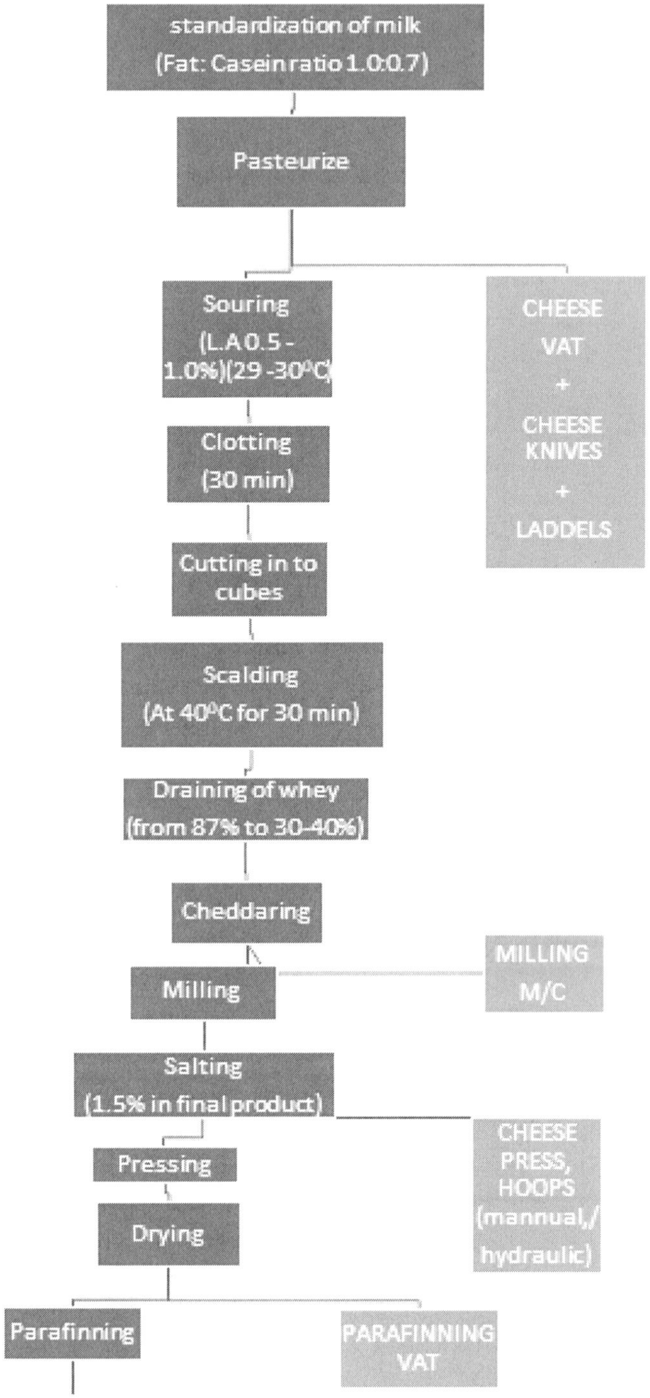

(L.A. = Lactic Acid; Clotting is by addition of Enzyme)

Figure 23.1 Sequence of operations of Cheese making

8. Cheddaring is the process of piling and sequential handling of curd particles to obtain desired curd properties.
9. Milling is the process of cutting the large curd particles into uniform sized cubes to apply salt.
10. Salting of the curd, in some of the cheeses to bring about the taste and further aid in ripening.
11. Loading of curd blocks in hoops and pressing cheese hoops is to expel further the whey.
12. Surface drying to make it suitable for applying paraffin.
13. Paraffining to keep away the curd particles from contamination and aid in fermentation in unhindered state during cold storage.

The basic design of a cheese vat is shown in Figure 23.2 (a). It is an open rectangular vat, curved at the ends, with provisions for agitating, heating, and draining of the whey. It may or may not be insulated. It is of low height to facilitate various operations on the cheese curd obtained after coagulation. For larger capacities the vat, may have mechanized agitators and curd cutting devices for reducing labour. To improve hygiene further, horizontal vats are completely enclosed, with provisions for agitation and curd cutting.

The cheese curd after milling and salting, is loaded into square or round hoops (Figure 23.3), and pressed at a fixed pressure and period to further expel the whey. The hoops are designed in such a way, that the top and bottom plates (which have holes) collapse in a fixed side rim plate, as the draining of whey progress over time. Usually, the draining period is overnight in smaller capacity hoops, where the pressure is by spring. The applied pressure varies as the whey is drained from the hoops. In larger capacities, the pressure is maintained by air operated piston, at a constant pressure.

Mechanized Cheese making Equipment

In meeting the demands of increased cheese consumption, the manual and batch wise operations were found to be tedious and laborious. Various basic steps of cheese making were initially mechanized. Such approach was not feasible for different types of cheeses as their manufacture requires verity of intermediate steps. The operations of agitation and cutting of cheese blocks were first adopted for mechanized operations.

The mechanized cheese vat has a horizontal gear all along the length of the vat (Fig 23.4). This shaft is positioned above the vat. Two vertical shafts are connected through gears and drive mechanism to the horizontal shaft. The vertical shafts rotate at slow speed. The vertical shafts operating at slow speed can move along the overhead horizontal gear. The shafts when fitted with agitators will agitate the milk or cut the curd when fitted with horizontal or vertical curd cutting devices. These curd cutting devices are basically rectangular frames with a connecting shaft

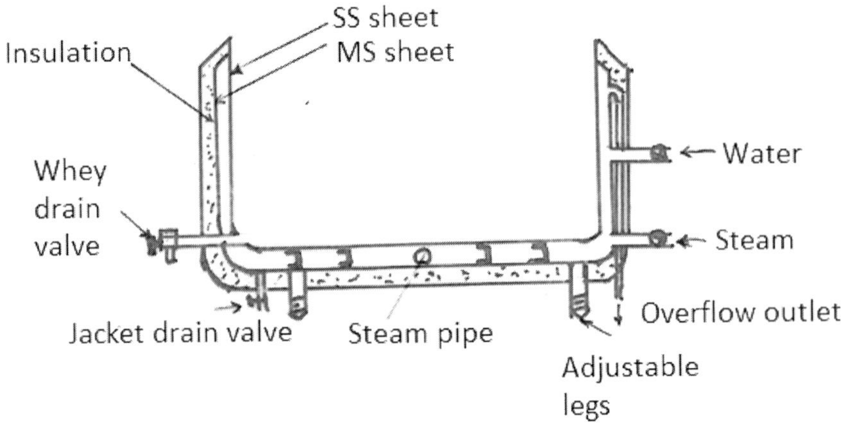

(a) Cheese Vat cross sectional view

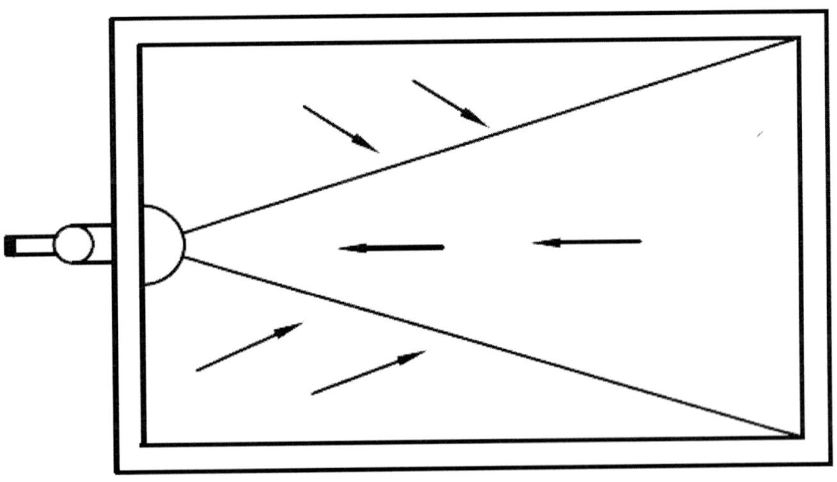

(b) The cheese vat bottom profile showing slope (top view)

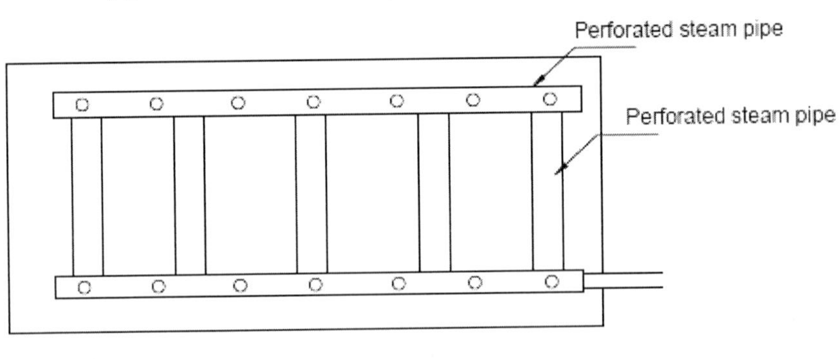

(c) Steam pipe distribution lines at bottom of the cheese vat

Figure 23.2. Cheese Vat and its Details

(a) Cheese Hoops (b) Pneumatic Cheese Press

Figure 23.3. Cheese Hoops and its Pressing System

provided at the top. Stainless-steel wires are fixed to the frame both horizontally and vertically. The constructional details of mechanized cheese vat are shown in Figure 23.4.

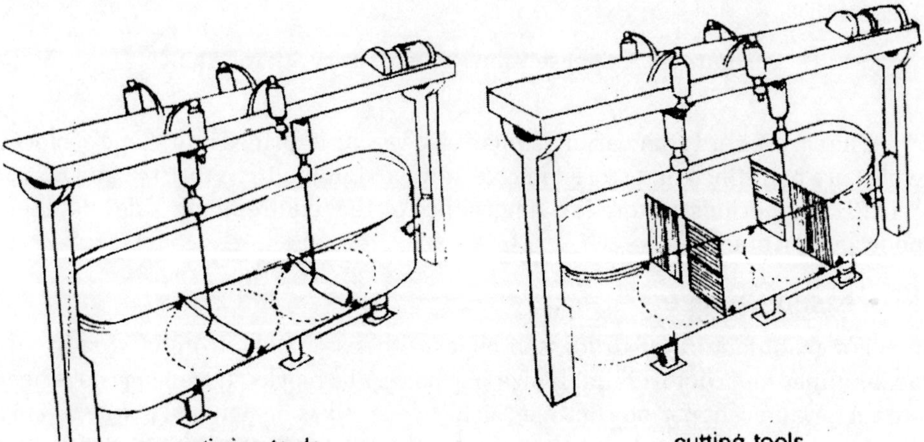

(Courtesy of Alfa - Laval, Sweden)

Figure 23.4. Mechanized Vat for Cheese making

An improved design of cheese vat is Oscillating stirred tank cheese vat in which the basic structure is a horizontal enclosed cylindrical tank, with a large shaft fitted concentrically to the horizontal axis of the tank. It is powered by motor with arrangement to oscillate in a certain arc. This shaft is fitted with agitator and blades in the upper and lower segments so that when the shaft oscillates in an arc, it either agitates or cuts the curd, depending on radial segment of the shaft. The tank has jacket for heating by hot water or steam. Other fittings are man hole door, ball feet. The design of this vat is shown in Figure below (Pasilac Horizontal cheese making tank):

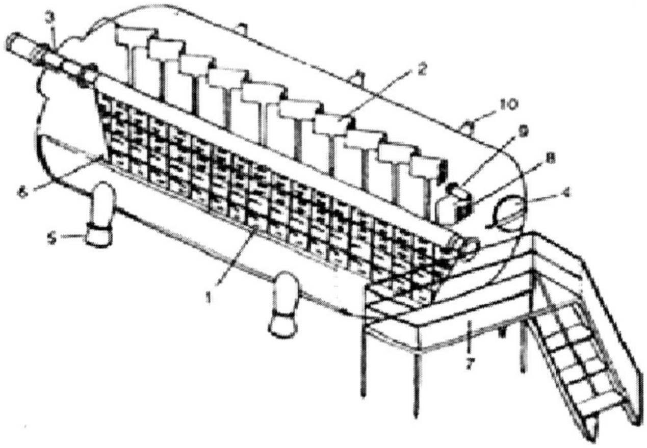

1. Cutting device; 2. Stirring pads; 3. Motor and drive; 4. Manhole; 5. Support fitted with pneumatic bellows; 6. Level gauge; 7. Service platform; 8. Control panel; 9. Tank light; 10. Spray ball.

Figure 23.5. Cheese Vat with Oscillatory Stirred Tank

The cheese curd after various steps of processing is formed into shape of blocks, which are paraffined and stored in cold store maintained at 8 to 10 °C and 85 per cent RH. The periods of storage is determined by the requirements of development body and flavour.

Processed Cheese

For manufacture of processed cheese, cheese blocks from cold store are deparaffined and cut to a suitable size (Fig 23.6). The blocks chosen may have been stored for different periods of time. Such cheese mix is heated and processed in a Process cheese vat to arrest further enzymatic activity. The final product is packed in various sizes. The Process cheese vat is an enclosed jacketed vessel with provision to mix and heat the cheese mix which is highly viscous. The product is mixed with anchor type agitator to enhance the heat transfer. The vat has a provision to incorporate culinary steam, to enhance uniform heating. A mechanically operated lid is provided for closing or opening and lifting. Once the process is completed the lid is lifted and the vat is tilted mechanically to unload the processed cheese, into containers. This is done while the product is warm and free flowing. The major components of the process cheese are shown in the Figure 23.7.

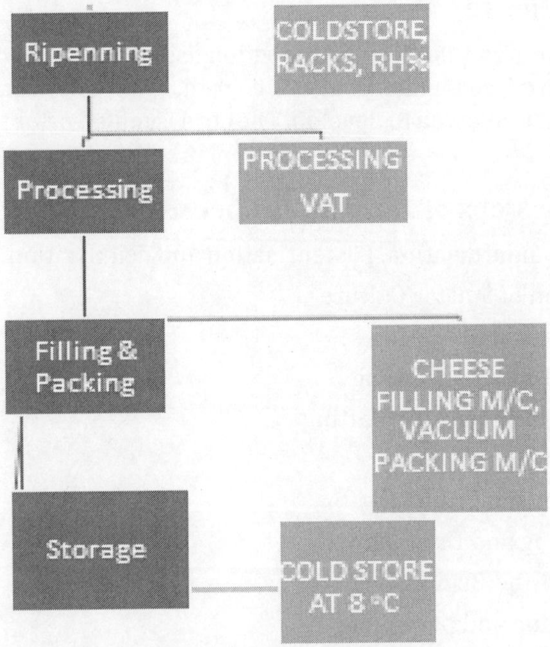

Figure 23.6. Flow Chart for Manufacturing Process Cheese from Ripened Cheese

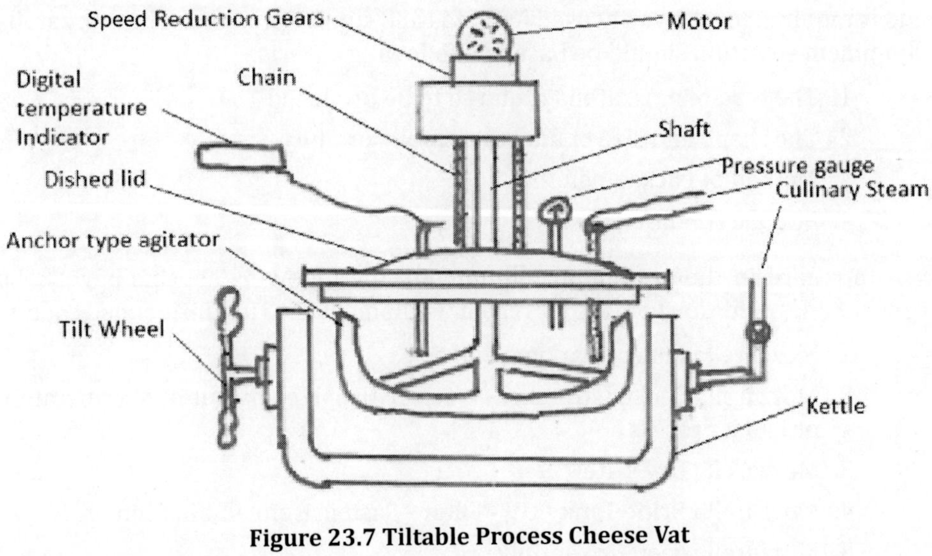

Figure 23.7 Tiltable Process Cheese Vat

The cheese after unloading from the process cheese vat is packed in verity of sizes and shapes and wrapped in foil. It is also filled in cans in suitable quantities and sealed.

Mozzarella Cheese

Traditional mozzarella cheese production uses standardized and pasteurized milk that is cultured, coagulated, cut and cooked. After whey drainage, the curd is matted while acid is allowed to develop. The curd is milled prior to being conveyed to the cooker.

Manufacturing Steps of Mozzarella Cheese

- Milk Standardization, Pasteurisation and acidification
- Addition of Starter Culture
- Renneting and Cutting
- Incubating and Scalding
- Whey Draining/Dry Matting
- Milling
- Cooking and Stretching
- Moulding and Dry Salting
- Cooling and Brining
- Shredding and packaging

Mozzarella Processing Equipment

The manufacture of this type of cheese differs from other steps in stages of Cooking and Stretching. The development of stretchability is an essential property and is required for the verity used in Pizzas and similar products. The mozzarella equipment selection should be based on following criteria:

1. The type of mozzarella required to be produced
2. The shape and size of final mozzarella product
3. Method of Packaging.
4. Storage conditions

In regard to the options for equipment, the following aspects have to be considered, in addition to those for regular equipment used for unripenned cheese:

1. Mozzarella Cheese Vat type and size
2. Mozzarella Cooker stretcher - Water bath or steam – automatic operation or batch process
3. Mozzarella Dry salter
4. Mozzarella Brine Tank - UV - Filter - Pasteuriser combination or
5. Mozzarella pasteurizer only
6. Mozzarella static mixer if acid is added
7. Draining table

8. Dry matting conveyor
9. Mozzarella moulder/block former
10. Shredder/dicer/shaping into cubes

The critical processes of cooking, stretching (or kneading) and moulding are all combined in same equipment. The operations of kneading and stretching require equipment for mechanized production. At higher pH, the kneading operation is required for longer time.

Stretching and Moulding Equipment

This equipment facilitates the production of stretched cheese. The curd blocks are fed to the upper hopper either manually or through a conveyor. The curd blocks are then crushed and transferred into a rotating drum made for stretching the dough. The drum has peripheral exit holes from which Mozzarella balls are ejected. Stretching machines of capacity varying from 20 to 200 kg/hr are available.

Figure 23.8. Stretching and Moulding Equipment

Continuous Mozzarella Cheese making equipment

The equipment is illustrated in Figure 23.9. The cheese curd is made in the

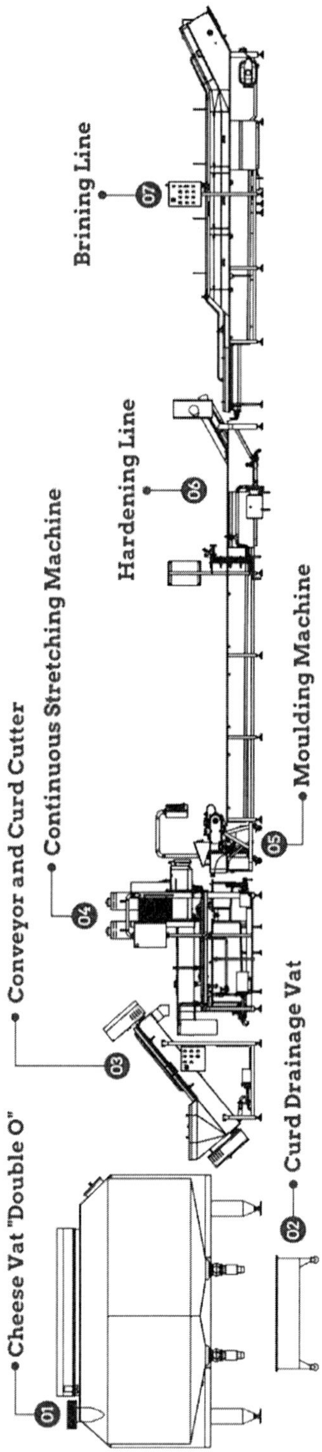

Figure 23.9. Continuous Mozzarella Cheese making Equipment (*Courtesy*: COMAT Equipment Manufacturing Co.)

regular cheese vat, or other types mentioned earlier. The mozzarella cheese curd enters the cooker where the plasticizing and working process takes place in hot water. The cooking process imparts the fibrous stretching structure for the mozzarella cheese. The hot plastic curd is extruded into moulds of the desired shape and weight. The cheese is then immersed in chilled brine to absorb salt and to cool. Dry salting can also take place during the kneading process. The hardening line allows the curd to become firm and retain its shape. Standardising the fat to Casein ratio of the milk increases the product yield, improves the body and texture.

Chapter 24

Membrane Processing

Introduction

Separation of solutes from a solution underwent many developmental changes, going down to the separation at molecular level.

Separation mechanisms: The most common separation method in dairy industry is centrifugation as applied to milk in clarification and separation. The other methods are Separation of dissolved components of differing molecular weights which are broadly termed as Fractionation, Demineralization, Purification *etc.* The removal of water would lead to Concentration of the solute components.

Membrane Processing

It is a familiar knowledge that a semi permeable membrane, allows diffusion of solvent across the membrane but not the solute. This transfer of solvent can be reversed by application of higher pressure on concentrated fluid side than the liquid pressure. This leads to expelling of moisture and increasing the concentration of the liquid. This process finds ready application in Dairy Industry where concentration of milk prior to drying to manufacture of milk powder, is needed, with least thermal exposure to heat sensitive milk components.

Similarly, different components of any liquids similar to milk can be separated by choosing the size of this membrane pores and operational methods. Depending on the sizes of the components to be separated, the membranes can be chosen. A membrane process allows the selective transfer of some species of molecules in a solution from one bulk phase to the other bulk phase. The various aspects of such operations are classified under the broad category Membrane Processing.

Membrane separation processes are based on the ability of membranes of the appropriate physical and chemical nature to discriminate between molecules primarily on the basis of molecular weight. The Membrane processing is further categorized as follows depending on the size of molecules to be separated.

1. Microfiltration (MF)
2. Ultrafiltration (UF)
3. Nanofiltration (NF)
4. Reverse Osmosis (RO)

Further, in addition to the method of application of pressure across the filtration membrane, an electric charge may also be used as a driving force to transport electrically charged particles across an ion selective membrane. This process is termed as Electro Dialysis.

The size of Molecules, that are retained classify the membrane operations, as shown in Figure in 24.1.

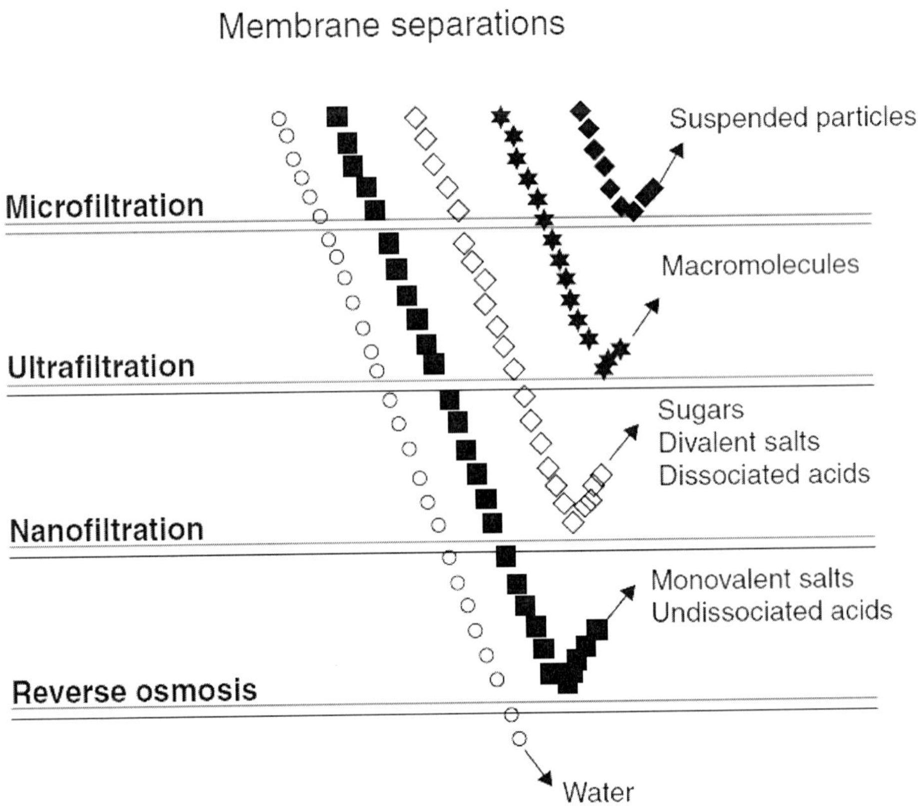

Figure 24.1. Membrane Separations (Source: Munir Cheryan)

Membrane processing depends on permeability of a porous membrane. The membrane permits passage of certain components called as "Permeate" and retains or rejects the other, called as "Retentate". The molecules smaller than pore size of the membrane pass through and those larger are retained. The feed solution is under pressure in the membrane. Two streams are formed from the feed by the membrane. The general layout of the process is shown in Figure in 24.2

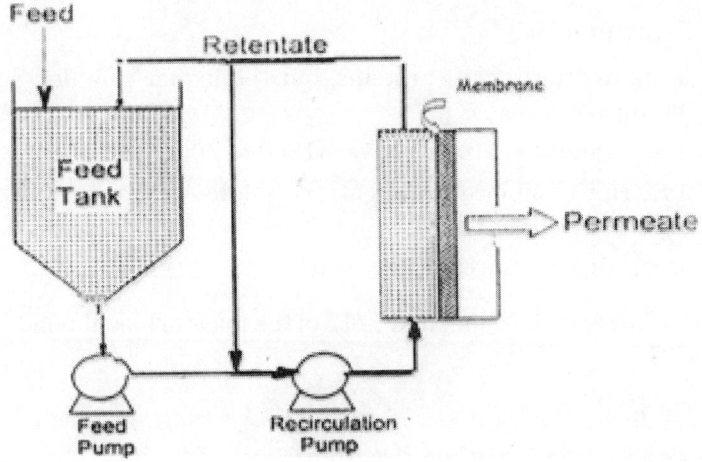

Figure 24.2. Membrane Separation Process Hookup

The advantages of membrane processing are: 1) The energy requirement is low, as there is no phase change and energy is used is for pumping only 2) The process can be carried out at ambient temperature 3) As heat transfer is absent the equipment is simple.

Specific use of membrane processing is:

1. Changing the constituents of milk stressing the individual needs such as lactose intolerance
2. Whey, which is a byproduct in cheese manufacture and has high BOD value can be concentrated for recovery of proteins. Further, the lactose in the permeate solution can also be recovered by RO.
3. Cheaper method of concentration than evaporation and can be applied as a step prior to drying of milk.
4. Possibility of incorporating whey proteins recovered through ultrafiltration into cheese
5. Environmental pollution from whey can be reduced
6. Milk can be concentrated without causing damage to protein or changing flavor, unlike in evaporation and drying.

The limitations are:

1. There is limitation on viscosity of the product during concentration process
2. As the concentration increases pumping costs increase

The essential features of different membrane separation processes are as follows:

Micro Filtration (MF)

1. It allows passage of proteins, but retains fat globules, somatic cells, bacteria *etc.*
2. Membrane pore size is between1000 to 2000 nm
3. In terms of Molecular weight, the solutes that are retained are above 2,00,000 Daltons*
4. Operating pressures is 0.3 to 2.0 bar.

*Dalton is a unit of mass, defined as 1/12 of the mass of an unbound neutral atom of carbon-12.

Applications

1. Coarse filtration of bacteria from Water, milk, whey, cheese brine *etc.*
2. Delipidisation, or removal of fat with whey
3. MF applied before processing of whey or milk by UF, increases the permeation rate two times

In the application of Bactofuge and Bactotherm, MF is used in combination in the UHT processing for long shelf-life milk.

Ultrafiltration (UF)

It is a method of simultaneously purifying, concentrating and fractionating macromolecules or fine colloidal suspensions. It is generally known as "Sieving Process". It retains proteins, fats, suspended particles. Its Membrane pore size is 1 – 50 nm and Molecular weight cut off is 1,000 – 2000 Daltons. The range of operating pressure is 1 – 15 bar.

Applications

1. Skim milk concentration
2. In concentration of milk in Cheese making to reduce volume of whey
3. Concentration of whey for recovery of proteins

Reverse Osmosis

It is also known as "Hyper filtration". It is essentially a dewatering technique. The membranes for this process have pores in the range 0.5 – 2 nm and Molecular weight cut off is around 100 Daltons. The operating pressure is high, in the range 15 – 75 bar.

Applications

1. Concentrating of milk, whey, in combination with UF and evaporators
2. Removal of water from milk to facilitate its bulk transportation
3. Water demineralization for potable supply
4. Purification of sea water

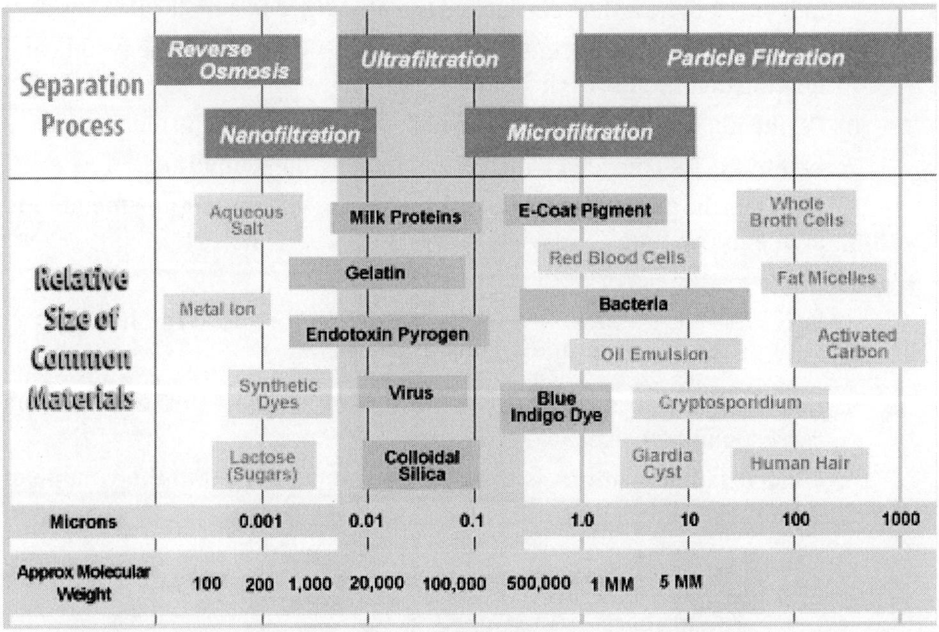

Figure 24.3. Relative Sizes of common materials and Membrane Process

Membrane Technology Terminology

The Membrane technology is different from normal process of filtration. The direction of flow of feed with respect to membrane is parallel instead of perpendicular. It is called Cross flow filtration. Further, not all the filterable material passes through the membrane. The feed is recirculated, certain percentage of the filterable material passes through the membrane at a certain rate. All these processes, have terminology, which are defined below:

1. Semipermeable Membrane: This is the most important component of the process. It permits passage of certain components of the feed or solution and retains the others on the basis of size/molecular weight.
2. Module or Cartridge: Used to refer either to completely assembled membrane or the external housing without the membrane subassembly.

3. Asymmetric Membrane: Normally, a membrane is of single material. But, due to their thin layer, they would need certain supporting structure, within its matrix. This type of membrane is a composite one, with a dense skin layer on the top of a spongy support layer below it.
4. Molecular weight cut off: It is molecular weight of the smallest test macromolecule that is retained by the membrane. It must be noted that molecular weight and the size are not directly proportional. For example, water with a molecular weight of 18 is has diameter of 0.3 nm, while Lactose with molecular weight of 342 has a diameter of 0.8 nm. It is an indication of how densely packed the atoms are in a molecule.
5. Feed: It is the material under process which pumped into the system and flowing on one side of the membrane.
6. Permeate: It is part of the feed which passes through the membrane.
7. Retentate: It is the feed which is retained by the membrane.
8. Flux: It is the permeate flow rate expressed as volume per unit membrane area per unit time.

$$\text{Flux} = \frac{\text{Volume of permeate}}{\text{Area of membrane X Time}} = \text{The units are liters/m}^2\text{ h}$$

9. Relative Flux: It is the ratio of the flux of solution to the flux of the corresponding permeate
10. Concentration factor: It is the ratio of volume of feed to the volume of retentate.

$$\text{CF} = \frac{\text{Volume of feed}}{\text{Volume of retentate}}$$

11. Transmembrane Pressure (TMP): It is the difference between mean of inlet pressure of feed (P_i) and outlet pressure of retentate (P_o) and pressure of permeate (P_p). This is the pressure that drives the filtration process.

$$\text{TMP} = \frac{P_i + P_o - P_p}{2}$$

12. Diafiltration: It is a process in which water equal to the quantity of permeate is added to the retentate and the process of UF is continued.
13. Concentration Polarization: It is the phenomena of accumulation of rejected solids on the surface of membrane, causing a large concentration gradient of solute within the boundary layer. It is the result of steady state reached when convective movement of solids towards membrane and the diffusive movement of solids away from the membrane balance each other.

It is due to this reason that flux becomes independent of pressure. Flux is controlled by the boundary layer thickness, and rate of back transfer of polarized molecules by diffusion.

The transport of concentrated solids back into the feed is governed by Fick's law of diffusion, which is governed by the following relationship.

$$J_{AB} = - D_{AB} (dc_A/dz) \quad \text{(Eq. 24.1)}$$

where,

J_{AB} = Molar flux of component A of mixture of AB in the direction of Z which is perpendicular to direction of flow (kg mol of A/m^2 s)

D_{AB} = Molecular diffusivity of component A in mixture of A and B (m^2/s)

c_A = Concentration of A (kg mol/m^3)

z = distance(m)

14. Membrane Fouling: It is the accumulation or adsorption of macromolecules or colloidal particles, such as proteins, lipids, macro-organisms and/or inorganic salts, on the membrane surface. It may also be caused due to precipitation of permeable solutes, such as sugars and salts, due to clogging within the membrane pores.

    While the Concentration Polarization is time-independent and a reversible process; membrane fouling is an irreversible phenomenon. During, fouling, the flux drops with operating time, usually rapidly in the initial stages and slowly with increasing feed concentration.

15. Retention factor: This parameter specifies the ability of a membrane to retain the molecule. It depends on the amount of solute passing through the membrane with permeate. Concentration factor (R) is given by.

    $$R = (C_f - C_p)/C_f \quad \text{(Eq. 24.2)}$$

    where,

    C_f = Concentration of molecules in feed (kg mol/m^3)

    C_p = Concentration of molecules in permeate (kg mol/m^3)

16. Rejection factor: It is same as above and expressed as

    $$R = 1- (C_p/C_f) \quad \text{(Eq. 24.3)}$$

    This relationship is used in water treatment.

    Further this relationship is not applicable to multi component system.

17. Separation Factor: It is the measure of performance of the membrane separating solute from solvent. This concept is more appropriate to water purification than milk concentration.

$S = C_f/C_p$ (Eq. 24.4)

where,

C_f, C_p are mole fractions of solute in solvent for feed and permeate respectively

The expression for change in concentrations in a Batch Process is different as R varies with time.

$\ln (C/C_o) = R \ln (Vo/V)$ (Eq. 24.5)

where,

C_o = Initial concentration at volume V_o (kg mol/m^3)

C = Concentration at any other volume V (kg mol/m^3)

Materials for Membrane Construction

Introduction

The range of membranes offered by the various manufacturers appear confusing since several materials are used to make membranes and are available under an array of trade names. In reality, relatively few materials are actually used and only a few basic membrane types form the bulk of the membranes sold and used.

Membrane Structure

Membranes have a tight top layer in contact with the product. This layer is also called the skin layer. It is thin, typically <0.1 micron. The membrane itself is 150 - 250 micron thick; the bulk of the membrane provides structural support for the skin layer.

Following materials are used for the membrane construction:

1. **Cellulose Acetate (CA)**

- ☆ Contains 38 – 40 per cent of acetyl group by weight.
- ☆ Membranes are strong and flexible.
- ☆ Used for RO, NF and UF applications.

Advantages

- ☆ Low price
- ☆ Less prone to fouling due to its hydrophilic nature.

Limitations

- ☆ Low resistance to pH changes.
- ☆ Maximum operating temperature is 35°C.
- ☆ Low resistance to chemicals used for cleaning and sanitization (*e.g.* chlorine)
- ☆ Can be decomposed by microorganisms.

2. Synthetic Polymers

a. *Ploysulphones (PSO)*

- ✰ Used for UF and MF membrane since 1975.

Advantages

- ✰ Exceptional resistance to temperature and pH
- ✰ These membranes have least tolerance oil, grease, fat and polar solvents.
- ✰ Aromatic Polysulphones are resistant to oxidation, remain dimensionally stable, resistant to acids, alkali, salt solutions and detergents even at elevated temperatures or moderate pressure.

b. *Polyvinylidenefluoride*

- ✰ These are resistant to hydrocarbons and oxidizing environments.
- ✰ These have excellent resistance to abrasion, temperature and chemicals.

c. *Polyacrylonitrile (PAN)*

d. *Polytetrafluoroethylene (PTFE)*

This material has extraordinary chemical resistance and is used for wide range of temperature.

e. *Aliphatic and aromatic polyamides*

3. Composite Membranes

- ✰ Made to replace cellulose acetate RO membranes.
- ✰ The main advantages are relatively high flux and salt rejection.
- ✰ They also have good resistance to temperature and pH.
- ✰ High tolerance for oxidizing environments.
- ✰ Composite membranes are made in two-layer and three-layer designs. The precise composition is proprietary.
- ✰ Around 1980, FilmTec marketed two-layer design which immediately became the industry standard for water desalination.
- ✰ In the mid-1980s Desalination Systems, Inc. (DSI) began making composite membranes with three-layer design.
- ✰ The three-layer design composite membrane is available for RO and NF and is the best choice for treating vast array of difficult process streams.

4. Other Membrane Materials

- ✰ **ABS: Useful** material in water desalination.
- ✰ **Delrin:** Also called POM. It has limited resistance in low pH environments.
- ✰ **PE:** Polyethylene has good chemical stability but has limited temperature stability.

☆ **PP** Polypropylene is a polymer resistant to chemicals. The temperature stability is limited, and it has a tendency to creep.

☆ **PVC:** PVC is mostly used for low pressure piping. It is inexpensive, but has severe temperature limitations. It can only be used for less demanding applications such as desalination of ground water.

☆ **C-PVC** Chlorinated PVC has significantly better temperature stability than ordinary PVC. Environmental and public health concerns may preclude PVC from being an acceptable construction material in the future.

☆ **PVDF** (Polyvinylidene fluoride): An excellent but rather expensive material. It has good heat stability and is chemically as resistant as Teflon.

☆ **FRP** Glass fiber reinforced polyester is widely used for housings. It has become the standard in water desalination. Corrosion resistance is its prime advantage, closely followed by low price.

☆ Epoxy A very special type of material made by Membratec, Germany. The housings are thin walled and can withstand very high pressure.

Selection of Membrane Materials

The Table 24.1 shows the choice of membrane materials based on chemical resistance.

Table 24.1. Tolerance of Material of Construction to various Chemicals

	Composite	CA	PSO	PVDF	PAN	SiO_2	Cellulose
3 < pH < 8	✔	✔	✔	✔	✔	✔	✔
pH<3 or pH>8	✔	✖	✔	✔	✔	✔	✔
Temp >35°C	✔	✖	✔	✔	✔	✔	✔
Humic acid	(✔)	✔	✖	✖	(✔)	✖	✔
Proteins	✔	(✔)	✔	(✔)	(✔)	✔	✔
Polysaccharides	(✔)	✖	✔	✖	(✔)	✔	✖
Textile waste	✔	✖	✔	(✔)	✔	✖	✖
Aliphatic hydrocarbon	✖	✖	✖	(✔)	✔	✔	✔
Aromatic hydrocarbon	✖	✖	✖	✔	✖	✔	(✔)
Oxidizers	✖	(✔)	✔	✔	(✔)	✔	(✔)
Ketones, Esters	✖	✖	✖	✔	✖	✔	(✔)
Alcohol	✔	✖	✔	✔	✔	✔	✔

✔ means high resistance

(✔) means either that the information is based on theory or that practical results have proved to be dubious

✖ means high resistance

Chapter 25

Applications of Membrane Processes

Applications of Membrane Processing in Dairy Industry

Membrane processing is extensively used in the dairy industry. The most of the uses are in the large-scale application of RO systems for concentration of milk for long distance transportation. Many of the other applications use a combination of membrane processes. For example, in the usage of membranes for producing pure water, the raw water is first treated with MF and then with the RO membranes, so that the later process is more effective. The Figure 25.1 gives the possible uses of membrane processes in dairy industry.

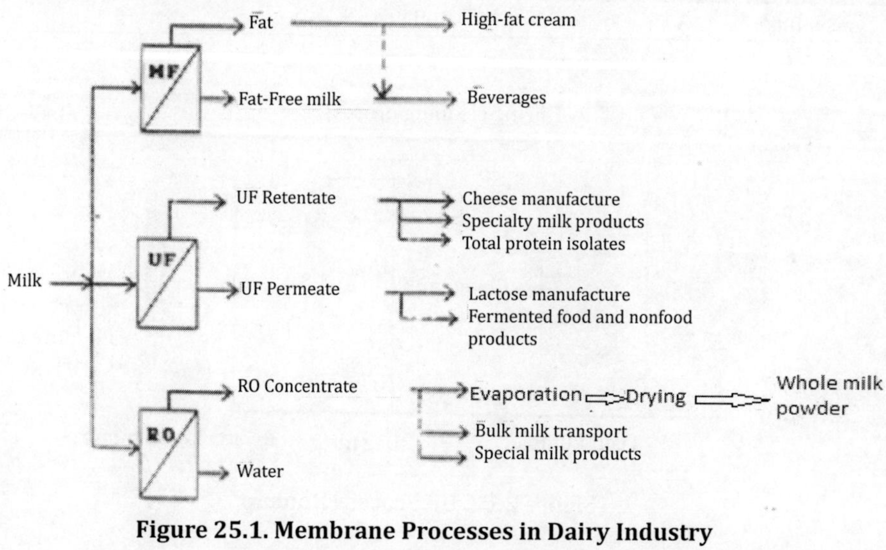

Figure 25.1. Membrane Processes in Dairy Industry

Ultrafiltration

The process of UF will be discussed more in detail in this chapter, as it finds important applications in dairy and food industries.

Ultrafiltration can be defined as a pressure driven membrane process that can be used in the separation and concentration of liquids having solutes with molecular weights between 10^3 – 10^6 Dalton.

In UF process high molecular weight components, such as proteins, and suspended solids are retained, while all low molecular weight components, lower than cut-off value of the membrane, pass through freely. The solutes such as mono and disaccharides, salts, amino acids, organics, inorganic acids or sodium hydroxide pass through the membrane with the permeate. The Figure 25.2 shows general hookups of the UF membrane system.

Characteristics of the UF Membranes

Table 25.1. UF Membrane Specifications

Membrane structure	Asymmetrical
Thickness	150 – 250 μm
UF Thin film	1 μm
Pore size	0.2 – 0.02 μm
Rejection (Retention)	Macro molecules, proteins, polysaccharides
Membrane materials	Ceramic, Poly Sulphones, Poly Vinylidene fluoride, Cellulose Acetate
Membrane module	Tubular, hollow fiber, spiral wound plate-and-frame
Operating pressure	1-10 bar
Mechanism of membrane retention	Molecular screening
Typical flux	30 – 300 lit/m²h

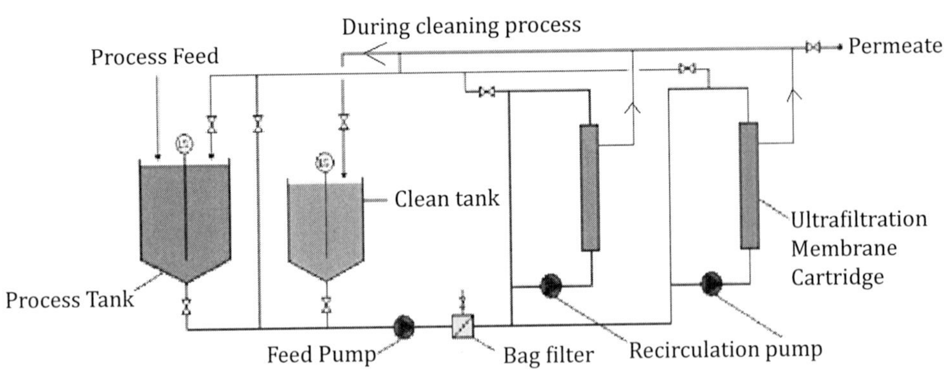

Figure 25.2. UF Process Hookup

Membrane Modules

Following are the common types of membrane modules:

1. Tubular Module

In tubular module with ceramic membrane, the filter element is a ceramic filter. The thin walls of the channels are made of fine-grained ceramic and constitute the membrane. The support material is coarse grained ceramic. Referring to Figure 25.3, the construction details are given below:

- ☆ 18 x12.5 mm perforated stainless steel tubes
- ☆ Tubes are assembled in a shell-and-tube like construction and connected in series.
- ☆ A replaceable membrane insert tube is fitted inside each of the perforated stainless steel pressure support tubes.
- ☆ Permeate is collected on the outside of the tube bundle in the stainless steel shroud.

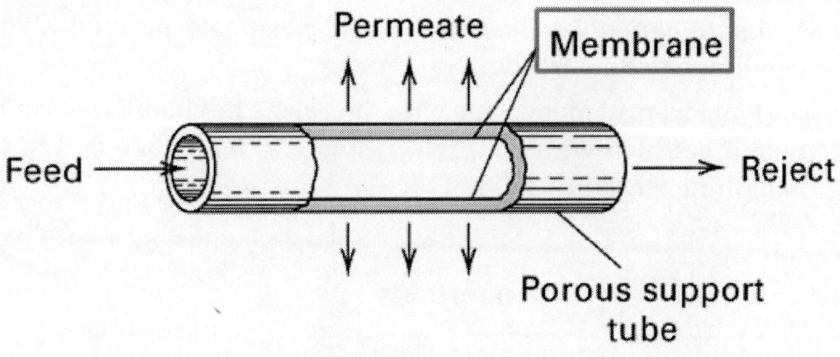

Figure 25.3. Tubular Module

2. Hollow Fibre (Figures 25.4 a,b)

- ☆ Cartridges each having bundles of 45 to 3000 hollow-fibre elements.
- ☆ The fibres are oriented in parallel.
- ☆ Fibers are potted in a resin at their ends and enclosed in the permeate-collecting-tube of epoxy.
- ☆ The membrane has an inner diameter ranging from 0.5 to 2.7 mm.
- ☆ The active membrane surface is on the inside of the hollow fibre.
- ☆ The outside of the hollow-fibre wall, has a rough structure and supports the membrane.
- ☆ The feed stream flows through the inside of these fibres, and permeate is collected outside and removed at the top of the tube.

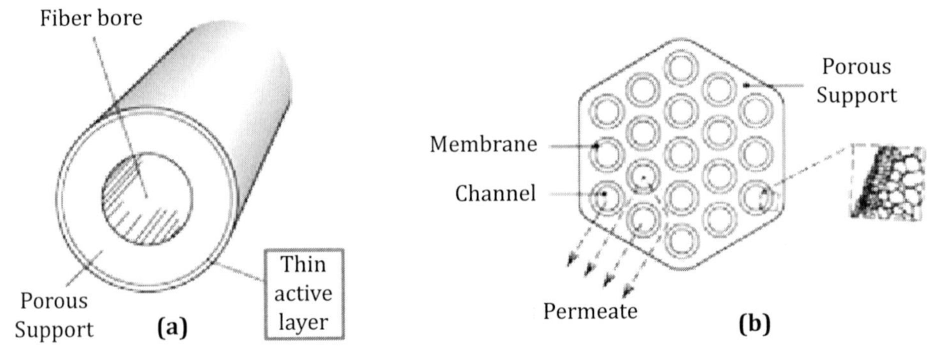

Figure 25.4. Hollow Fibre Module

3. Spiral Wound (Figure 25.5)

- ☆ Contains one or more membrane envelopes, each of which contains two layers of membrane separated by a porous material for permeate flow.
- ☆ *Permeate channel spacer* allows the permeate passing through the membrane to flow freely.
- ☆ The two layers of membrane with the permeate channel spacer between them are sealed with adhesive at two edges and one end to form the membrane envelope.

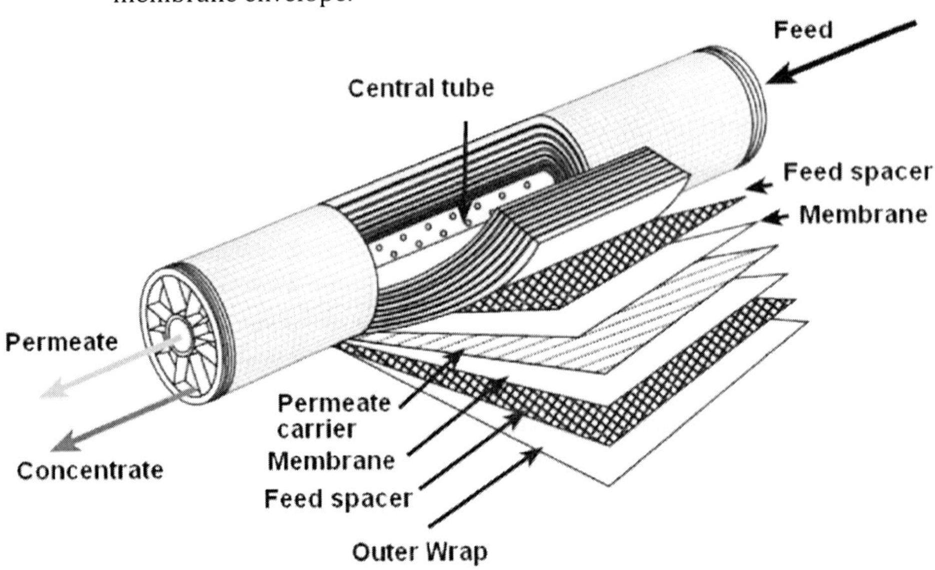

Figure 25.5. Spiral Wound Module

- ☆ The open end of the envelope is connected and sealed to a perforated permeate collecting tube
- ☆ The feed channel spacer is placed in contact with one side of each membrane envelope.

4. Plate and Frame Design (Figure 25.6)

- ☆ It consists of membranes sandwiched between membrane support plates arranged in form of stacks.
- ☆ The liquid feed is forced through the narrow channels which may be configured for parallel flow or as a combination of parallel and serial flow.

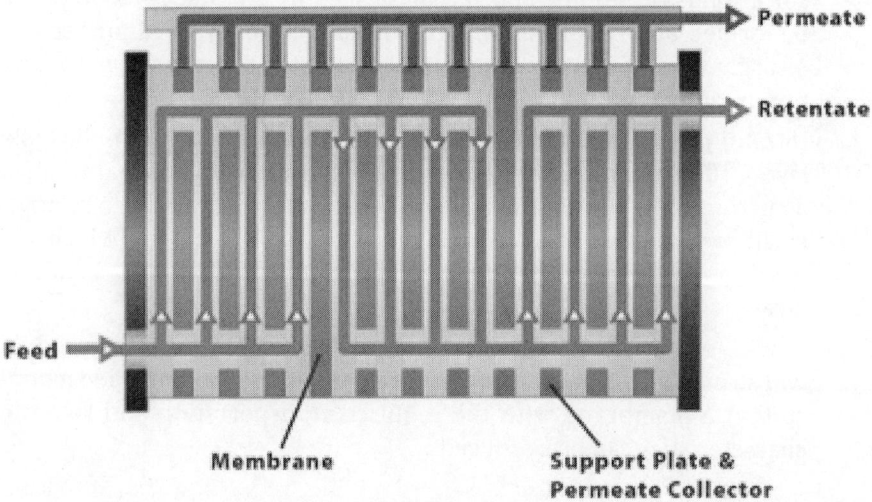

Figure 25.6. Plate and Frame Module

Materials used in Membrane Manufacturing

The materials used in membrane manufacturing are Cellulose Acetate, Polyamide, Polysulfone which are polymers and Ceramics.

Fouling of Membrane

Fouling of membrane causes decline in flux rate when all operating parameters like pressure, flow rate, temperature and feed concentrations are kept constant. It can be avoided by: 1) Pretreatment of feed 2) By maintaining minimum axial velocity 3) By keeping dynamic pressure of flow higher 4) By keeping pH of proteinaceous feeds far from iso-electric point.

U.F. Membrane preparation Methods

- Phase inversion: This is a technique, where, a polymer plus solvent (polymer solution) is cast in a proper supporting layer and then submerged in a coagulation bath containing non solvent. Due to the solvent and nonsolvent exchange, precipitation takes place, and a membrane is formed.
- Thermal inversion: It involves quenching a solution to a temperature regime, where the polymer precipitates.
- Dynamic membrane: which is also called as Secondary membrane, is formed on the underlying support material when the filter solution contains fine particles. The membrane itself may no longer necessary, since solids rejection will be accomplished by the cake layer which can be formed and re-formed in situ. These are mostly used for recovery of organic matter from domestic wastewater to enhance net energy production.
- Ultrathin composite membranes: They are comprised of a thin, dense skin layer (100 nano meter or below) performing molecular separation and a microporous support providing mechanical strength. Here a relatively small pore membrane is coated by even more selective layer (which often has its own skin) applied by a technique resulting in a very thin separating layer.
- Track – etched membranes: This is a high-precision manufactured film whose pore size, shape and density can be varied in a controlled manner so that a membrane with the required transportation and retention characteristics can be produced.

Disadvantages

- As the surface of membrane is not smooth, accumulation of scale promotes bacterial growth.
- The membrane voids provide space for growth of micro-organisms.
- Cellulose acetate membranes have high sensitivity to several cleaning and sanitizing solutions.

Applications

1. Separation and fractionation of individual milk proteins from other components in milk like lactose and minerals.
2. Enzyme recovery in various operations like recovery of lactase in lactose hydrolysis
3. Fractionation of cheese whey

4. Pre – concentration of milk in cheese manufacture
5. Sugar refining
6. Vegetable protein processing. *e.g.* soy proteins
7. Concentration of egg white
8. Animal products industry. *e.g.* gelatin
9. Biotechnology applications
10. Fruit juices and other beverages

Reverse Osmosis

Introduction

Reverse Osmosis (RO) employs tightest possible membrane in liquid/liquid separation. Water in principle is the only material passing through the membrane; essentially all dissolved and suspended materials are retained. The more open types of RO membranes are sometimes confused with nanofiltration (NF). In Reverse osmosis molecules and ions (molecular weight less than 1000 and molecular size less than 0.001μm) are separated from the solvents. The principle of RO is explained in Figure 25.7.

A pressure driven membrane system is used for dewatering of liquid foods.

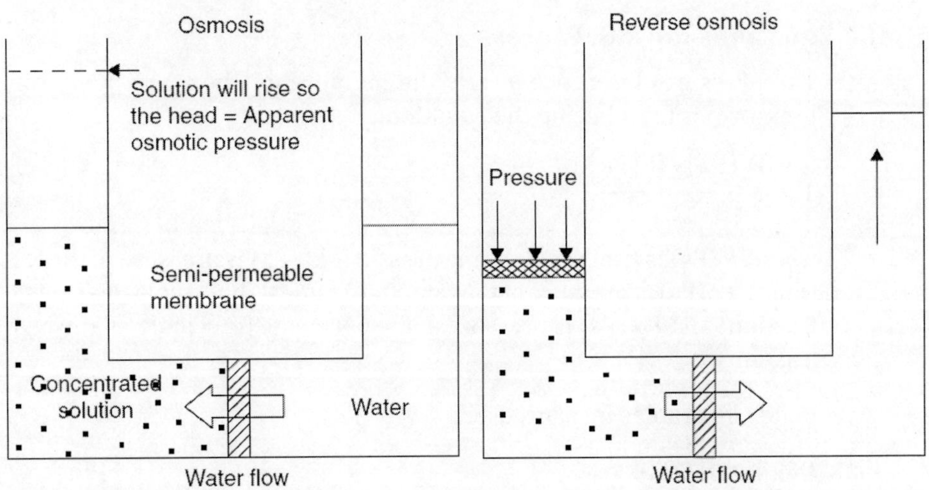

(a) Principle of Osmosis (b) Principle of RO

Figure 25.7. Principle of RO Process

Characteristics of the RO Membranes

Characteristics of the RO membranes show in Table 25.2.

Table 25.2. RO Membrane Specifications

Membrane	Asymmetrical
Thickness	150 μm
Thin film	1 μm
Pore size	<0.002 μm
Rejection (Retention)	High molecular weight components, low molecular weight components, NaCl, glucose, amino acids
Membrane materials	Cellulose Acetate, thin film
Membrane module	Tubular, spiral wound plate-and-frame
Operating pressure	15-150 bar
Transport mechanism through membrane	Diffusive transport
Typical flux rate	3 – 30 lit/m²h

Materials used in Membrane Manufacturing

The materials used for R.O.membranes are Cellulose acetate, Polymers (Polysulphones, polyamides, PVC, polystyrene, polycarbonates, and polyethers). Composite or ceramic membranes (porous carbon, zirconium oxide, alumina).

Useful Equations for R.O. Process

- The pressure difference across the membrane (the trans-membrane pressure) is found using the equation:

$$P = \frac{P_f(Pa) + P_r(Pa)}{2} - Pp \qquad \text{(Eq.25.1)}$$

where, P (Pa) is trans-membrane pressure, *P*f (Pa) is pressure of the feed (inlet), *P*r (Pa) is pressure of the retentate (outlet) (high molecular weight fraction) and *P*p (Pa) is pressure of the permeate (low molecular weight fraction).

- Water flux rate is measured as:

$$J = kA(\Delta P - \Delta \Pi) \qquad \text{(Eq. 25.2)}$$

where, *J* (kg/h) is flux rate, *k* (kg m^{-2} h^{-1} Pa^{-1}) is mass transfer coefficient, *A* (m^2) is area of the membrane, ΔP (Pa) is applied pressure and $\Delta\Pi$ (Pa) is change in osmotic pressure.

- Osmotic pressure Π = MRT (Eq. 25.3)

where, *T* (K) is absolute temperature of the solution, *R is* universal gas constant, *M is* molar concentration and Π(Pa) is osmotic pressure.

Advantages of RO

The advantages of R.O. are 1) The removal of water is accomplished without a change in phase or state of the solvent. 2) The process can be operated from ambient to 50°C temperature. Thus, thermal degradation of nutrients in the concentrated product is minimum. 3) There is negligible loss of volatiles and food quality. 4) Lower labour and operating costs.

Limitations

1. Variation in the product flow rate increasing concentration of the feed
2. Higher capital costs than evaporation equipment
3. A maximum concentration to 30 per cent total solids is attainable
4. Fouling of the membrane which reduces the permeate flux rate and increases operating time.

Applications

1. Concentration of milk, whey and UF permeate
2. Preparation of indigenous dairy products like khoa, chakka *etc.*
3. Concentration and purification of fruit juices, enzymes, fermentation liquors and vegetable oils
4. Concentration of wheat starch, citric acid, egg white, coffee, syrups, natural extracts and flavours
5. Clarification of wine and beer
6. Demineralization and purification of water from boreholes, rivers or and desalination of sea water

Electrodialysis

Introduction

Electro Dialysis (ED) is a membrane process in which ions are transported through a semi permeable membrane, under the influence of an electric potential. Electrodialysis is an electrically driven membrane separation process that is capable of separating, concentrating, and purifying selected ions from aqueous solutions (as well as some organic solvents).

Modern electrodialysis became a practical reality with the development of the first reliable ion-exchange membranes having both good electrolyte conductivity and ion-perm selectivity. Electrodialysis was first commercially exploited for the desalination of brackish water. In the early 1980's a completely new area for the application of electrodialysis was opened up with the introduction of bipolar membranes for the recovery of acids and bases from the corresponding salts.

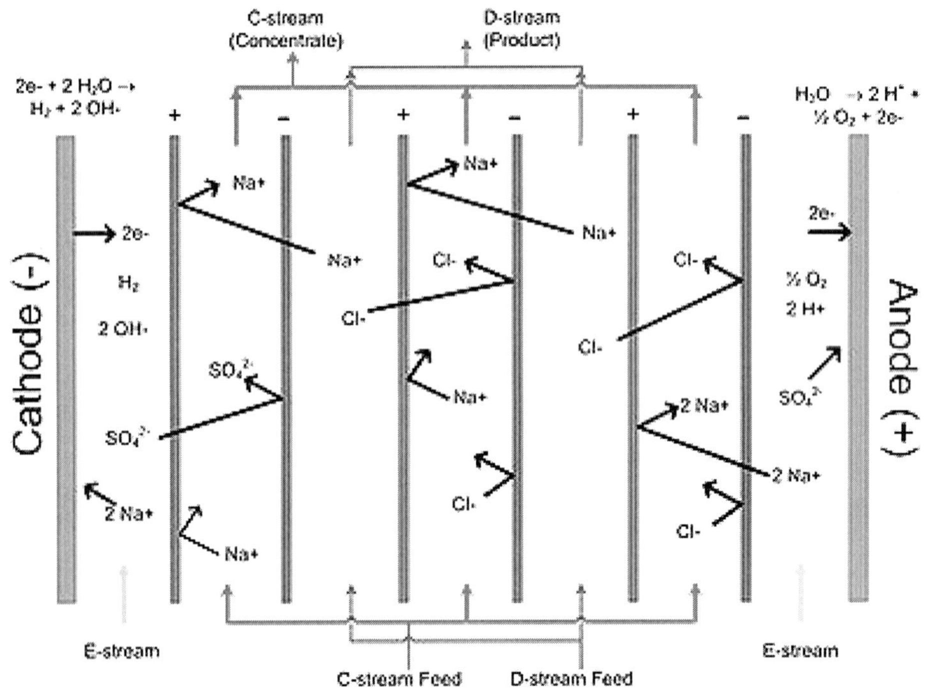

Figure 25.8. Electrodialysis Process

Principle

The principal components of this system (Figure 25.8) are: 1) Dilute Feed Stream (D) 2) Concentrate Stream(C) 3) Electrode Stream (E) 4) Cell compartments formed by ion exchange membranes and 5) Electrodes

The basic principle of the process is: Electrodes, when a potential difference is created across them attracts oppositely charged ions depending on their polarity. Cations (positively charged ions) are attracted toward cathode (-) and Anions (negatively charged ions) are attracted toward anode (+). Anion exchange membrane (positively charged) allows anions to pass through but checks cations. Similarly, cation exchange membrane (negatively charged) allows cations to pass through but checks anions. Equal number of cations and anions get collected in C stream (transferred from D stream). Thus, overall charge balance is maintained. The overall result of the electro dialysis process is increase in ion concentration in the concentrate stream with depletion of ions in the dilute solution feed stream.

Materials and Specifications

1. Ionic mobility: The ionic mobility is directly proportional to the specific conductivity and inversely proportional to number of solute molecules in the solution, which is about 3-6 x 10^2 mm/sec.

2. Membranes: Electrodialysis membranes are comprised of polymer chains - styrene-divinyl benzene and are made anionic with quaternary ammonium groups and made cationic with sulphonic groups.
3. Operating voltage: 1-2 V DC is applied across each pair of membranes.
4. Membrane arrangement:
    a. Anion and cation exchange membranes are arranged alternately in parallel between an anode and a cathode.
    b. Distance between membranes is 1 mm or less.
    c. A plate and frame arrangement similar to a plate heat exchanger or a plate filter is used.

Electrodialysis using Bipolar Membranes: A bipolar membrane in an electric field generates hydrogen ion on one side and hydroxyl ion on the other side. Therefore electrodialysis with bipolar membrane can be used to produce acid and base from salt.

Current Efficiency

Current efficiency is a measure of how effectively ions are transported across the ion exchange membranes for a given current. Current efficiency is generally a function of feed concentration and is given by the following relationship:

$$\xi = \frac{zFQ_f(C^d_{inlet} - C^d_{outlet})}{NI} \qquad \text{(Eq.25.4)}$$

where,

ξ = current utilization efficiency

z = charge of the ion

F = Faraday constant, 96,485 Amp-s/mol

Q_f = dilute stream flow rate, L/s

C^d_{inlet} = dilute stream ED cell inlet concentration, mol/L

C^d_{outlet} = dilute stream ED cell outlet concentration, mol/L

N = number of cell pairs; and I = current, Amps.

Advantages

1. ED system separates the salts from solution without phase change, which results in relatively low energy consumption.
2. When brackish water is desalted by ED system, the product water needs only limited pre-treatment. Typically for disinfection of product water only chlorination is required.
3. As ED system removes only ionized species, it is particularly suitable for separating non-ionized from ionized components.

4. Since Osmotic pressure is not an influencing factor in ED process, pressure can be used for concentrating salt solutions to 20 per cent or higher.

Limitations

1. Organic matter, colloids and SiO_2 are not removed by the ED system.
2. Pretreatment of Feed water is necessary to prevent fouling of ED stacks.
3. Elaborate controls are required, and keeping them at optimum condition can be difficult.
4. Selection of the materials of construction for membranes and stack is important to ensure compatibility with the feed stream.

Applications

1. Removal of salts from liquid foods.
2. Demineralization of milk products and whey for infant formula and special dietary products.
3. Concentration of salts, acids and bases.
4. Adjustment of pH in foods and wine.
5. Large scale brackish and seawater desalination.
6. Drinking water production
7. Pre-demineralization (for process water, chemical manufacturing, boiler makeup water)
8. Glycol desalting
9. Glycerin purification

Different Types of Membranes

Ion Permeable Membranes

Membranes for electrodialysis are typically hydrocarbon films with ion exchange functional groups attached to the polymer chains. Hydrocarbon membranes are usually categorized as homogeneous and heterogeneous.

For heterogeneous membranes, the film-forming polymer is usually polyethylene or polyvinylidene fluoride, but other polymers are also be used. These membranes are thick, opaque and mechanically strong, but they tend to have higher resistance than homogeneous membranes.

Typical homogeneous membranes have a polymer matrix of styrene cross linked with divinylbenzene (DVB) and ion exchange functional groups of sulfonic acid or quaternary amines. Higher crosslinking improves selectivity and membrane stability by reducing swelling, but it increases electrical resistance (Figure 25.9).

High charge density reduces resistance and increases selectivity, but promotes swelling and thus necessitates higher crosslinking. A compromise between selectivity, electrical resistance, and dimensional stability is achieved by proper adjustment of crosslinking and fixed-charge densities.

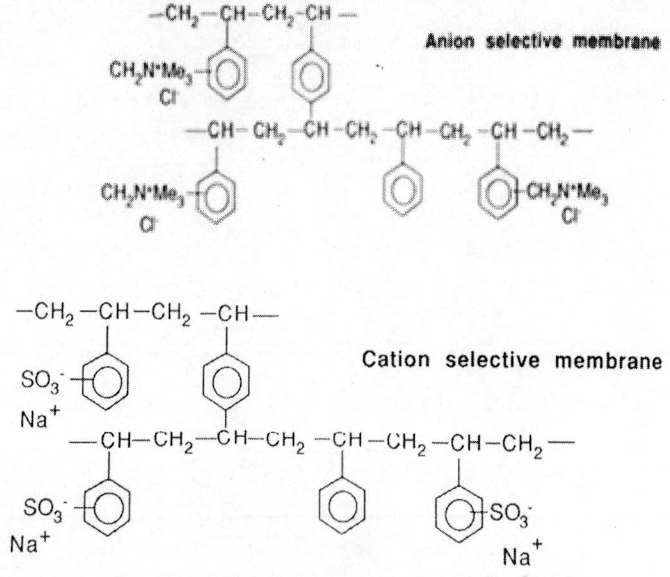

Figure 25.9. Chemical Structure of Membrane

Bipolar Membranes

Bipolar membranes consist of an anion-permeable membrane and a cation permeable membrane laminated together. Multiple bipolar membranes along with other ion permeable membranes can be placed between a single pair of electrodes in an electrodialysis stack for the production of acid and base from a neutral salt solution. When this composite structure is oriented such that the cation-exchange layer faces the anode it is possible, by imposing a potential difference across the membrane, to split water into proton and hydroxyl ions. This results in the production of acidic and basic solutions at the surfaces of the bipolar membranes. A Potential difference of 0.8 V is necessary for modern bipolar membrane for this process.

Annexure I

Steam Pipeline Capacities at Specific Velocities

Pressure (Bar)	*Velocity (m/s)*	*kg/hr*									
		Nominal Bore of Pipe in mm									
		20	*25*	*32*	*40*	*50*	*65*	*80*	*100*	*125*	*150*
3	15	37	60	93	127	345	385				
	25	56	100	152	225	425	632				
	40	87	157	250	357	595	1025				
5	15	49	87	126	187	352	526				
	25	81	195	211	308	548	865				
	40	131	225	338	495	855	1350				
6	15	59	105	153	225	425	832	925	1555		
	25	97	162	253	370	659	1065	1520	2530		
	40	157	270	405	395	1025	1620	2270	4210		
8	15	70	125	190	285	475	800	1125	1990	3025	4540
	25	122	205	320	465	810	1260	1870	3240	6220	7120
	40	192	327	510	730	370	2065	3120	5135	8395	12470
10	15	95	155	250	372	626	1012	1465	2495	3995	5860
	25	145	257	405	562	990	1530	2206	3825	8295	8995
	40	216	408	615	910	1635	2545	3600	6230	9880	14390

Pressure (Bar)	*Velocity (m/s)*	*kg/hr*									
		Nominal Bore of Pipe in mm									
		20	*25*	*32*	*40*	*50*	*65*	*80*	*100*	*125*	*150*
14	15	121	205	310	465	810	1270	1870	3220	5215	7390
	25	195	331	520	740	1375	2080	3120	5200	8500	12560
	40	305	555	825	1210	2195	3425	4735	9610	13060	18630
Recommended steam velocity: 25 m/s											

Source: NDDB.

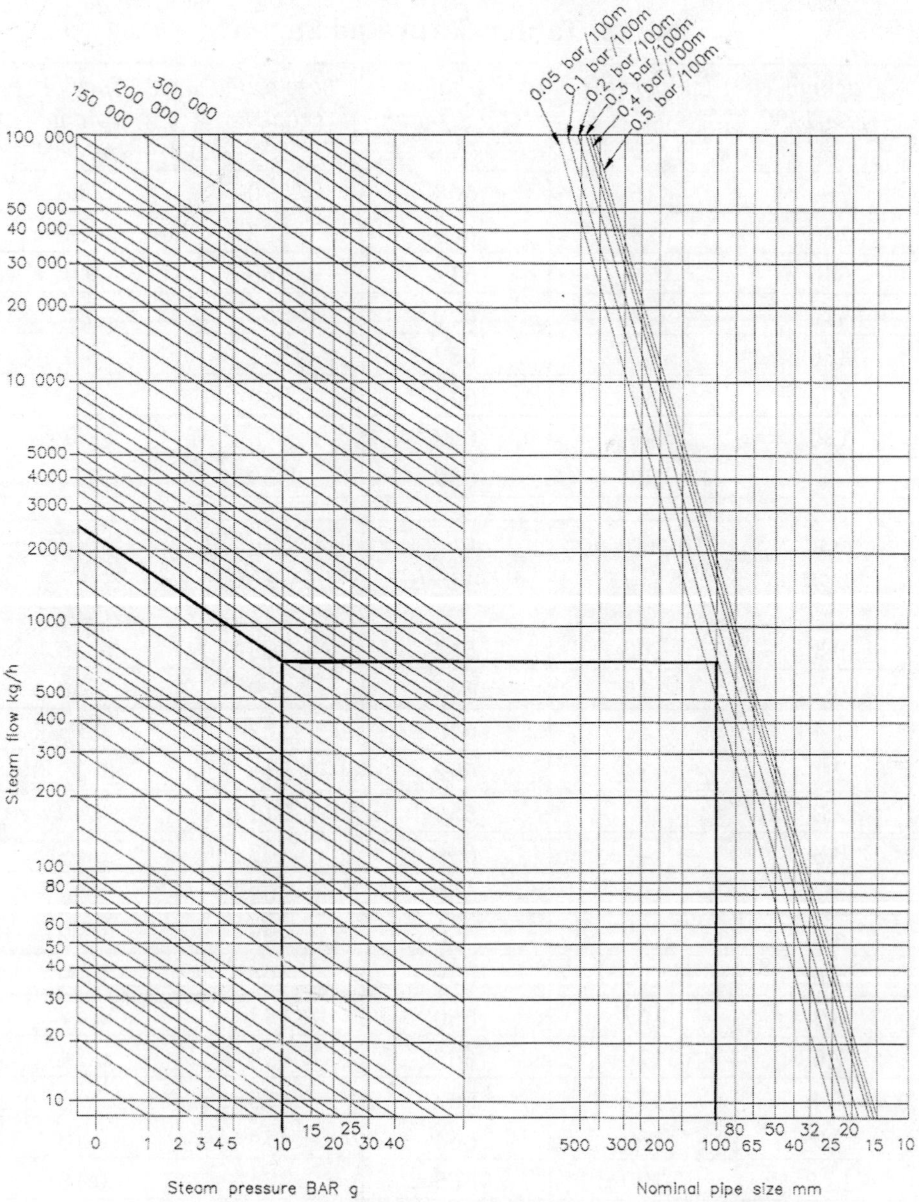

Steam Pipeline Capacities along with Pressure Losses (*Source*: Seppo Ralli).

Annexure 2

Steam Table: Saturated Steam

Vacuum (mmHg)	*Temperature (°C)*	*Enthalpy (kCal/kg)*	*Latent Heat (kCal/kg)*	*Specific Volume (m^3/kg)*
750	16	603	590	78.6
735	25	609	582	41.0
640	56	612	565	9.0
510	72	628	556	4.7
330	82	632	551	3.2
200	89	635	546	2.4
130	95	637	543	2.0
0	100	639	539	1.7

Pressure (kg/cm^2)	*Temperature (°C)*	*Enthalpy (kCal/kg)*	*Latent Heat (kCal/kg)*	*Specific Volume (m^3/kg)*
0.3	108	642	533	1.30
0.7	115	645	529	1.00
1.4	126	648	522	0.75
2.1	134	652	516	0.29
2.8	142	654	511	0.49
3.5	147	656	507	0.45
4.2	153	657	503	0.36
4.9	157	658	499	0.32
5.6	162	659	496	0.29
6.3	167	660	492	0.27
7.0	170	661	490	0.24
8.4	177	662	485	0.21
9.9	183	664	479	0.18
11.2	189	665	474	0.16
12.7	193	666	470	0.15
14.1	197	667	466	0.13

Properties of Saturated Steam Table (in SI units)

Temp. °C T	Press. kPa P	Specific Volume m^3/kg Sat. Liquid v_f	Sat. Vapor v_g	Internal Energy kJ/kg Sat. Liquid u_f	Evap. u_{fg}	Sat. Vapor u_g	Enthalpy kJ/kg Sat. Liquid h_f	Evap. h_{fg}	Sat. Vapor h_g	Entropy kJ/kg·K Sat. Liquid s_f	Evap. s_{fg}	Sat. Vapor s_g
0.01	0.6113	0.001 000	206.14	.00	2375.3	2375.3	.01	2501.3	2501.4	.0000	9.1562	9.1562
5	0.8721	0.001 000	147.12	20.97	2361.3	2382.3	20.98	2489.6	2510.6	.0761	8.9496	9.0257
10	1.2276	0.001 000	106.38	42.00	2347.2	2389.2	42.01	2477.7	2519.8	.1510	8.7498	8.9008
15	1.7051	0.001 001	77.93	62.99	2333.1	2396.1	62.99	2465.9	2528.9	.2245	8.5569	8.7814
20	2.339	0.001 002	57.79	83.95	2319.0	2402.9	83.96	2454.1	2538.1	.2966	8.3706	8.6672
25	3.169	0.001 003	43.36	104.88	2304.9	2409.8	104.89	2442.3	2547.2	.3674	8.1905	8.5580
30	4.246	0.001 004	32.89	125.78	2290.8	2416.6	125.79	2430.5	2556.3	.4369	8.0164	8.4533
35	5.628	0.001 006	25.22	146.67	2276.7	2423.4	146.68	2418.6	2565.3	.5053	7.8478	8.3531
40	7.384	0.001 008	19.52	167.56	2262.6	2430.1	167.57	2406.7	2574.3	.5725	7.6845	8.2570
45	9.593	0.001 010	15.26	188.44	2248.4	2436.8	188.45	2394.8	2583.2	.6387	7.5261	8.1648
50	12.349	0.001 012	12.03	209.32	2234.2	2443.5	209.33	2382.7	2592.1	.7038	7.3725	8.0763
55	15.758	0.001 015	9.568	230.21	2219.9	2450.1	230.23	2370.7	2600.9	.7679	7.2234	7.9913
60	19.940	0.001 017	7.671	251.11	2205.5	2456.6	251.13	2358.5	2609.6	.8312	7.0784	7.9096
65	25.03	0.001 020	6.197	272.02	2191.1	2463.1	272.06	2346.2	2618.3	.8935	6.9375	7.8310
70	31.19	0.001 023	5.042	292.95	2176.6	2469.6	292.98	2333.8	2626.8	.9549	6.8004	7.7553
75	38.58	0.001 026	4.131	313.90	2162.0	2475.9	313.93	2321.4	2635.3	1.0155	6.6669	7.6824
80	47.39	0.001 029	3.407	334.86	2147.4	2482.2	334.91	2308.8	2643.7	1.0753	6.5369	7.6122
85	57.83	0.001 033	2.828	355.84	2132.6	2488.4	355.90	2296.0	2651.9	1.1343	6.4102	7.5445
90	70.14	0.001 036	2.361	376.85	2117.7	2494.5	376.92	2283.2	2660.1	1.1925	6.2866	7.4791
95	84.55	0.001 040	1.982	397.88	2102.7	2500.6	397.96	2270.2	2668.1	1.2500	6.1659	7.4159
	MPa											
100	0.101 35	0.001 044	1.6729	418.94	2087.6	2506.5	419.04	2257.0	2676.1	1.3069	6.0480	7.3549
105	0.120 82	0.001 048	1.4194	440.02	2072.3	2512.4	440.15	2243.7	2683.8	1.3630	5.9328	7.2958
110	0.143 27	0.001 052	1.2102	461.14	2057.0	2518.1	461.30	2230.2	2691.5	1.4185	5.8202	7.2387
115	0.169 06	0.001 056	1.0366	482.30	2041.4	2523.7	482.48	2216.5	2699.0	1.4734	5.7100	7.1833
120	0.198 53	0.001 060	0.8919	503.50	2025.8	2529.3	503.71	2202.6	2706.3	1.5276	5.6020	7.1296
125	0.2321	0.001 065	0.7706	524.74	2009.9	2534.6	524.99	2188.5	2713.5	1.5813	5.4962	7.0775
130	0.2701	0.001 070	0.6685	546.02	1993.9	2539.9	546.31	2174.2	2720.5	1.6344	5.3925	7.0269
135	0.3130	0.001 075	0.5822	567.35	1977.7	2545.0	567.69	2159.6	2727.3	1.6870	5.2907	6.9777
140	0.3613	0.001 080	0.5089	588.74	1961.3	2550.0	589.13	2144.7	2733.9	1.7391	5.1908	6.9299
145	0.4154	0.001 085	0.4463	610.18	1944.7	2554.9	610.63	2129.6	2740.3	1.7907	5.0926	6.8833
150	0.4758	0.001 091	0.3928	631.68	1927.9	2559.5	632.20	2114.3	2746.5	1.8418	4.9960	6.8379
155	0.5431	0.001 096	0.3468	653.24	1910.8	2564.1	653.84	2098.6	2752.4	1.8925	4.9010	6.7935
160	0.6178	0.001 102	0.3071	674.87	1893.5	2568.4	675.55	2082.6	2758.1	1.9427	4.8075	6.7502
165	0.7005	0.001 108	0.2727	696.56	1876.0	2572.5	697.34	2066.2	2763.5	1.9925	4.7153	6.7078
170	0.7917	0.001 114	0.2428	718.33	1858.1	2576.5	719.21	2049.5	2768.7	2.0419	4.6244	6.6663
175	0.8920	0.001 121	0.2168	740.17	1840.0	2580.2	741.17	2032.4	2773.6	2.0909	4.5347	6.6256
180	1.0021	0.001 127	0.194 05	762.09	1821.6	2583.7	763.22	2015.0	2778.2	2.1396	4.4461	6.5857
185	1.1227	0.001 134	0.174 09	784.10	1802.9	2587.0	785.37	1997.1	2782.4	2.1879	4.3586	6.5465
190	1.2544	0.001 141	0.156 54	806.19	1783.8	2590.0	807.62	1978.8	2786.4	2.2359	4.2720	6.5079
195	1.3978	0.001 149	0.141 05	828.37	1764.4	2592.8	829.98	1960.0	2790.0	2.2835	4.1863	6.4698
200	1.5538	0.001 157	0.127 36	850.65	1744.7	2595.3	852.45	1910.7	2793.2	2.3309	4.1014	6.4323
205	1.7230	0.001 164	0.115 21	873.04	1724.5	2597.5	875.04	1921.0	2796.0	2.3780	4.0172	6.3952
210	1.9062	0.001 173	0.104 41	895.53	1703.9	2599.5	897.76	1900.7	2798.5	2.4248	3.9337	6.3585
215	2.104	0.001 181	0.094 79	918.14	1682.9	2601.1	920.62	1879.9	2800.5	2.4714	3.8507	6.3221
220	2.318	0.001 190	0.086 19	940.87	1661.5	2602.4	943.62	1858.5	2802.1	2.5178	3.7683	6.2861
225	2.548	0.001 199	0.078 49	963.73	1639.6	2603.3	966.78	1836.5	2803.3	2.5639	3.6863	6.2503
230	2.795	0.001 209	0.071 58	986.74	1617.2	2603.9	990.12	1813.8	2804.0	2.6099	3.6047	6.2146
235	3.060	0.001 219	0.065 37	1009.89	1594.2	2604.1	1013.62	1790.5	2804.2	2.6558	3.5233	6.1791
240	3.344	0.001 229	0.059 76	1033.21	1570.8	2604.0	1037.32	1766.5	2803.8	2.7015	3.4422	6.1437
245	3.648	0.001 240	0.054 71	1056.71	1546.7	2603.4	1061.23	1741.7	2803.0	2.7472	3.3612	6.1083
250	3.973	0.001 251	0.050 13	1080.39	1522.0	2602.4	1085.36	1716.2	2801.5	2.7927	3.2802	6.0730

Annexure 3

Some Useful Conversion Factors

Length	*Area*
1 cm = 0.3937 inch	1 cm^2 – 0.155 $inch^2$
1 m = 3.2808 feet	1 m^2 = 10.764 $feet^2$
1 feet = 0.3048 m	1 $feet^2$ = 0.092903 m^2
1 mile = 1.6093	
Volume	*Mass*
1 m^3 = 35.315 $inch^2$	1 kg = 2.2051 lbm
1 litre = 0.001 m^3	1 lbm = 0.45359 kg
1 $feet^3$ = 28.316 litre = 0.028316 m^3	1 ton = 2000 lbm = 907.18 kg
Density	*Acceleration*
1 g/cm^2 = 1000 kg/m^3 = 1000 g/L	1 m/s^2 = 3.2808 ft/s^2
1 lbm/ft^3 = 16.018 kg/m^3	1 g = 32.174 ft/s^2 =9.8067 m/s^2
	= 0.016018 g/cm^3
Volume Flow	*Mass Flow*
1 m^3/min = 35.315 ft^3/min	1 kg/s = 132.28 lbm/min
= 16.667 L/s	1lbm/h = 0.45359 kg/h
Pressure	*Angle*
1 atm = 101.325 Pa	1 radian (rd) = 57.296 deg (°)
= 14.696 lbf/in^2 or PSI	1° = 3600 sec = 60 min
= 1.0133 bar	= 0.017453 rad
1 lbf/in^2 = 6894.7 Pa = 144 lbf/ft^2	
1 torr = 1 mmHg at 32 °F	
Heat and Energy	*Heat Flow and Power*
1 Joule (J) = 1 N. m = 0.23885 g.cal	1 Watt(W) = 1 J/s =0.01433 k.Cal/min
1kWh = 860 k.Cal/h	1 hp = 1055.06 J = 745.70 W
	1 boiler hp = 9809.5 W
	1 Ton of Refrigeration = 12000 Btu/h = 3516.8 W
Thermal Conductivity	*Heat Transfer Co-efficient*
1 W/(m K) = 0.8599 k.Cal/(h m °C)	1 W/(m^2 K) =85986 kg. Cal/h m^2 °C

Thermal Resistivity	*Thermal Resistance*
1 m.K/W = 100 cm °C/W	1 h. ft^2 °F/Btu = 0.17611 m^2 K/W
	1 m^2 K/W = 1.163 h m^2 °C/kg Cal
Thermal Diffusivity	*Temperature*
1 m^2/s = 3600 m^2/h = 10.764 ft^2/s	T°C=(T °F – 32)/1.8
	T °F = 1.8 (T°C+32)
Viscosity (kinematic)	*Viscosity (Dynamic)*
1 centistoke = 0.0036 m^2/h	1 centipoise (cP)=0.001 Pa.s
	1 Pa.s = 100 cP =10 Poise

Annexure 4

Properties of some of the Milk and Milk Products (*Source*: NDDB)

Product	*Density* (kg/m^3)	*Specific Heat* (*k.Cal/ kg - °C m*)	*Thermal Conductivity*		*Freezing Point °C*	*Viscosity*	
			Temp °C	*k.Cal/ h-m²-°C m*		*Temp. °C*	*Viscosity (cP)*
Milk	1032	0.935	15	0.43	-0.54	20	2.12
Skim Milk	1035	0.95	15	0.47	–	25	1.4
Cream,35 per cent fat	999	0.895	15	0.30	–	15	9.0
Icecream, 100 per cent OR	540	0.600	–	–	–	-8	$20x10^9$
Ice cream mix	1080	0.800	–	–	-2.5	10	42
Butter (unsalted)	999	0.555	5	0.17	0	–	–
Condensed Whole Milk (50 per cent TS)	1160 1160	0.940 0.940	2080	0.26 0.31	-3.5	–	–
Sweetened Cond. milk	1285	0.590	–	0.31	-15	–	–
Milk powder	570	0.770	–	0.37	-12.9	–	–

Approximate Milk Equivalent of Milk Products (*Source*: NDDB)

Product	*Kg Product per 100 kg Milk*		*Kg Milk for 1 kg Product*	
	BM (6 per cent F, 9 per cent SNF)	*CM (4 per cent F, 8.5 per cent SNF)*	*BM*	*CM*
Skim milk	85.2	90.2	1.2	1.1
Butter milk	7.6	5.1	13.2	19.6
Cream 40 per cent fat	14.8	9.8	6.8	10.2
Butter	7.2	4.7	13.9	21.0
Ghee	5.7	3.8	17.5	26.3
Dahi	91.0	77.0	1.1	1.3
Chhana	24.3	20.0	4.1	5.0
Khoa	21.7	18.0	4.6	5.6
Conc. Milk,50 per cent TS	30.0	25.0	3.3	4.0
Skim milk powder	9.5*	9.0*	10.5**	11.0**
W. milk powder	15.5	12.9	6.5	7.8
Cheese,Cheddar	13.6	10.4	7.4	9.6
Cheese,Cottage	6.9*	6.2*	14.8**	16.1**
Casein, raw	7.5*	6.8*	13.6**	14.8**
Whey	86.4	89.6	1.2	1.1

* Kg per 100 kg of skim milk; ** kg skim milk required.

Annexure 5

Convention in Use of SI Units

Symbols are written in lower case, unless they are from the name of a person *e.g.* Hertz, Newton, Kelvin, Watt (*e.g.* N for Newton and W for Watt)

Symbols do not change in the plural and never written with a period except at the end of a sentence.

For numbers having 5 or more digits, the digit should be placed in groups of three separated by spaces (instead of commas) counting both to the left and right to the decimal point.

[7 589 000; 0.012 55; 30 x 10^6]

A dash is to be used to separate units that are multiplied together (*e.g.* N – m) while mN is a millinewton.

Prefixes Recommended in SI Units

Multiples	*Prefixes*	*Symbols*	*Sub-multiples*	*Prefixes*	*Symbols*
10^{12}	Tera	T	10^{-1}	Deci	d
10^{9}	Giga	G	10^{-2}	Centi	c
10^{6}	Mega	M	10^{-3}	Milli	m
10^{3}	Kilo	k	10^{-6}	Micro	μ
10^{2}	Hecto	H	10^{-9}	Nano	n
10	Deka	da	10^{-12}	Alto	p

No	*Convention*	*Wrong*	*Right*
1	Double prefixes are not allowed	μkg	Mg
2	Prefix symbol and unit symbol should be together	1 n m	1 nm
3	Unit and prefix are raised to the same power	$10^{-9}m^3$	$10^{-27}m^3$
4	Symbols do not change with plural	10 kgs	10 kg
5	Symbols are not written with period except at the end of sentence	10 kg.	10 kg
6	Space must be left between numerals and the fist letter of the symbol	10kg	10 kg
7	If a numerical value is less than one, a zero should precede the decimal point	.6	0.6

No	*Convention*	*Wrong*	*Right*
8	Solidus (/) only between the symbols in the numerator and those in denominator	m/s/s	m/s^2 or $m.s^{-2}$
9	Coined names should not be used	Kilo	Kilogram
10	When names are written out, division is indicated by the word 'per'	9 m/ second	9 m per second

Index

Agitation Equipment 138
- Agitator types 138

Asceptic 80
Aseptic filling
- Atomizing devices 230
- Automated 35

Bactofugation 67
Bactotherm 68
- Bag Filters 250

Ball making machine 285
- Batch Butter churn 264
- Batch fluidized bed 258
- Batch Ice cream freezer 290
- Batch process time taken 87
- Boiling point elevation 155

Boiling 149
Bottle Washing
- Bucket type ice cream freezer 289
- Bulk filling 133

Butter making machines
- Butter melting vat 273
- Butter trolley 271

Can Washing
- Capacity 12
- Care of pipelines 14
- Carton filling 126

Centrifugal separation 49
- Cheese hoops 299

Cheese making equipment 295
- Cheese Vat 298

CIP
- Classification of drying 216
- Classification of Indian dairy products 277
- Classification of method of separation 248
- Classification of separators 57
- Classification of Spray Dryers 225

Classification 106

Classification 106

Comparison 122

Components 163

Condenser Design 187

Constant drying rate period 210

Container 123

Continuous butter making machine 267

Continuous Ice cream freezer 291

Continuous khoa making equipment 281

Continuous mozzarella cheese equipment 304

Continuous sterilizer 107

Control of fat 64

Controls 242

Controls 30

Controls 93-94

Cream Separation

Cyclone Separators 249

Cyclone 38

Direct heating sequence 116

Direct line standardization 65

Direct Steam Injector 168

Draft tubes 140

Drive mechanisms 60

Dryer efficiency 241

Drying

Dust collection 38

Electrical power required 78

Electrodialysis 326

Electrostatic precipitators 254

Evaporators

Factors affecting bulk density 237

Falling film 152

Falling rate period 211

FDV 95

Filtration

Fires and Explosion 245

Fluidization 255

Fluidized bed driers 259

Fluidized bed flow regimes 256

Forced Circulation 155

Form Fill Sealing Machine 132

Fouling of membrane 321

Free and Bound Water 206

Freezing Point 288

Fritz Process 268

Gas operated khoa making equipment 280

Gasket material

General points 11

Ghee boiler 275

Ghee making equipment 272

Glass 10

Gravity Settling 47

Hermetic separators 58

Hollow fibre module 320

Homogenizer

Horizontal 20

Hot air system 227

HTST 89

Hydraulic drive mechanism 265

Hydrolock 108

Ice Cream making equipment 287
Inclined SSHE 282
Installation 14
Internal fluidized bed 261
Line setup 83
LMTD Correction Factor 98
Maintenance 97
Materials for membrane 314
Materials used in dairy 6
Mechanical Vapour Recompressor 193
Mechanization of Indian Dairy Products
Mechanized cheese making equipment 299
Membrane Processes in dairy industry 317
Membrane processing 307
Membrane separation process hookup 309
Membrane separations 308
Method of sterilizing Carton 127
Micro Filtration 310
Milk Storage Tank
Mozzarella Processing equipment 302
Multiple effect evaporators
Oscillatory Stirred tank 300
Pasteurizer
Pasteurizers 166
Plastic Tubing 12
Plate and frame module 321
Plate arrangement 91
Plate type 113
Pneumatic atomization 236
Pneumatic conveying 238
Polymers used in dairy, HDPE,CAB, LDPE 9
Porosity 256-257
Power consumption 142
Power required to rotate cylinder 53
Preheaters 164
Pressure nozzle 233
Prestratification tank 273
Prevention of losses through exhaust air 244
Principles 41
Processed cheese vat 301
Rate of 42
Rate of Evaporation 218
Reconstitution 24
Regenerative heaters 167
Reuse 35
Reverse Osmosis 323
Roller drier 217
Rotary Atomization 230
Safety interlocks 243
Sanitary design 3 – 5
Sanitary features 1
Sanitary pipelines
Selection of membrane materials 316
Self desludging 61
Separation
Separation
Sequence of operations of cheese making 296

Silos 23

Size of pipe lines 15

Soaker type 32

Solid-Gas separation 247

Specific requirements for manufacture 2

Spiral wound module 320

Spray Drying

Stages of drying 209

Stainless Steel pipes

Stainless 8

Steam Infusion 121

Steam Injection 119

Steaming block 28

Steel 7

Sterilization

Stoke's Law 47

Straight through 29

Stretching and Molding equipment 303

Support of pipes 16

Temperature rise 79

Temperatures 13

Thermo Vapour Recompressor 191

Three stage khoa making equipment 284

Tiltable khoa pan 278

Time-Temperature relationship 117

Tub 28

Tubular module 319

Tubular 114

Twin cylinder SSHE 283

Two stage drying 239

Two-stage dryer 261

Types of Feed arrangements 178

Types 150

Types 34

UF process hookup 318

UHT processing \

Ultrafiltration 310

Unbound Moisture 207

Vacuum controlling devices 201

Vacuum Pan 151

Vacuum producing devices 196

valve profiles 74

Valves wearing off 81

Vapour Separators 172

Venturi scrubber 39

Vertical plate 44

Vertical 22

Viscous force 48

Wet scrubbers 252

Wetting of heat transfer area 175